AF182569

To my daughter Lara, who gives meaning to my world,

and her father Claudio, the love of my life.

Contents

Preface

To the memory of Mario Castagnino:
I still miss him very much.

When I first met Mario Castagnino at the end of 2000, he was already an internationally recognized senior physicist, but he had never worked in the field of philosophy. Our first collaborations focused on a topic that interested us both at the time: irreversibility and the arrow of time. Then we began to analyze the phenomenon of decoherence as a particular case of irreversibility, and in this way we got closer to quantum mechanics. I remember that I immediately told him: "Doc, we can work on any topic, except on the interpretation of quantum mechanics, because on that subject everything has been said." But, as the drunk who found Pedro Navaja lying on a New York street said, "La vida te da sorpresas": the surprising fact here is that, after many years, I now find myself writing the prologue of a book that deals precisely with the Modal-Hamiltonian Interpretation of quantum mechanics, which I began to develop together with "Doc" Castagnino almost twenty years ago, and whose first version was published in 2008.

Since then, the interpretation was articulated and enriched in many works that, in general, I published in co-authorship. Thus, deep down, the content of this book is almost a collective work in which many people have participated: my students, my collaborators, and other researchers from whom I learned a lot. This is the opportunity to explicitly thank each one of them.

First of all, I would like to thank the young people who have always injected me with enormous doses of energy and vitality. They are the members of my group, who, in general, started as my students and gradually became my collaborators and independent researchers. In strict alphabetical order, they are: Hernán Accorinti, Fiorela Alassia, Juan Sebastián Ardenghi, Bruno Borge, Mariana Córdoba, Sebastian Fortin, Manuel Herrera, Federico Holik, Jesús Jaimes Arriaga, Juan Pablo Jorge, Mariano Lastiri, Cristian López, Marcelo Losada, Juan Camilo Martínez González, Matías Pasqualini, Ignacio Rojas, Leonardo Vanni, and Alfio Zambon. Many of these names will appear throughout the book, when I include references to the works we developed in collaboration on the topics discussed here. And the names that are not included in this book appear, nevertheless, in co-authored publications on other topics, such as philosophy of information, philosophy of chemistry, or questions related to the problem of realism. Precisely because names and topics are intertwined in multiple publications, I can speak of a strongly cohesive research group, of which I am deeply proud. All these people are for me a parallel family: I am bound to them by much more than purely academic ties. Among them, I want to thank particularly Cristian López and Bruno Borge for their detailed and critical reading of some chapters of the book, and Sebastian Fortin and Juan Sebastian Ardenghi for their meticulous revision of the mathematical developments of Chapter 12.

Regarding the senior scholars from whom I learned so much, there are some people who marked my intellectual development. First of all, I would like to mention Dennis Dieks, through whom I entered the world of the modal interpretations of quantum mechanics, and who trusted me enough to invite me to co-author the entry "Modal Interpretations of Quantum Mechanics" for the *Stanford Encyclopedia of Philosophy*. My exchange with Dennis was not only highly fruitful philosophically, but it was also essential from a personal point of view: being a woman, with an Italian surname, with a poor fluency in English, trained far from the most relevant academic centers, and working in an underdeveloped country in the south of the southern hemisphere, the fact that a philosopher of physics of the stature of Dennis Dieks believed in my value as a researcher allowed me to overcome my insecurities and dedicate myself seriously to my work. I will never be able to express clearly enough the magnitude of my gratitude.

This is also a place to remember Newton da Costa, who encouraged me to develop the ontological part of the interpretation. Our collaborative work was the basis for the building of the ontological structure of the quantum world proposed by the interpretation, which in recent years has begun to attract the interest of some metaphysicians of physics. Besides having been an immeasurable honor to have participated in these publications, that work gave me the opportunity to know closely the endearing "Professor Newton," his joy, his perennial smile, and his masterful way of encouraging his collaborators by turning the complex into simple. I am just one of the many people who miss him.

I would especially like to mention here the remarkable physicist-mathematician Manuel Gadella, who many years ago had the patience to fill four blackboards of a classroom at the University of Valladolid to teach me the mathematics of rigged Hilbert spaces, of which he is the leading international specialist. That first contact has led to some collaborative publications. But over time, my relationship with Manolo has become a close friendship that we have continued to cultivate to this day. That friendship allowed me to take advantage of his time to ask him to review the mathematical chapter of this book, which he did with infinite patience. So I thank Manolo in general, for his teachings and for his friendship and, in particular, for his revision work. I hope we will work together again very soon.

I am very happy to say that, recently, the Modal-Hamiltonian Interpretation began to attract the attention of the remarkable logician and philosopher of science Décio Krause. I convinced him that a theory of bundles of properties makes sense, and he convinced me of the importance of having an adequate metalanguage to talk about the ontology I propose. This exchange has already resulted in a collaborative publication (together with Fede Holik and Juan Pablo Jorge) and announces new and interesting works. But perhaps my greatest satisfaction is that this collaboration has led to a strong personal friendship, favored by geographical and linguistic proximity, and irrigated with considerable amounts of beer. Thank you, Décio, for being my friend!

While we are on the subject of personal relationships, academic activity has given me other friends who brighten up my life: Roberto Laura, who taught me to add joy to physics, Ana Rosa Pérez Ransanz, with whom we share so many talks about our deepest feelings, Lucía Lewowicz, whose strength to face life never ceases to amaze me, and Chérif Matta, from whom I learn not only quantum chemistry. I also extend my thanks to all of them.

In connection with the preparation of this book, I would like to thank Otávio Bueno and Yanis Pianko: their critical reading of my works allowed me to clarify several aspects of my proposal as presented in this book. I also want to acknowledge in general all the anonymous reviewers of the articles that shaped the Modal-Hamiltonian Interpretation, but in particular, I would like to express my gratitude to the four anonymous reviewers of the proposal for this book, whose favorable opinions allowed it to be accepted for publication by such a prestigious publishing house as Oxford University Press. I also thank Oxford University Press, in particular Sonke Adlung (Senior Acquisitions Editor, Physical Sciences) and Jodie Keefe (Project Editor) for their patience in guiding me throughout the editorial process.

Last but not least, I want to deeply thank my closest family: my daughter Lara, simply for existing, and my husband Claudio, because I would not be who I am without him.

O.L., March 2025.

Introduction

The problem of the interpretation of non-relativistic quantum mechanics is certainly one of the most discussed topics in the philosophy of physics, which has kept philosophers and scientists busy for more than a century. During a long period after the first formulations of the theory, the orthodox interpretational framework was tied to an instrumentalist reading: quantum features were discussed only in terms of possible measurement results. However, in the last decades, the traditional instrumentalist perspective has begun to lose its original strength, and several realist interpretations of quantum mechanics have been proposed. In one of its many meanings, realism is the philosophical position according to which the aim of scientific theories is to describe reality. From this realist viewpoint, the purpose of an interpretation of a scientific theory is to say how reality would be if the theory were true. The present proposal is framed in this realist stance.

It is important to emphasize from the outset that this is not a text where quantum mechanics is explained: certain basic knowledge about the theory is presupposed. Nor is it a book whose main purpose is to present and analyze different approaches to quantum mechanics. While other positions will be discussed when necessary, the ultimate goal is to introduce the Modal-Hamiltonian Interpretation (MHI), arguing in its favor from different perspectives.

The MHI is a realist non-collapse interpretation belonging to the family of Modal Interpretations of quantum mechanics. By contrast to the traditional Modal Interpretations, which make the preferred context of definite-valued observables dependent on the instantaneous state of the system, the MHI endows the Hamiltonian with a determining role both in the definition of systems and subsystems and in the selection of the preferred context.

From a scientific point of view, it will be proven that the MHI can account for the measurement problem both in its traditional version and in the most recent measuring scenarios. Furthermore, it will be shown that the application of the interpretation leads to conclusions that are in agreement with the everyday practice of physics, both regarding many concrete physical models and concerning the role of Galilei-invariance in the interpretation. These considerations about the symmetry of the theory will also pave the way for the extrapolation of the interpretation to relativistic quantum field theory.

From a philosophical point of view, a description of the structure of the quantum ontology proposed by the MHI will be offered. The ontology will be designed by taking into account many central concepts of traditional metaphysics, in many cases to challenge them. The result will be an ontology of properties, in which the ontological category of object is absent, so that quantum systems are non-objectual bundles of properties. Although this is a metaphysically heterodox proposal, those who understand it will gain a coherent and meaningful picture of the quantum ontology and will be able to make inferences about its structure.

The two aspects, although conceptually distinguishable, are not independent. On the contrary, both are closely connected to provide a unified picture of quantum reality that offers consistent answers to all the conceptual and ontological challenges posed by the theory. In addition, this unified picture furnishes foundational elements to take a position on specific physical issues, such as the interpretation of the concept of quantum decoherence or the problem of the entanglement of indistinguishable particles.

On this basis, the book is organized into thirteen chapters—which proves that I am not superstitious.

The first three chapters are introductory. Chapter 1 includes a reflection on what we are talking about when we speak of "quantum mechanics," and states my position as to why the interpretation of the theory deserves further discussion. Chapter 2 briefly introduces the algebraic formalism of quantum mechanics, which is adopted as the formal starting point of the proposed interpretation, and explains the role of the Galilei group in endowing the theory with physical content. In Chapter 3, the roots of the Modal Interpretations in the works of Bas van Fraassen are recalled, and the common features of the members of the modal family as well as the differences between them are presented.

The following five chapters are devoted to introducing the MHI from different perspectives. In Chapter 4, the MHI will be described in terms of interpretative postulates, which establish the ontological reference of the terms of the formalism and identify the preferred context of the definite-valued observables of a quantum system. Since a "good" interpretation should show its agreement with the orthodox practice of physics, Chapter 5 is devoted to arguing for the physical relevance of the MHI by applying it to very well-known models and experimental results. Chapter 6 deals with the famous quantum measurement problem: it explains how the MHI offers a solution to the problem for both ideal and non-ideal measurements, and how, in the non-ideal case, the interpretation distinguishes between reliable and non-reliable measurements; this chapter also shows how the MHI accounts for consecutive measurements and provides a novel, more physical understanding of the very idea of quantum measurement. After clarifying the concepts of covariance and invariance, Chapter 7 considers in what sense quantum mechanics is covariant and under which conditions it is invariant under the transformations of the Galilei group; this conceptual framework provides the tools for formulating an explicitly Galilei-invariant version of the MHI. Closing this second group of chapters, the aim of Chapter 8 is to make a case for a closed-system approach that leads to a top-down view of quantum mechanics; this view is in conceptual agreement with the ontology of properties to be developed in the following group of chapters.

The next three chapters contain the ontological component of the MHI. Chapter 9 provides the metaphysical preliminaries that serve as the basis for the discussions of the two following chapters: the specificity of the category of object, the difference between universals and tropes, the relation between determinable and determinate properties, the different types of modality, the difference between actualism and possibilism, and the notion of metaphysical indeterminacy. On the basis of these technical metaphysical resources, Chapter 10 sets out in detail the structure of the ontology proposed by the MHI: an ontology of properties without objects, in which systems are non-objectual bundles of instances of universal determinable properties and possibility is understood in possibilist terms that support quantum indeterminacy. Since arguing in favor of a certain quantum ontology over others requires showing how fruitful it is in offering reasonable solutions to interpretative problems, Chapter 11 is devoted to show the advantages of the proposed picture for dealing with the ontological challenges of quantum mechanics—contextuality, non-separability, and indistinguishability—and to emphasize how this ontology of properties provides the philosophical background to the top down view introduced in Chapter 8.

Although the MHI solution to the measurement problem was discussed in Chapter 6, in the last decades, more complex measurement scenarios have caught the attention of philosophers and physicists alike, some as old as the so-called Wigner's friend experiment, and others recently designed to test the theoretical predictions of quantum mechanics. For this reason, Chapter 12 revisits the measurement problem from the perspective of the challenges that those scenarios pose to the interpretation, analyzing in particular the Frauchiger-Renner argument and the so-called encapsulated measurements.

Finally, Chapter 13 is a "miscellanea" chapter, which addresses some issues that should not be neglected, but are not closely related to each other: the relations between the MHI and other interpretations and quantum theories, the emergence of an event-time according to the MHI, and a brief discussion about the possibility of extrapolating the MHI to relativistic quantum field theory. Although the first three sections of this chapter can be read in any order, the last one makes sense as a closing section since it explains my general philosophical position, which accounts for my interpretative strategy regarding quantum mechanics.

The book closes with some Final Remarks, where certain arguments in favor of the reasonability and fruitfulness of the MHI are added to those already set forth throughout the book. Finally, it is emphasized that the research on the MHI is a lively project, which can be and indeed is being developed in different directions, in particular, the metaphysics of science, the work on quasi-set theory as a metalanguage for the interpretation, and the extrapolation of the MHI to quantum field theory, among others. Perhaps the ultimate purpose underlying the writing of this book is the hope that other philosophers of science will become involved in this project for its further development.

1

Theories and Interpretations

The purpose of this book is to present an interpretation of quantum mechanics. But what are we talking about when we speak of "quantum mechanics?" Are we talking about a single theory or about many related theories? And what do we mean by "interpretation" here?

In this chapter, I will address these questions in an introductory way. Of course, I do not intend to take part in the extensive discussions about what a scientific theory is, what is meant by interpreting a theory, in what situation we encounter equivalent theories, and many other questions related to or derived from these. My aim here is to present some preliminary ideas in a way that, I hope, will be sufficient to avoid misunderstandings in the following chapters. But, above all, I will state my position as to why, after so many years and so much literature on the subject, the interpretation of quantum mechanics deserves further discussion.

1.1 What identifies a scientific theory?

The question of what a scientific theory is dominated much of the twentieth-century philosophical reflection on science. And although during the last decades the philosophy of science turned to other topics, particularly the role played by models in science, that discussion continues to appear repeatedly in the literature.

Traditionally, two approaches regarding scientific theories are distinguished: the syntactic view and the semantic view.

It is often said that, for the *syntactic view*, theories are axiomatic collections of sentences cast in a logical language. This position was so dominant during the first half of the twentieth century that Hilary Putnam (1962) dubbed it the "received view." But from the 1960s onward, criticisms of this view multiplied (see, e.g., Putnam 1962, van Fraassen 1970, Suppe 1974). As Hans Halvorson (2016, 587) states, "it seems that between 1975 and 2010, beating up on RV [the received view] was the favorite pastime of many philosophers of science." Considered in perspective, it is now recognized that some of those criticisms were appropriate, but others less so (see, e.g., Lutz 2012, 2017, Halvorson 2016).

According to the *semantic view*, a theory is a class of models or, at least, it is identified by the class of its models. Thus, Bas van Fraassen states that "if the theory as such is to be identified with anything at all—if theories are to be reified—then a theory should be identified with its class of models" (1989, 222). Soon, the semantic view became the orthodox approach to the nature of scientific

The Modal-Hamiltonian Interpretation of Quantum Mechanics. Olimpia Lombardi, Oxford University Press.
© Olimpia Lombardi (2026). DOI: 10.1093/9780198951407.003.0001

theories (Frigg 2006), and only a few authors were concerned to point out its limitations (e.g., Cartwright, Shomar, and Suarez 1996, Morrison 1999, Halvorson 2012).

The first point I will emphasize here is that I am not going to engage in this discussion, since it would divert me from my particular objective. On the one hand, here I am not interested in knowing how to characterize the very concept of scientific theory in a general way, but only in having some criteria to identify a theory, to decide when we are confronted with versions of the same theory, and to distinguish between different theories. Moreover, I tend to agree with Halvorson (2016, 597) that "the syntactic and semantic views are attempts to capture aspects of scientific theorizing, not the essence of scientific theorizing." On the other hand, the debate on the syntactic versus the semantic view seems to be conceived as independent of the particular scientific discipline and applicable to any science. Besides that I am rather skeptical that such a broadly applicable concept of scientific theory can finally be formulated—indeed, I believe that in certain disciplines, such as chemistry, the concept of theory is not central at all, and the leading role is played by models—, here I am only interested in physics, and within it, in one particular field, that of quantum mechanics.

If one asks physicists what a scientific theory is, most will answer that it is a formalism plus an interpretation (see, e.g., Cushing 1994, Bauman and Wolf 2018). However, this does not settle the issue, as both elements involve subtleties.

The *formalism* of a physical theory is a certain mathematical language (with its own rules of deduction) and certain statements, formulated in that language, which play the role of the basic postulates of the theory. The formalism of Newtonian mechanics is a traditional example: it is the language of infinitesimal calculus plus three basic postulates formulated in that language. As physics is a highly mathematized discipline, formalism is an ever-present element in physical theories. However, this does not mean that a formalism identifies a theory: two different formalisms do not necessarily lead to different theories. For example, Heisenberg's matrix mechanics and Schrödinger's wave mechanics rely on different formalisms, but in general they are not considered to be distinct physical theories. Physicists often talk about "*mathematical equivalence*" between different formalisms, in which case two supposedly different theories with mathematically equivalent formalisms would actually be different formulations of the same theory. But this depends on what we mean by "mathematical equivalence." The term itself suggests that it is an entirely rigorous concept. However, as David Wallace (2022) points out, it is not so easy to characterize such a relation in a completely precise way. The naive idea of mathematical equivalence as isomorphism does not work, since certain non-isomorphic formalisms are not considered in practice to be formalisms of different theories, but rather different mathematical presentations of the same theory (Barrett and Halvorson 2016). This is precisely the case of Heisenberg's matrix mechanics and Schrödinger's wave mechanics, whose formalisms are not isomorphic but which are not conceived as two different theories (see Halvorson 2012).

Despite the absence of a fully general and precise characterization of "mathematical equivalence," in many cases the practice of physics is quite clear about when the relation is satisfied: for example, with respect to general relativity, the formalism in terms of forces (Weinberg 1972) is mathematically equivalent to the formalism in terms of space-time curvature (Misner, Thorne, and Wheeler 1973). In other cases, by contrast, this is a hotly debated issue. For example, while the formalisms of Hamiltonian and Lagrangian mechanics have traditionally been regarded as mathematically equivalent, this position is now being questioned (North 2009, Curiel 2014, Hunt, Carcassi, and Aidala 2025), and it has been pointed out that the conclusion on this issue depends on the standard of equivalence that is adopted (Barrett 2019). In this sense, Sebastian De Haro (2021) points out that, in this particular case, it is reasonable to take the relevant isomorphism to be the Legendre transformation, which, as usually explained in textbooks, carries one back and forth between the Lagrangian and Hamiltonian frameworks.

Whereas mathematical equivalence is often interpreted as indicating that we are dealing with the same theory, this does not imply that conclusions about a physical theory can be drawn by looking only at the formalism. Indeed, the same formalism can "speak" about very different subject matters (see Weatherall 2019, Wallace 2022). For instance, the so-called wave equation is a second-order linear partial differential equation that can describe sound waves, seismic waves, electromagnetic waves, or many other wave phenomena. And this brings us directly to the second element of a physical theory: interpretation.

Since mathematics is a formal discipline, the formalism, taken at face value, in principle says nothing about the natural world (see Ruetsche 2011, Muller 2015): its *interpretation* is what endows it with semantic content. According to the standard position, interpreting a formalism amounts to assigning extra-linguistic references to the formal terms (see De Haro 2021; for a representationalist conception of interpretation, see Wallace 2022). On the one hand, perhaps inspired by Bryce DeWitt's (1970, 160) famous dictum, "The mathematical formalism of the quantum theory is capable of yielding its own interpretation," there are those who claim that the formalism of a theory determines its interpretation. This is obviously false, because the same formalism admits different interpretations: it is a fact of logic that any consistent formalism has multiple interpretations, in general in an infinite number. On the other hand, if a physical theory is an interpreted formalism, then, strictly speaking, *different interpretations of the same formalism* lead to different theories. However, in general this case is regarded as corresponding to one and the same theory with different interpretations. Below, I will point out that the interpretation of a formalism is manifested at two different levels; nevertheless, at either level in which there is an interpretative divergence of the same formalism, in the practice of physics the situation is usually understood as referring to a single theory with different interpretations.

In turn, two interpretations may be different because they apply to different formalisms. When the formalisms are not only different but mathematically non-equivalent, then it is easy to state that we are dealing with two different theories.

However, the conclusion is not so straightforward when the formalisms are considered mathematically equivalent.

It is quite natural to think that, if two formalisms are mathematically equivalent, the corresponding theories say the same about reality: in other words, what Bas van Fraassen (1980, 14, 57) calls the "picture of the world drawn by the theory" is the same in the two formalisms. Then, the interpretative divergences are viewed as different ways of expressing the same ontology. If this were the case, we would be dealing not only with mathematically equivalent formalisms, but also with equivalent theories. However, I believe that this stance does not distinguish between two levels of interpretation, which, to my knowledge, have not been sufficiently addressed in the literature.

On the one hand, a formalism is interpreted in terms of physical items, such as "system," "observable," "state," "evolution," etc., and, given a particular formalism, there is usually no disagreement among physicists regarding this, let us call it, *physical interpretation*. If two formalisms are considered equivalent but, precisely because they are different, are endowed with different physical interpretations, then it is assumed that there is a way to translate the terms of one physical interpretation into the terms of the other. For example, if the mathematical equivalence between the formalisms of Hamiltonian and Lagrangian classical mechanics is accepted, as De Haro proposes, it is also accepted that there is a physically meaningful translation between the languages corresponding to the two physical interpretations; in this sense they would be providing the same picture of the world. Another example is the relation between the Hilbert space formalism and the algebraic formalism of Standard Quantum Mechanics (SQM): although different, they are considered mathematically equivalent and there is a physically meaningful way of switching from the language of one physical interpretation to the language of the other. In the case of theories with different formalisms but meaningfully inter-translatable physical interpretations, one could say with fair certainty that they are *physically equivalent* theories.

On the other hand, not all the elements of the formalism of a theory, when interpreted physically, are empirically accessible. In general, experimental results only supply values to some of these formal elements, whereas the remaining ones, although endowed with the semantic content provided by the physical interpretation, acquire their particular "values" in an indirect way. For example, in quantum mechanics it is accepted that we have experimental access to the values of the observables, but not to the wavefunction or the quantum state. The part of the physical interpretation that assigns semantic content to the physically interpreted formal elements that are experimentally accessible can be called *minimal interpretation*, and it is what allows the theory to be empirically tested. Such empirical testing also requires that the theory has *operational content*, which consists of certain prescriptions necessary to identify, at least, the systems under study, the preparation procedures on such systems, and the appropriate measuring apparatuses to carry out the experiments corresponding to the quantities to be measured. If two theories, physically equivalent or not, are equivalent with respect to the experimental results involved in the minimal interpretation and obtained through their shared operational content, then they are said to be *empirically equivalent*. An influential example is that proposed by Clark Glymour (1980) in the classical framework. The Newtonian theory of gravitation was originally formulated in terms of bodies that accelerate in the presence of a gravitational field, which in turn depends on the distribution of matter. However, according to another formulation of the theory, there is no gravitational field: analogously to the case of general relativity, bodies follow geodesics in curved space-time, whose curvature depends on the distribution of mass. Despite the difference between the two theories, they are empirically equivalent in the sense that, for any model of the original formulation, there is a model of the non-standard formulation with the same matter distribution in which bodies follow the same trajectories as in the original one. In addition to illustrating the case of empirical equivalence, this example allows us to move on to the second level of interpretation, which I want to emphasize.

It is debatable whether the two classical theories of gravitation compared by Glymour are merely empirically equivalent or also physically equivalent in terms of a precise but sufficiently broad notion of "definitional equivalence" (see discussion in Weatherall 2016, 2019). What is more difficult to deny is that both theories describe different realities. According to James Owen Weatherall (2019, 5), Glymour "wished to emphasize that theories could be empirically equivalent to one another, but nonetheless theoretically inequivalent, in the sense of making substantially different claims about the world." In other words, it is necessary to distinguish the physical interpretation of a formalism from its *ontological interpretation*, which tells us what is the structure of the ontology, i.e., what are the ontological categories of the items that constitute the domain referred to by the theory. Confronted with a theory, two physicists may agree on the formalism and even on the physical interpretation, but they may disagree on the ontological interpretation. For example, in quantum field theory, leaving aside the discussion on the standard Lagrangian formalism versus the algebraic formalism and considering exclusively either of them, disagreements persist as to whether the ontological referents of the theory are particles or fields (see discussion in Kuhlmann 2023).

As I said above, a formalism does not fix its interpretation, neither the physical nor the ontological one. However, this does not imply that formalisms are ontologically neutral: different formal systems, even if mathematically equivalent, may suggest different ontologies. A typical example is the theory of natural numbers, which can be formulated on the basis of Peano's axioms or in terms of Russell's set-theoretic construction: although mathematically equivalent, the two formulations have different ontological connotations. From Peano's perspective, natural numbers admit a realistic, Platonic interpretation, according to which they exist as abstract entities. Russell's formulation, on the contrary, is more favorable to a nominalist interpretation, according to which reality is populated by individuals and classes of individuals, but not by natural numbers. An example from physics is, again, the case of Hamiltonian and Lagrangian mechanics: regardless of whether they are considered mathematically equivalent, they evoke different ontological pictures. Hamiltonian mechanics offers a dynamical picture: it describes temporal processes which, starting from an initial state, all evolve in the same temporal direction under the guidance of a dynamical law. Lagrangian mechanics, on the contrary, suggests a static picture with a frozen time: there are two states at once, and the trajectory between them is given by a variational principle, in particular, the principle of least action, which selects one among all possible trajectories. Some prominent physicists have even endowed Lagrangian mechanics with a certain teleological significance, regarding it as the reintroduction of final causes into physics (see the discussion in Ben-Menahem 2018, in particular, the quote from Planck on page 150; see also Glick 2023, Veldman 2024). But even if this teleological reading is rejected, it is not difficult to realize that the two theories suggest different realities with respect to the very conception of time and dynamics (do not miss the short story "Story of Your Life" by Ted Chiang 2002).

The aim of this section was not to discuss particular cases of scientific theories and their relations. I only tried to present, in a simple and necessarily schematic way, certain concepts about how to identify a physical theory, its versions, and eventually its relations with other theories of the same "family." I hope that, although one may disagree about the specific examples used here, the concepts presented in this section were clear enough to be applied to the field of quantum mechanics, where all of them are plainly illustrated.

1.2 Quantum mechanics: theory or theories?

Although quantum mechanics is often spoken of as if it were a single physical theory with multiple interpretations, the discussion in the previous section shows that this way of speaking is not accurate and requires several clarifications.

Let us begin with the so-called *Standard Quantum Mechanics* (SQM). Despite being "standard," there are different ways of conceiving it. For some, SQM is identified exclusively by its operational content; something along these lines is what Wayne Myrvold (2022a) is saying when he talks about the "operational core" of quantum mechanics, which consists of recipes for calculating probabilities of outcomes of experiments performed on systems subjected to certain state preparation procedures. However, this still does not tell us what the formalism of SQM is: is it the language of Hilbert spaces or can it be another mathematically equivalent language, such as the algebraic one? Nor does it tell us which are the statements, expressed in the corresponding language, that properly interpreted play the role of basic postulates of the theory: for example, is the projection postulate part of the formalism of the SQM, or is it exclusively part of the operational content of the theory? Some authors refer to SQM as the theory presented in textbooks (see, e.g., Egg 2021; for a criticism, see López 2024). But, besides the fact that didactic presentations of the theoretical bases of SQM vary greatly among textbooks (compare, for example, the texts by Asher Peres 1993 and by Leslie Ballentine 1998), this view does not answer the questions just raised.

These remarks make it clear that identifying the theory we are talking about is not so simple even when we refer to its standard version. Nevertheless, I consider that a discussion about what is the "true" characterization of SQM lacks philosophical relevance, like all discussions that attempt to find ultimate definitions or truths. I propose, then, to identify SQM in this work as follows. The formalism of SQM is given by (i) the language of Hilbert spaces or the algebraic language, conceived as mathematically equivalent, together with (ii) the statements which, physically interpreted, play the role of postulates of the theory (these postulates will be introduced in the next chapter). Now, even if this proposal is accepted, strictly one should distinguish two versions of the SQM, depending on whether the projection postulate is included in the formalism or not: say, SQM_1 if it is included, and SQM_2 if it is not included and, at most, collapse is understood as a mere recipe belonging to the operational content of the theory. In both cases, the formalism is endowed with the usual physical interpretation, part of which is the minimal interpretation that allows the theory to be experimentally tested. SQM also includes what I called above "operational content," which provides the heuristic tools for experimental testing. In principle, SQM does not have an ontological interpretation of its own: this is one of the aspects that open the possibility for what are usually called its different "interpretations."

When talking about SQM, many authors implicitly assume SQM_1, and this immediately leads to the so-called *Copenhagen Interpretation*, which many conceive as the orthodox interpretation of SQM, originated in the works of the founding fathers of quantum mechanics, mainly Niels Bohr, Werner Heisenberg, and Max Born. However, the deep differences between the approaches of these three authors, and even the many variations in Bohr's own writings, make it impossible to characterize the interpretation univocally (see Faye 2019). Certainly, Don Howard (2004, 680) is correct when he states that "until Heisenberg coined the term in 1955, there was no unitary Copenhagen interpretation of quantum mechanics." In general, however, the Copenhagen Interpretation is identified by the following assumptions: the inherent indeterminism involved in the collapse of the wavefunction, the complementarity between corpuscular and wave pictures, and the statistical interpretation of the wavefunction. In principle, it seems reasonable to regard the Copenhagen Interpretation, in its different forms, as an ontological interpretation of SQM, in particular of SQM_1, which assigns a precise content to the projection postulate: the postulate refers to the collapse of the wavefunction, which is conceived as a real and essentially indeterministic physical process.

Another ontological interpretation of SQM, also in its SQM_1 version, is that known as "*QBism*" (short for "quantum Bayesianism"), originally proposed by Carlton Caves, Christopher Fuchs, and Rüdiger Schack (2002a,b), although not yet under that name. According to this interpretative approach, quantum mechanics does not describe states of affairs in the world, but it is concerned with our knowledge and beliefs about the world. In particular, a quantum state is conceived as representing the epistemic state of an agent, and probabilities are interpreted in a subjectivistic way, as proposed by Bruno de Finetti (1937). Thus, the Born Rule is not a physical law but an empirically motivated norm of rationality, and the projection postulate represents the process by which agents update their knowledge on the basis of new information (see Healey 2022). Although QBism is often considered an interpretation of SQM, in particular of SQM_1, strictly speaking it is a different theory from that arising from the Copenhagen Interpretation, insofar as the two involve strongly divergent ontological assumptions. Nevertheless, in principle they are empirically equivalent theories: indeed, the original underlying aim of QBism was to clarify the obscurities of the Copenhagen Interpretation and not to question the empirical success of SQM_1 on which the Copenhagen Interpretation is based.

An interpretation of SQM_1 that has sparked great interest in the community of the philosophy and the foundations of quantum mechanics in recent times is the *Relational Quantum Mechanics* proposed by Carlo Rovelli (1996; for an updated version, see Laudisa and Rovelli 2021). Although initially devised as a refinement of the "textbook" Copenhagen Interpretation, Relational Quantum Mechanics relies on particular theses that endow it with its interpretative specificity. According to this interpretation, quantum variables take values only at interactions, and the values the variables acquire in one of the systems are only relative to the other system affected by the interaction. As a consequence, a quantum-mechanical description cannot be taken as an "absolute" representation of reality, even if the relative nature of the description does not involve any kind of subjectivity. From this perspective, the wavefunction—and, in general, the quantum state—is not realistically interpreted: it is a theoretical tool designed to facilitate the computation of the probabilities of future interaction events on the basis of certain given knowledge. As in the case of QBism, Relational Quantum Mechanics does not call into question the empirical success of SQM_1 and, as a consequence, it is also empirically equivalent to SQM_1.

Other ontological interpretations of SQM, but now in its SQM_2 version, are those comprising the family of *Everett-Like Interpretations*, which originated in the works of Hugh Everett III (1957, 1973) and whose main feature is their rejection of the collapse hypothesis. To this family belong (i) Everett's own *Relative State Interpretation*, which introduces the difference between absolute and relative state and a measure of typicality over relative states (see Barrett 2023), (ii) the *Many Worlds Interpretation*, originally proposed by DeWitt (1970), according to which each of the terms of a superposition corresponds to a coherent world, and all of the worlds are equally real (see Vaidman 2021), (iii) the *Many Minds Interpretation*, proposed by David Albert and Barry Loewer (1988), that distinguishes between the observer's physical state, which always evolves in a deterministic way, and the observer's mental state, which randomly jumps in each measurement interaction. All these interpretations, while sharing formalism and minimal interpretation, lead to very different ontological pictures. Therefore, although they are often regarded as different interpretations of the same physical theory, in a strict sense they should be considered as different physical theories. However, they are all empirically equivalent to each other. The interesting question is whether these theories are empirically equivalent to SQM_1—with any of its interpretations—, which includes the projection postulate as part of the formalism and the collapse of the wavefunction as a non-unitary and indeterministic physical process. Traditionally, such equivalence has been implicitly assumed, to the point that the issue is not commonly discussed in the usual presentations of the Everett-Like Interpretations. However, this assumption has now been revised in the light of the so-called encapsulated measurements. Indeed, Veronika Baumann and Stefan Wolf (2018) have shown that, in the case of that type of measurement, the theories coming from Everett-Like Interpretations and those coming from SQM_1 lead to different probabilistic predictions (I will come back to encapsulated measurements from a general viewpoint in Section 12.4 of Chapter 12).

Another family of ontological interpretations of the SQM, also in its SQM_2 version, is the family of the *Modal Interpretations*, which finds their roots in works of Bas van Fraassen (1972, 1974) of the 1970s. That original proposal relied on the distinction between what he called the "dynamical state" and the "value state" of a system at any instant: the *dynamical state* determines what may be the case, whereas the *value state* represents what actually is the case, that is, all the system's physical properties that are precisely definite. Van Fraassen's proposal is "modal" because the quantum state is the dynamical state and, in general, only tells us what is *possible* and not what is actual. In the 1980s several authors proposed interpretations that can be regarded as modal in the above sense, leading to a family of Modal Interpretations (see Dieks and Vermaas 1998, Lombardi and Dieks 2024) to which the following versions belong: (i) the *Biorthogonal-Decomposition Modal Interpretation* (Kochen 1985, Dieks 1988, 1989), (ii) the *Spectral-Decomposition Modal Interpretation* (Vermaas and Dieks 1995), (iii) the *Atomic Modal Interpretation* (Bacciagaluppi and Dickson 1999), (iv) the *Perspectival Modal Interpretation* (Bene and Dieks 2002), and the *Modal-Hamiltonian Interpretation* (Lombardi and Castagnino 2008). Since the central topic of this book will be the Modal-Hamiltonian Interpretation (MHI), I will not dwell here on Modal Interpretations because I will return to the subject in much more detail in the following chapters. Here I only point out that all these modal versions, although regarded as different interpretations of the same physical theory—as in the case of the Everett-Like Interpretations—, are properly different theories since they include different interpretative postulates for the selection of the observables that acquire precise actual values.

Up to this point, I have introduced cases that, strictly speaking, are different theories (see Acuña 2021), but which are traditionally treated as interpretations of the same theory, SQM, in some of its versions. However, there are other cases of theoretical proposals in which the difference with respect to SQM is clearly manifested at the level of formalism. Although originally these cases were also thought of as "interpretations"—and even in some works they continue to be presented as such—, it is now recognized that they are different theories. The best-known examples of this situation are Bohmian Mechanics and Spontaneous Collapse Theory.

Bohmian Mechanics was inspired by the so-called de Broglie-Bohm Theory, which was initially proposed by Louis de Broglie (1928) and reformulated by David Bohm (1952). In its contemporary version, it can be found in the works of Detlef Dürr, Sheldon Goldstein, and Nino Zanghì (1992, 2013). Bohmian Mechanics is a completely different theory from SQM since it is formulated in configuration space: Hilbert space is not essential to the proposal. Moreover, the Schrödinger equation is complemented with the Guiding Equation, which is a first-order evolution equation for the positions of the particles. Nevertheless, the formalism includes a part that admits the same minimal interpretation as SQM, which makes it possible to experimentally compare the two theories and conclude that they are empirically equivalent (for a discussion about this point, see Das and Dürr 2019). From the point of view of the ontological interpretation, Bohmian Mechanics proposes an ontology of particles with definite positions and momenta at any time, which in principle can be distinguished from each other on the basis of their deterministic trajectories: probabilities represent ignorance about the precise evolution of the particles of the system. However, this ontological picture leaves room to complete the ontological interpretation in different directions, in particular regarding the way the wavefunction and the quantum potential are understood (for a detailed discussion, see Solé 2013, Matarese 2023).

The *Spontaneous Collapse Theory* finds its origin in a work of GianCarlo Ghirardi, Alberto Rimini, and Tullio Weber (1986), and is often known as GRW theory after the authors' names. The basic idea is that the collapse of the wavefunction is a spontaneous, random process occurring at random times. To describe this process, the deterministic and linear Schrödinger equation of SQM is modified by adding stochastic and nonlinear terms. Thus, although the same mathematical language as that of SQM is used, the statement that—properly interpreted—describes the dynamical evolution of the quantum system is different, and this introduces a difference in the formalism itself. Shortly thereafter, the theory was reformulated giving rise to a new version, the *Continuous Spontaneous Localization* theory (Pearle 1989, Ghirardi, Pearle, and Rimini 1990), in which the discontinuous jumps of the original version are replaced by a continuous stochastic evolution in Hilbert space. The physical interpretation of the Spontaneous Collapse Theory, in its two versions, matches that of the SQM in the parts they share. However, the Spontaneous Collapse Theory introduces new terms d and f, which are physically interpreted as the localization accuracy and the average localization frequency. Moreover, the two theories are not empirically equivalent since the Spontaneous Collapse Theory makes predictions that differ slightly from those of the SQM (see Ghirardi and Bassi 2024). From an ontological viewpoint, the Spontaneous Collapse Theory has been endowed with two different ontological interpretations. According to the matter density ontology, the wavefunction describes a continuous field of matter density, which varies at different points in the three-dimensional physical space and changes over time (see Ghirardi, Grassi, and Benatti 1995). According to the flash ontology, on the contrary, each jump of the wavefunction—a flash—represents an event occurring at a point in physical space, and the time evolution of the wavefunction represents the probability of occurrence of future flashes (Tumulka 2006).

Given the substantial differences between the different characters of this conceptual map, the question arises: what do they all have in common to be subsumed under the label "quantum mechanics?" Here, Myrvold's idea of "operational core," as what is shared by the different proposals and which allows all of them to be empirically contrasted, becomes relevant. Such a core consists of the statements to which the minimal interpretation applies—the part of the interpretation on which all quantum physicists agree—plus what was called here "operational content," which is defined by scientific practice itself. It is this shared basis that allows all those involved in these matters to recognize themselves as physicists working in the field of quantum mechanics.

This brief review of the different theoretical formulations of quantum mechanics is undoubtedly schematic and necessarily incomplete. The aim has not been to provide an exhaustive overview of the field, but to show how the range of formulations that appear in this landscape may differ in their formalisms, in their empirical predictions, and/or in their interpretations, whether physical or ontological. Thus, the proposal that will engage us in this book, the MHI, finds its position in this complex network, so that in the following chapters we will be able to establish the similarities, differences, and relations with other formulations. However, the question about the significance of this work still remains.

1.3 Why is it worth insisting on interpretation?

As seen in the previous section, in the context of this "map of madness" (for a much more complete version, see Cabello 2017), most formulations are empirically equivalent to SQM, and in some cases the major divergences appear in the field of ontological interpretation. So, what is the point of carrying on the debate about the "interpretation of quantum mechanics?" Someone might say that predictive divergences will only have physical significance when the experimental tests will be effectively achievable and that the ontological divergences are completely sterile and only serve to entertain philosophers. It is quite clear that, if I am in the process of writing a book on an interpretative proposal for quantum mechanics, I do not agree with this assessment. Now it is time to explain why.

I am certain that it is worth insisting on the interpretation of quantum mechanics because only at the interpretative level is a proper understanding of the subject achieved. Of course, the problem now consists in elucidating what is meant by *understanding.*

For much of the twentieth century, with the preeminence of logical empiricism, the interest of philosophers of science focused on scientific prediction and explanation. In turn, explanation was conceived in terms of "legal coverage," relegating understanding to the realm of the idiosyncratic and subjective, so that it would be nothing but psychological acclimation (see, e.g., Hempel 1966), or equating the notion of understanding to the concept of explanation (Friedman 1974). Over time, however, the hegemony of logical empiricism gradually faded away; with this process, while the concept of explanation broadened considerably, the notion of understanding began to come back from ostracism. Thus, the literature on scientific understanding has grown remarkably in the last decades (see references in de Regt 2017, Gordon 2017, Grimm 2021). It is not my aim here to comment on that literature nor do I intend to give a precise characterization of what "understanding" means (I could do neither). All I will attempt here is to give a couple of notes on what I mean by "understanding," why I consider that an interpretation of quantum mechanics provides understanding, and what the cognitive and pragmatic advantage of reaching such an epistemic state is.

Understanding is an epistemic state that entails more than mere knowing: I may know what the fall of the Western Roman Empire was, but, at the same time, I may not understand that historical process. As Henk de Regt claims, "not only knowledge of principles and background conditions but also particular skills of the subject are involved in the process of achieving understanding." (de Regt 2004, 101). I believe that this plus of understanding over knowledge involves two features: *projectability* and *transferability*. On the one hand, subjects understand a certain knowledge better to the extent that they can project it further. For example, they can use it successfully to obtain new knowledge, they can infer causal connections from it, they can draw conclusions on its basis, they can transfer it to cases not originally considered and even strongly novel, they can even imagine counterfactual situations that fit it. On the other hand, subjects understand a certain knowledge better to the extent that they can transmit it to other subjects, even those who are not familiar with it. For example, the fact that a scientist can "explain" their topic of study to a layman is a symptom of their understanding, without requiring that the layman achieves a high level of comprehension. Of course, if the topic is very technical, in this process the "explainer" has to appeal to multiple resources, such as the use of images, metaphors, analogies, etc.; but it is precisely their capacity to generate such resources that shows their degree of understanding. According to a well-known anecdote, Richard Feynman, who prided himself on being able to devise ways to explain even the most complex ideas to beginning students, was once asked by a Caltech faculty member to explain why spin one-half particles obey Fermi-Dirac statistics. Rising to the challenge, he said, "I'll prepare a freshman lecture on it." But a few days later he told his colleague, "You know, I couldn't do it. I couldn't reduce it to the freshman level. That means we really don't understand it." (Goodstein and Goodstein 1996).

These notes make clear why understanding is much more than familiarity and psychological acclimatization. No matter how much we get used to accepting a certain knowledge, this is not enough to be able to project and transmit it. In turn, it is clear that understanding is different from explaining: understanding carries a pragmatic component that requires a certain type of personal elaboration; it involves different kinds of skills of the person who understands. In this sense, de Regt (2004, 103) speaks of the "epistemically relevant, pragmatic sense of understanding which, while it is not purely objective, should also not be equated with the subjective, highly individual 'feels right' experience." That pragmatic component is what "allows for the possibility that two people may disagree about whether a phenomenon is understood, while they both possess the same theoretical and background knowledge" (de Regt 2004, 103). The latter is precisely what happens in quantum mechanics.

On the basis of these notes of what I conceive as understanding, it is easy to see why I consider that an interpretation of quantum mechanics, particularly an ontological interpretation, provides understanding. If we only possess a formalism with a minimal interpretation, we do not reach an understanding of the matter since we hardly have the necessary tools to project and transfer our knowledge. The interpretation of the formalism is what allows us to extend the quantum knowledge to new situations, to make inferences about which facts are in accordance with the picture provided by the theory, and to evaluate how well the world described by the theory fits into the ontological categories that are at the basis of our traditional conception of reality. And it is only in terms of an interpretation that we can attempt to convey an image of quantum reality through ordinary language.

But even accepting what has been said so far about the notion of understanding, someone with instrumentalist temperament might insist: if the "operational core" of quantum mechanics has been so overwhelmingly successful in empirical terms, why do we need to understand it through interpretation? My answer is that understanding in the field of quantum mechanics has consequences for scientific practice itself. Certain interpretations lead to theories that may be more akin to other theoretical domains of physics, such as quantum field theory, or to other scientific disciplines, such as chemistry (I will come back to these issues in the following chapters). Furthermore, the instrumentalist spirit is perhaps one of the factors that discourages young people who seek to understand how the world works and that drives them away from physics. But I believe that the answer to the instrumentalist challenge depends primarily on how each of us conceives of the scientific enterprise. If science is understood exclusively as a means to produce technology in the form of concrete or abstract objects to be injected into consumer society and/or to improve the material lives of human beings, then not only do the discussions about the interpretation of quantum mechanics make no sense, but almost the entire reflection in the philosophy of science is utterly sterile. But those of us who believe that science is an end in itself because it is essential to the intellectual development of human beings will continue to think that the understanding provided by the interpretations of quantum mechanics has intrinsic value.

In summary, only through some interpretation can we achieve understanding of quantum reality and can we conceive what the world would be like according to the theory. In other words, only this interpretative work allows us to make sense of the quantum world.

2
Standard Quantum Mechanics: Formalism and physical content

Since the Modal-Hamiltonian Interpretation (MHI) is based on the formalism and the minimal interpretation of Standard Quantum Mechanics (SQM), this chapter may seem superfluous: it could be argued that everyone knows what SQM is. However, as remarked in the previous chapter, the identification of SQM is not as immediate as is usually assumed. Therefore, if an interpretation of quantum mechanics is to be precisely developed, the underlying theoretical commitments must be clear from the very beginning. The second reason why this chapter is not superfluous is that the MHI differs from the traditional Hilbert space formulation of SQM by adopting the algebraic formalism as its departing point. Although this might seem a minor formal detail, it will play a central role in the ontological picture arising from the MHI interpretation, as will be seen in Chapter 10.

2.1 Algebraic formalism and physical interpretation

The so-called *algebraic formalism* of quantum mechanics is a formal language for SQM that abstracts and generalizes the traditional Hilbert space formalism. The algebraic approach finds its roots in the work of John von Neumann (1932). Nevertheless, at present it is mainly based on the theory of C*-algebras, which was initially formulated by Israel Gelfand and Mark Naimark (1943) and later developed by Irving Segal (1947a,b) and Rudolf Haag (Haag and Kastler 1964; see also Haag 1992). Although originally proposed to provide a mathematically rigorous framework for systems with an infinite number of degrees of freedom, in particular in quantum field theory, at present the algebraic formalism is also employed in SQM (see, e.g., Landsman 2007, Earman 2015). Here I will offer a very simple and not completely rigorous sketch of the formalism, but which will suffice for the purposes of this book (for a concise but rigorous presentation, see Landsman 2009; an updated and thorough presentation of the topic can be found in Takesaki 2003).

A *C*-algebra* $\mathcal{A}$ is an associative algebra over the field $\mathbb{C}$ of the complex numbers, such that:

- It is complete in a norm $\|\cdot\|$ satisfying $\|ab\| \leq \|a\| \; \|b\|$, for all $a, b \in \mathcal{A}$.
- It has an involution operation $a \to a^*$ such that $\| a^* a \| = \| a \|^2$; this is the abstract notion of adjointness for elements that are not defined by their action on a Hilbert space.

The Modal-Hamiltonian Interpretation of Quantum Mechanics. Olimpia Lombardi, Oxford University Press.
© Olimpia Lombardi (2026). DOI: 10.1093/9780198951407.003.0002

- It is assumed to be closed under products, linear combinations, and the involution.
- It contains an identity $I \in \mathcal{A}$ that satisfies $I^* = I$ and $Ia = a$ for all $a \in \mathcal{A}$.

In this formal context, an *algebraic state* on a C*-algebra $\mathcal{A}$ is defined as a linear functional $\omega : \mathcal{A} \to \mathbb{C}$ such that:

- It is normalized in that $\omega(I) = 1$.
- It is positive in that $\omega(a^*a) \geq 0$, for all $a \in \mathcal{A}$.

An algebraic state ω on a C*-algebra $\mathcal{A}$ is said to be *pure* if it cannot be written as a non-trivial convex combination $\omega = \lambda_1\omega_1 + \lambda_2\omega_2$, with $0 < \lambda_1, \lambda_2 < 1$, $\lambda_1 + \lambda_2 = 1$, and where ω_1, ω_2 are also algebraic states on $\mathcal{A}$; otherwise ω is said to be *mixed*.

The algebraic language has a standard physical interpretation: a quantum system corresponds to a C*-algebra $\mathcal{A}$, its observables correspond to the self-adjoint elements of $\mathcal{A}$, and its quantum state corresponds to an algebraic state ω. Although in the algebraic formalism the algebra $\mathcal{A}$ is usually called 'algebra of observables,' this terminology is not entirely precise because not all the elements of the algebra, but only the self-adjoint ones, represent observables.

The abstract algebraic language is related to the Hilbert space language by the Gelfand–Naimark–Segal (GNS) construction (Gelfand and Naimark 1943, Segal 1947a), which proves that, for any algebraic state ω on a C*-algebra $\mathcal{A}$, the algebra $\mathcal{A}$ has a *mathematical representation* π_ω on a Hilbert space $\mathcal{H}_\omega$: ω is mathematically represented by a normed trace (density) operator ρ on $\mathcal{H}_\omega$, and the elements of $\mathcal{A}$ are mathematically represented by operators O on $\mathcal{H}_\omega$. When the state ω, represented by the density operator ρ, is pure, there is a vector $|\Psi\rangle \in \mathcal{H}_\omega$ such that $\rho = |\Psi\rangle\langle\Psi|$.

For simplicity, here I will remain in the context of the C*-algebra case and its Hilbert space mathematical representation. Nevertheless, it must be stressed that the algebraic formalism is not confined to the Hilbert space representation. In fact, C*-algebras do not contain unbounded operators, so other *-algebras are necessary for treating them. In the case of different *-algebras, the conditions for the representation of the algebra have to be established. For instance, by means of a generalization of the GNS theorem, it can be proved that certain nuclear *-algebras can be represented by rigged Hilbert spaces (Iguri and Castagnino 1999; for applications of rigged Hilbert spaces to quantum mechanics, see Bohm and Gadella 1989).

Although, strictly speaking, the abstract algebraic language and the language of its mathematical representation are different, for simplicity in what follows I will use the language of the Hilbert space representation to endow it with a physical interpretation: a quantum system $\mathcal{S}$ is represented by a space $\mathcal{O}$ containing the self-adjoint elements of the von Neumann algebra $\mathcal{B}(\mathcal{H})$ of linear bounded operators on the Hilbert space $\mathcal{H}$, and its quantum states correspond to density operators ρ, whose role is to yield the expectation values of all the observables corresponding to the operators belonging to $\mathcal{O}$: $\rho(O) = Tr(\rho O) = \langle O \rangle_\rho \in \mathbb{R}$, for all $O \in \mathcal{O}$.

But beyond the mathematical details and regardless of whether the abstract or the representational language is used, the conceptually relevant feature of the algebraic approach is that it emphasizes the difference between observables and states, whose corresponding spaces may not even be the same; this is precisely the case in those nuclear algebras represented by rigged Hilbert spaces. In the algebraic framework, observables are the basic elements of the theory, and states are secondary elements, defined in terms of the basic ones. To the extent that quantum states correspond to linear functionals on the algebra representing the system, their physical "nature" is exhausted in the task of computing the expectation values of the system's observables. In other words, from the conceptual viewpoint suggested by the algebraic approach, states must not be confused with observables: they must not be understood as any kind of property of the quantum system. As John Earman stresses, one should never forget "the mantra of the algebraic approach: a system state is an expectation value functional on the system algebra." (Earman 2015, 324). This point will be essential to the MHI ontological interpretation to be developed in Chapter 10.

2.2 The postulates of the theory

On the basis of the above brief presentation of the algebraic approach, the SQM as considered in the present work will be fixed by three *representative postulates* RP and two *quantum postulates* QP, all of them endowed with the physical interpretation introduced in the previous section.

The first representative postulate gives the mathematical characterization of a quantum system:

> **RP1**: A *quantum system* $\mathcal{S}$ is represented by a pair $(\mathcal{O}, H)$, $\mathcal{S}:(\mathcal{O}, H)$, where: (i) the *observables* of $\mathcal{S}$ are represented by the operators $O \in \mathcal{O}$, where $\mathcal{O}$ is the space containing the self-adjoint elements of the von Neumann algebra $\mathcal{B}(\mathcal{H})$ of linear bounded operators on the Hilbert space $\mathcal{H}$, and (ii) $H \in \mathcal{O}$ represents a particular observable called '*Hamiltonian*' of $\mathcal{S}$.

The second representative postulate mathematically characterizes the possible values of the observables of a quantum system:

> **RP2**: Given a quantum system represented as $\mathcal{S}:(\mathcal{O}, H)$, and the observable represented by $O = \sum_i o_i \Pi_i \in \mathcal{O}$—with integral in the continuous case—, where $o_i \neq o_j$ for $i \neq j$ and the $\Pi_i \in \mathcal{O}$ are projection operators—eigenprojectors of O—, each eigenvalue o_i of O represents one of the *possible values of the observable* represented by O.

The third representative postulate introduces the quantum state of a quantum system:

> **RP3**: Given a quantum system represented as $\mathcal{S}:(\mathcal{O}, H)$, its *quantum state* is represented by a linear functional ρ on the space $\mathcal{O}$, such that (i) $Tr(\rho)=1$ (normalization condition), and (ii) $\langle u|\rho|u\rangle \geq 0$ for any $|u\rangle \in \mathcal{H}$ (non-negativeness condition).

Once the representative postulates establish how the physical items "system," "observable," and "state" are mathematically represented, the quantum postulates can be introduced. The first one is the *Dynamical Postulate*, which states how the initial state of the quantum system evolves in time:

> **QP1 (*Dynamical Postulate*)**: Given a quantum system represented as $\mathcal{S}:(\mathcal{O}, H)$ with initial state represented by ρ_0, the state of $\mathcal{S}$ at time t is represented by a linear functional $\rho(t)$ on the space $\mathcal{O}$ given by the Schrödinger equation (in its von Neumann version): $\rho(t) = e^{-iHt/\hbar} \rho_0 \, e^{iHt/\hbar}$.

The second quantum postulate is the *Born Rule*, which introduces the probabilistic component in the theory. According to the instrumentalist readings of quantum mechanics, the rule assigns probabilities to the results of measurements. As repeatedly noted, this requires that "measurement" be conceived as a primitive concept of the theory. By contrast, here a measurement is considered an ordinary physical interaction that must be accounted for by the theory, that is, by the formalism and its interpretation. Therefore, the Born Rule plays the role of a postulate that assigns probabilities to the facts corresponding to the different values of the observables. If the symbol $A: a_k$ is used to denote the fact that an observable represented by the operator A has the value represented by the eigenvalue a_k of A, the Born Rule can be expressed as follows:

> **QP2 (*Born Rule*)**: Given a quantum system represented as $\mathcal{S}:(\mathcal{O}, H)$ in a state represented by ρ, and considering an observable represented by the operator $A = \sum_i a_i \Pi_i \in \mathcal{O}$, where the $a_i \neq a_j$ for $i \neq j$ are the eigenvalues of A and the $\Pi_i \in \mathcal{O}$ are the eigenprojectors of A, the *Born probability* P_B of the occurrence of the fact $A: a_k$ in state ρ can be computed as
>
> $$P_B(A: a_k, \rho) = Tr(\rho \, \Pi_k),$$
>
> where $\Pi_k \in \mathcal{O}$ is the eigenprojector of A corresponding to a_k.

In particular, since any projection operator $\Pi \in \mathcal{O}$ represents an observable with two values, represented in turn by the eigenvalues $\pi_1 = 1$ and $\pi_2 = 0$ of Π, the Born probability of the fact $\Pi: 1$ in the state ρ is computed as

$$P_B(\Pi: 1, \rho) = Tr(\rho \, \Pi). \tag{2-1}$$

If the Born probability P_B is computed over all the projection operators $\Pi \in \mathcal{O}$, the resulting function does not satisfy the Kolmogorov definition of probability: such a definition applies on a Boolean algebra, whereas the set of the facts corresponding to the projectors Π does not have a Boolean structure. For this reason, some authors define a generalized probability function (non-Kolmogorovian) over the orthoalgebra of quantum facts (see, e.g., Hughes 1989, Cohen 1989). In the present proposal, the Kolmogorovian character of probability will be preserved by defining *different probability functions* on the Hilbert space $\mathcal{H}$, as explained below.

Let us define a *context* as a *complete set of orthogonal projectors* (*CSOP*) $\{\Pi_\alpha\}$, with $\Pi_\alpha \in \mathcal{O}$ and $\alpha = 1$ to M, such that

$$\sum_{\alpha=1}^{M} \Pi_\alpha = I \quad \text{and} \quad \Pi_\alpha \Pi_\beta = \delta_{\alpha\beta} \Pi_\alpha, \tag{2-2}$$

where I is the identity operator in $\mathcal{O}$. If $\{|\alpha\rangle\}$ is a basis of the corresponding Hilbert space, the set $\{\Pi_\alpha = |\alpha\rangle\langle\alpha|\}$ is a CSOP. All the projection operators Π_B of the form $\Pi_B = \sum_J \Pi_J$, where the $\Pi_J \in \{\Pi_\alpha\}$ and $J \in \mathcal{J} \subset \{1,\ldots,M\}$, define a Boolean algebra $B^\alpha = (\mathcal{P}^\alpha, \cap, \cup, 0, 1)$, where $\mathcal{P}^\alpha = \cup \{\Pi_B\}$. Then, through the Born Rule QP2, a state, represented by the functional ρ on $\mathcal{O}$, defines a probability function $f_\rho^\alpha : \mathcal{P}^\alpha \rightarrow [0,1]$ on $\mathcal{P}^\alpha$ that satisfies the axioms of Kolmogorov:

- $f_\rho^\alpha(\mathcal{U}) = 1$
- $f_\rho^\alpha(\varnothing) = 0$
- If $X, Y \in \mathcal{P}^\alpha$ and $X \cap Y = \varnothing$, then $f_\rho^\alpha(X \cup Y) = f_\rho^\alpha(X) + f_\rho^\alpha(Y)$

As a consequence, given a CSOP $\{\Pi_\alpha\}$ and the projection operators of the form $\Pi_B = \sum_J \Pi_J$:

- The probability $f_\rho^\alpha(\Pi_B : 1)$ ascribed to the fact denoted by $\Pi_B : 1$ is computed with the Born Rule as

$$f_\rho^\alpha(\Pi_B : 1) = P_B(\Pi_B : 1,\ \rho) = Tr(\rho\, \Pi_B). \tag{2-3}$$

- For any observable represented by an operator $A = \sum_i a_i \Pi_{Bi}$, the probability $f_\rho^\alpha(A : a_k)$ ascribed to the fact denoted by $A : a_k$ is identified with the probability $f_\rho^\alpha(\Pi_{Bk} : 1)$ ascribed to the fact $\Pi_{Bk} : 1$:

$$f_\rho^\alpha(A : a_k) = f_\rho^\alpha(\Pi_{Bk} : 1) = Tr(\rho\, \Pi_{Bk}). \tag{2-4}$$

Summing up, since Booleanity is retained in each context, it can be said that the state represented by ρ defines *a different probability function on each context*, but each one of these probability functions satisfies the definition of Kolmogorov. In other words: regarding probability, one definition, many functions.

Given the above presentation, it is clear that I am referring to the version of the SQM that, in the previous chapter, I called SQM_2: the projection postulate is not included in the formalism and, consequently, the collapse hypothesis is not adopted as another interpretative postulate. This is so because, as will be seen in the next chapter, the members of the family of Modal Interpretations, to which the MHI belongs, are non-collapse interpretations.

In order to preserve the precision of the presentation, up to this point I have distinguished between the *physical language*—to which the terms 'quantum system,' 'observable,' 'value of an observable,' and 'quantum state' belong—and the *mathematical language*—to which the terms 'space of operators,' 'operator,' 'eigenvalues of an operator,' and 'linear functional' belong. On the basis of this distinction, I stressed that the physical items are *represented* by mathematical objects belonging to the formal domain. However, preserving this distinction all along the text would make the reading very dense and heavy without adding anything from a conceptual point of view. Therefore, under the assumption that the distinction has been understood, both languages will be mixed hereafter, as is in fact usual in the textbooks on quantum mechanics. Thus, I will speak of the system $\mathcal{S}:(\mathcal{O}, H)$, where $\mathcal{O}$ is a space of observables; I will say that the observable $O \in \mathcal{O}$ has eigenstates $|o_i\rangle$ and eigenvalues o_i; I will say that the initial state ρ_0 evolves unitarily according to the Schrödinger equation. However, when in Chapter 10 the proposed ontological interpretation will be introduced, I will recall again the distinction between physical and mathematical languages.

2.3 The Galilei group and the physical content of the theory

In the previous section I have presented a formalism for SQM with its physical interpretation. Nevertheless, as Leslie Ballentine (1998) points out, although such a physically interpreted formal structure is a necessary basis for the formulation of the theory, it has by itself very little empirical content. When concrete physical problems are to be solved, the relevant observables of the system, endowed with a clear empirical meaning, have to be identified. Those observables are closely related to space-time transformations.

A *continuous transformation* $\mathcal{T}$ with parameter s acts on observables A and state vectors $|\varphi\rangle$ as

$$A \longrightarrow \tilde{A} = U_s A\, U_s^{-1}, \tag{2-5}$$

$$|\varphi\rangle \longrightarrow |\tilde{\varphi}\rangle = U_s\, |\varphi\rangle\,, \tag{2-6}$$

where U_s is the family of unitary operators that describes the transformation $\mathcal{T}$. If s is allowed to become infinitesimally small, the unitary operator U_s can be expressed as

$$U_s = e^{iKs}, \tag{2-7}$$

where K is a self-adjoint infinitesimal operator independent of s, the *generator* of the transformation $\mathcal{T}$. In turn, when the state vector is represented as a function of space-time coordinates x, there is an inverse relation between transformations on function space and transformations on coordinates:

$$\varphi(x) = U_s\,\varphi(\tilde{x}). \tag{2-8}$$

As is well known, each physical theory has a corresponding *group of symmetry transformations*, in the sense that the dynamical law of the theory is covariant under all the transformations of the group. Here I am using the following notion of covariance. Let $X_1,\ldots,X_n$ be quantities that are functions of space-time coordinates, and $\tilde{X}_1,\ldots,\tilde{X}_n$ be the result of applying a certain group of transformations. A physical law $L\left(X_1,\ldots,X_n\right)=0$ is covariant under that group if it preserves its form under the transformations of the group: $L\left(\tilde{X}_1,\ldots,\tilde{X}_n\right)=0$ (I will come back to this point in detail in Chapter 7). On this basis, the symmetry group corresponding to non-relativistic quantum mechanics is the *Galilei group*.

Under the assumption that time can be represented by a variable $t \in \mathbb{R}$ and position can be represented by a variable $r=(x,y,z) \in \mathbb{R}^3$, the Galilei group $\mathcal{G}=\{T_\alpha\}$, with $\alpha=1$ to 10, is a group of continuous space-time transformations $T_\alpha : \mathbb{R}^3\times\mathbb{R} \to \mathbb{R}^3\times\mathbb{R}$ acting as

- Time-displacement: $t \to \tilde{t} = t+\tau$,
- Space-displacement: $r \to \tilde{r} = r+\rho$,
- Space-rotation: $r \to \tilde{r} = R_\theta r$,
- Velocity-boost: $r \to \tilde{r} = r+ut$,

where $\tau \in \mathbb{R}$ is a real number representing a time interval, $\rho=(\rho_x,\rho_y,\rho_z) \in \mathbb{R}^3$ is a triple of real numbers representing a space interval, $R_\theta \in \mathcal{M}^{3\times3}$ is a 3×3 matrix representing a space rotation by an angle θ, and $u=(u_x,u_y,u_z) \in \mathbb{R}^3$ is a triple of real numbers representing a constant velocity. Since the Galilei group $\mathcal{G}$ is a Lie group, the Galilei transformations T_α can be represented by unitary operators U_α over the Hilbert space, with the exponential parametrization $U_\alpha=e^{iK_\alpha s_\alpha}$, where s_α is a continuous parameter and K_α is the generator of the transformation T_α. Then, $\mathcal{G}$ is defined by ten group generators K_α: one time-displacement K_τ, three space-displacements K_{ρ_i}, three space-rotations K_{θ_i}, and three velocity-boosts K_{u_i}, with $i=x,y,z$. The generators of $\mathcal{G}$ form the Galilei algebra, that is, the Lie algebra of the Galilei generators. The combined action of all the transformations is given by

$$U_s = \prod_{\alpha=1}^{10} e^{iK_\alpha s_\alpha}. \tag{2-9}$$

In the case of quantum mechanics, strictly speaking, the symmetry group is the central extension of the Galilei group, where the central extension of a group is the smallest connected and simply-connected group extending the original group such that any unitary irreducible projective representation of the original group can be obtained from a linear unitary irreducible representation of the central extension. In the case of the Galilei group, the central extension is obtained as a semi-direct product between the Galilei algebra and the algebra generated by a central charge, which in this case corresponds to the mass operator $M = mI$, where I is the identity operator and m is the mass. The mass operator as a central charge is a consequence of the projective representation of the Galilei group. This means that the action of two transformations is equal to the product of the actions of the two transformations up to a phase; for example, $U_{\rho_i u_j} = exp\left(iK_{\rho_i}\right).exp\left(iK_{u_j}\right).exp\left(if(\rho_i, u_j)M\right)$: it is for this reason that a unitary representation of the Galilei group requires the introduction of the central charge $M = mI$ (see Weinberg 1995, Bose 1995). That phase is precisely what disappears in the central extension of the group. In order to simplify the presentation, from now on I will use the expression 'Galilei group' to refer to the corresponding central extension.

As a Lie group, the Galilei group is defined by the commutation relations between its generators:

$$\begin{array}{llll}
(a) & \left[K_{\rho_i}, K_{\rho_j}\right] = 0, & (f) & \left[K_{u_i}, K_{\rho_j}\right] = i\delta_{ij}M, \\
(b) & \left[K_{u_i}, K_{u_j}\right] = 0, & (g) & \left[K_{\rho_i}, K_{\tau}\right] = 0, \\
(c) & \left[K_{\theta_i}, K_{\theta_j}\right] = i\varepsilon_{ijk}K_{\theta_j}, & (h) & \left[K_{\theta_i}, K_{\tau}\right] = 0, \\
(d) & \left[K_{\theta_i}, K_{\rho_j}\right] = i\varepsilon_{ijk}K_{\rho_k}, & (i) & \left[K_{u_i}, K_{\tau}\right] = iK_{\rho_i}, \\
(e) & \left[K_{\theta_i}, K_{u_j}\right] = i\varepsilon_{ijk}K_{u_k}, & &
\end{array} \qquad (2\text{-}10)$$

where ε_{ijk} is the Levi–Civita tensor such that $i \neq k$, $j \neq k$, $\varepsilon_{ijk} = \varepsilon_{jki} = \varepsilon_{kij} = 1$, $\varepsilon_{ikj} = \varepsilon_{jik} = \varepsilon_{kji} = -1$, and $\varepsilon_{ijk} = 0$ if $i = j$. In quantum mechanics, when the system is free from external fields, the generators K_α represent the basic magnitudes of the theory (see Ballentine 1998): the energy $H = \hbar K_\tau$, the three momentum components $P_i = \hbar K_{\rho_i}$, the three angular-momentum components $J_i = \hbar K_{\theta_i}$, and the three boost components $G_i = \hbar K_{u_i}$. Then, by taking $\hbar = 1$, the commutation relations turn out to be

$$\begin{array}{llll}
(a) & \left[P_i, P_j\right] = 0, & (f) & \left[G_i, P_j\right] = i\delta_{ij}M, \\
(b) & \left[G_i, G_j\right] = 0, & (g) & \left[P_i, H\right] = 0, \\
(c) & \left[J_i, J_j\right] = i\varepsilon_{ijk}J_k, & (h) & \left[J_i, H\right] = 0, \\
(d) & \left[J_i, P_j\right] = i\varepsilon_{ijk}P_k, & (i) & \left[G_i, H\right] = iP_i. \\
(e) & \left[J_i, G_j\right] = i\varepsilon_{ijk}G_k, & &
\end{array} \qquad (2\text{-}11)$$

The rest of the physical magnitudes can be defined in terms of these basic ones: for instance, the three position components are $Q_i = G_i/m$, the three orbital angular momentum components are $L_i = \varepsilon_{ijk}Q_jP_k$, and the three spin components are $S_i = J_i - L_i$. The—central extension of the—Galilei group has three Casimir operators which, by definition, commute with all the generators of the group: they are the mass operator M, the spin-squared operator S^2, and the internal-energy operator $W = H - P^2/2m$. The eigenvalues of the Casimir operators label the irreducible representations of the group; so, in each irreducible representation, the Casimir operators are multiples of the identity: $M = mI$, where m is the mass, $S^2 = s(s+1)I$, where s is the eigenvalue of the spin S, and $W = wI$, where w is the scalar internal energy (the relevance of the Casimir operators to the present proposal will be discussed in Chapter 7).

In SQM, to each Galilei transformation T_α corresponds a transformation of states and observables such that (see eqs. (2-5) and (2-6))

$$|\varphi\rangle \rightarrow |\tilde{\varphi}\rangle = U_{s_\alpha}|\varphi\rangle = e^{iK_\alpha s_\alpha}|\varphi\rangle, \tag{2-12}$$

$$O \rightarrow \tilde{O} = U_{s_\alpha} O\, U_{s_\alpha}^{-1} = e^{iK_\alpha s_\alpha} O\, e^{-iK_\alpha s_\alpha}. \tag{2-13}$$

Moreover,

$$\tilde{O} = e^{iK_\alpha s_\alpha} O\, e^{-iK_\alpha s_\alpha} = O \quad \Leftrightarrow \quad [O, K_\alpha] = 0, \tag{2-14}$$

which means that the invariance of an observable O under a Galilei transformation T_α amounts to the commutation between O and the corresponding generator K_α.

2.4 The role of the Hamiltonian

The previous section explained how the Galilei group endows the formal skeleton of SQM with the empirical flesh that allows the theory to be applied to concrete physical situations. In this section, the space-time symmetries will serve to highlight the central role played by the Hamiltonian in the context of the theory.

As seen above, there is an inverse relation between transformations on function space and transformations on coordinates (see eq. (2-8)). In the case of time-displacement, the transformation on coordinates is $t_0 \rightarrow t_0 + \tau$ and, since the Hamiltonian H (strictly speaking, $H/\hbar$) is the generator of the transformation, U_s is $U_\tau = e^{i(H/\hbar)\tau}$. Therefore,

$$|\varphi(t_0)\rangle = e^{i(H/\hbar)\tau}\, |\varphi(t_0 + \tau)\rangle. \tag{2-15}$$

By making $t_0 = 0$ and $\tau = t$ in the above equation, we obtain

$$|\varphi(t)\rangle = e^{-i(H/\hbar)t}\, |\varphi(0)\rangle, \tag{2-16}$$

which has the form of the Schrödinger equation. However, it must be recalled that eq. (2-16) can be obtained only when the Hamiltonian H is *independent of* t: if H is a function of t, in general no simple closed form can be given for U_τ and, as a consequence, H cannot be conceived as the generator of time-displacements (see Ballentine 1989, 89). This means that the time-independence of the Hamiltonian is what endows the Schrödinger equation with a clear physical meaning, precisely, that of expressing time-displacements. In addition, the time-independence of the Hamiltonian is what makes the Schrödinger equation strictly applicable only to *closed systems*. This result is what implicitly supports the orthodox approach to quantum mechanics, where the quantum system is conceived as a closed, constant-energy system, unitarily evolving according to the Schrödinger equation; the Hamiltonian of the system, that is, its energy, only changes with time as the result of the interaction with other systems, described by means of an interaction Hamiltonian. Nevertheless, regarding this topic, non-orthodox approaches have been defended. For example, Nancy Cartwright (1983) proposed to consider a non-unitary generalized master equation for open quantum systems as the dynamical postulate of quantum mechanics instead of the Schrödinger equation. More recently, Michael Cuffaro and Stephan Hartmann (2024) formulated a strongly articulated open-system approach, according to which systems interacting with their environment are conceived of as fundamental, and the environment's influence is represented via the dynamical equations that govern the system's evolution (for a critical perspective, see Lombardi 2024a). The present proposal will not follow this approach: it will be based on the orthodox version of SQM with its closed systems view. I will come back to this point in Chapter 8.

Although in a—closed—quantum system the Hamiltonian H is time-independent and, then, invariant under time-displacements, it may have the remaining space-time symmetries or not. The fact that the Hamiltonian is invariant under a certain symmetry transformation with generator K and parameter s implies that H commutes with K (see eq. (2-14)):

$$\text{If } e^{iKs} H e^{-iKs} = H, \text{ then } [H, K] = 0. \tag{2-17}$$

This means that, when H is invariant under a certain continuous transformation, the generator of that transformation is a *constant of motion* of the system: each symmetry of the Hamiltonian defines a conserved quantity. For instance, the invariance of H under space-displacements in any direction implies that the momentum P is a constant of motion; the invariance of H under space-rotations in any direction implies that the total angular momentum J is a constant of motion. If, by contrast, H is invariant under space-displacements only in one direction, say x, only the component P_x of P is a constant of motion.

In turn, as is well known, the invariance under space-time transformations is closely related to the properties of space and time: invariance under time-displacements expresses the homogeneity of time, invariance under space-displacements expresses the homogeneity of space, and invariance under space-rotations expresses the isotropy of space. In addition, the fact that the Galilei group is the symmetry group of SQM means that the Schrödinger equation must be covariant under the full Galilei group. The covariance of the Schrödinger equation under time-displacements is guaranteed by the time-independence of the Hamiltonian, and this expresses the homogeneity of time. However, in non-relativistic SQM, fields are not quantized, but are treated as classical fields that act on the quantum system by breaking the homogeneity and/or the isotropy of space. Nevertheless, despite this, the dynamical law of the theory, $|\varphi(t)\rangle = e^{-iHt/\hbar}\,|\varphi(0)\rangle$, must remain covariant under space-displacements and space-rotations. Therefore, the breaking of the homogeneity and/or the isotropy of space resulting from the action of fields on the quantum system has to be "contained" in the form of the Hamiltonian: the non-homogeneity of space implies the non-invariance of H under space-displacements and, then, the fact that P (or one of its components) is not a constant of motion; the non-isotropy of space implies the non-invariance of H under space-rotations and, then, the fact that J (or one of its components) is not a constant of motion. In other words, the Hamiltonian H must embody the space asymmetries in each particular situation in order to preserve the covariance of the dynamical law of the theory (the interpretative relevance of this point will be discussed in more detail in Chapter 7).

Summing up, the physical content of SQM is provided by the physical observables that can be identified in terms of the Galilei group. However, not all of them have the same status in the light of the symmetry group. The Hamiltonian holds a privileged position because (i) it is the main character of the dynamical postulate of the theory, (ii) each one of its symmetries defines a constant of motion of the system, and (iii) it embodies the asymmetries resulting from external fields to keep the covariance of the dynamical postulate. These theoretical facts begin to outline the reasons for advocating an interpretation of SQM based on the priority of the Hamiltonian.

3
Modal Interpretations of Quantum Mechanics

Since the aim of this book is to introduce the Modal-Hamiltonian Interpretation (MHI) in a comprehensive way, I will begin by focusing on the family of Modal Interpretations of quantum mechanics, to which the MHI belongs. The purpose of this chapter is, then, to recall the common features of the Modal Interpretations as well as the differences between them (see Lombardi and Dieks 2024 for details). The presentation will be brief because its only purpose is to provide a conceptual map on which the MHI can find its place.

3.1 Bas van Fraassen and the origin of the modal approach

The roots of the Modal Interpretations can be found in the works of Bas van Fraassen (1972, 1974, 1991), who rejected the projection postulate and endowed quantum propositions with a modal content. Under the guidance of these ideas, he distinguished between the "dynamical state" and the "value state" of a system at each time:

- The *dynamical state* represents what *may be* the case: the values that all the physical properties may possess and the corresponding probabilities at the considered time.
- The *value state* determines what *actually is* the case: the values of all the physical properties that are definite valued at the considered time.

The dynamical state is what is commonly conceived as the quantum state of the system, which does not describe actual facts but only possible facts. It never collapses, so always evolves unitarily. The value state, by contrast, describes actual facts. Moreover, quantum systems have definite-valued properties at all times.

The concept of value state involves the idea that an observable of a quantum system may have a definite value even if the dynamical state of the system is not an eigenstate of that observable. This idea contradicts the so-called *eigenstate-eigenvalue link*, according to which a system possesses a definite value of an observable, represented by one of its eigenvalues, if and only if the quantum state is the corresponding eigenstate. Although it usually appears in textbooks of quantum mechanics, the eigenstate-eigenvalue link is not part of the formalism of SQM but an interpretative principle that, like the collapse hypothesis, may or may not be adopted (see Wallace 2008, 2019). Van Fraassen's approach accepts the "if" part, but denies the "only if" part of the link.

The Modal-Hamiltonian Interpretation of Quantum Mechanics. Olimpia Lombardi, Oxford University Press.
© Olimpia Lombardi (2026). DOI: 10.1093/9780198951407.003.0003

In other words, the definite values of the system's observables are not one-to-one represented by the dynamical state, but by the value state, which in the generic case differs from the dynamical state.

Van Fraassen called his proposal "modal" because the dynamical state in general only tells us what is possible and specifies probabilities for the various possibilities. According to him, this modal and probabilistic aspect is fundamental to quantum mechanics and should not be regarded as arising from an incompleteness of the description.

The apparently realistic aspect of van Fraassen's modal proposal may seem strange in the general context of his thinking. As is well known, he does not endorse scientific realism; from his "constructive empiricism," the aim of science is only empirical adequacy. Nevertheless, van Fraassen admits that a meaningful account of reality is necessary for a scientific theory to be intelligible (on this point, see Bueno 2014).

Empirical adequacy requires that the observable corresponding to the measuring apparatus' pointer is assigned a particular definite value when the system is at the end of a measurement process. In this case, the approach has to generate the Born probability measure over the possible measurement results. Van Fraassen's proposal was criticized by claiming that its characterization of measurement is too narrow to account for the processes usually regarded as measurements in laboratory practice (see Ruetsche 1996, Monton 1999). Nevertheless, these original ideas motivated a program directed to provide a more elaborate modal approach to quantum mechanics, which assigns definite properties to physical systems even when they are not subject to measurements.

3.2 General features of the modal family

On the basis of van Fraassen's original ideas, from the 1980s several authors proposed realist interpretations that, in retrospect, can be viewed as elaborations or variations on the quantum modal themes (for an overview and references, see Dieks and Vermaas 1998). Beyond the differences between the members of the modal family, all of them agree on the following points:

- The interpretations are based on the formalism of SQM in its SQM_2 version, that is, without the projection postulate. Thus, the quantum state always evolves unitarily according to the Schrödinger equation: there is no collapse—by contrast to Spontaneous Collapse Theory.
- The interpretations are realist in the sense that they aim at describing how reality would be if SQM were true—by contrast to Qbism.
- Quantum mechanics is a universal theory: it applies both to microscopic and to macroscopic systems. There is no Heisenberg cut—by contrast to Bohr's interpretation.

- Quantum mechanics applies to single systems: the quantum state does not refer to an ensemble of systems, but to a single system—by contrast to Leslie Ballentine's ensemble interpretation (1970).
- The quantum state of the system (pure or mixed) does not describe the actual properties of the system, but the probabilities corresponding to the values of all the physical properties that the system may possess.
- The Schrödinger equation gives the time evolution of probabilities, not of actual properties.
- A quantum measurement is an ordinary physical interaction. Therefore, during measurements the quantum state also evolves unitarily.
- Systems possess actual properties at all times, whether or not a measurement is performed on them.

Whereas the just mentioned features allow the different Modal Interpretations to be classified as belonging to a single family, the fundamental difference between them consists in the choice of the properties that take on definite values. In fact, the Kochen-Specker theorem (Kochen and Specker 1967) proves the impossibility of ascribing precise values to all the observables—properties—of a quantum system simultaneously, while preserving the functional relations between commuting observables. Therefore, any realist non-collapse interpretation is committed to selecting a subset of observables that can be consistently assigned definite values. Each modal interpretation thus supplies a *rule of definite-value ascription* or *actualization rule*, which picks out, from the set of all the observables of the system, the subset of definite-valued observables, which will be called *preferred context*. As will be seen in the next sections, such a subset may be time invariant or may change over time, depending on the particular interpretation considered.

Under the assumption that the actualization rule defines the subset of definite-valued observables exclusively on the basis of the dynamical state, in the mid-1990s, Rob Clifton (1995a, 1995b) and Michael Dickson (1995a, 1995b) concluded that the value states of the components of a two-part composite system should be fixed by the states occurring in the Schmidt—bi-orthogonal—decomposition of the dynamical state of the composite system, or by the projectors occurring in the spectral decomposition of the density matrices—obtained by partial tracing—representing the component systems. As will be explained in the next two sections, these results are at the core of the Biorthogonal-Decomposition Modal Interpretation and the Spectral-Decomposition Modal Interpretation.

The problem of the definite-valued observables was approached from a more general perspective by Jeffrey Bub and Clifton (1996; for an improved version, see Bub, Clifton, and Goldstein 2000). These authors started from the assumption that the subset of definite-valued observables is determined by the dynamical state $|\phi\rangle$ and/or a "privileged observable" R, representing a property that is always definite-valued irrespective of that state. From this premise, Bub (1997; see also 1992, 1994) drew the conclusion that Bohmian Mechanics is a modal interpretation, with position as the privileged observable R. Bub and Clifton (1996) considered the possibility of selecting the pointer of the measuring apparatus as the privileged observable R, in order to provide a straighforward solution to the measurement problem. However, as far as I know, before 2008 nobody conceived the possibility of selecting the Hamiltonian as the privileged observable, a fact that might appear surprising given its special theoretical position as explained at the end of the previous chapter. This would be precisely the strategy of the Modal-Hamiltonian Interpretation (I will come back to this point in more detail in the next chapter).

3.3 The Biorthogonal-Decomposition Modal Interpretation (BDMI)

In the Biorthogonal-Decomposition Modal Interpretation (BDMI), also known as "Kochen-Dieks modal interpretation" (Kochen 1985, Dieks 1988, 1989), the set of the properties that acquire definite values is determined by the quantum—dynamical—state of the system. In particular, the definite-valued observables are picked out by the biorthogonal (Schmidt) decomposition of the state.

> **Biorthogonal Decomposition Theorem**: Given a vector $|\psi\rangle$ of a Hilbert space $\mathcal{H}$ and a factorization of $\mathcal{H}$ in the tensor-product $\mathcal{H}^1 \otimes \mathcal{H}^2$, there exist bases $\{|a_i\rangle\}$ and $\{|p_i\rangle\}$ of $\mathcal{H}^1$ and $\mathcal{H}^2$, respectively, such that $|\psi\rangle$ can be written as a linear combination of terms of the form $|a_i\rangle \otimes |p_i\rangle$:
>
> $$|\psi\rangle = \sum_i c_i \, |a_i\rangle \otimes |p_i\rangle.$$
>
> If the absolute values (moduli) of the coefficients in this linear combination are all unequal, $|c_i| \neq |c_j|$ for $i \neq j$—non-degenerate case—, then the bases are unique (for a proof, see, for example, Schrödinger 1935).

The BDMI applies this theorem to quantum mechanics to state that, given a composite system $\mathcal{S}$, represented by the Hilbert space $\mathcal{H}$ and consisting of two components $\mathcal{S}^1$ and $\mathcal{S}^2$ respectively, represented by the Hilbert spaces $\mathcal{H}^1$ and $\mathcal{H}^2$, the state of $\mathcal{S}$ picks out the bases $\{|a_i\rangle\}$ and $\{|p_i\rangle\}$ (in the non-degenerate case, uniquely) for $\mathcal{H}^1$ and $\mathcal{H}^2$ respectively. According to this interpretation, these bases generate the properties that acquire definite values in the corresponding subsystems. In other words, the actualization rule of the BDMI states that, when the composite system $\mathcal{S}$ is in the pure state $|\psi\rangle$, the definite-valued observables of $\mathcal{S}^1$ and $\mathcal{S}^2$—which define the corresponding preferred contexts—are those that commute with all the projectors $|a_i\rangle\langle a_i|$ and $|p_i\rangle\langle p_i|$, respectively, corresponding to the basis vectors determined by the Biorthogonal Decomposition Theorem.

Since a measurement is an interaction between two systems, the actualization rule of the BDMI applies directly to the measurement situation. Let us consider an ideal measurement according to the standard von Neumann model, in which the system $\mathcal{S}$

to be measured interacts with a measuring apparatus $\mathcal{M}$. Before the interaction, $\mathcal{M}$ is prepared in a ready-to-measure state $|p_0\rangle$, eigenvector of the pointer observable P of $\mathcal{M}$, and the state of $\mathcal{S}$ is a superposition of the eigenstates $|a_i\rangle$ of an observable A of $\mathcal{S}$. The interaction introduces a correlation between the eigenstates $|a_i\rangle$ of A and the eigenstates $|p_i\rangle$ of P:

$$|\psi_0\rangle = \sum_i c_i \, |a_i\rangle \otimes |p_0\rangle \longrightarrow |\psi\rangle = \sum_i c_i \, |a_i\rangle \otimes |p_i\rangle . \tag{3-1}$$

The measurement problem consists in explaining why, being the state $|\psi\rangle$ a superposition of the $|a_i\rangle \otimes |p_i\rangle$, the pointer P acquires a definite value.

According to the actualization rule of the BDMI, in this case the preferred context of the measured system $\mathcal{S}$ is defined by the basis $\{|a_i\rangle\}$ of $\mathcal{H}^S$—the Hilbert space that represents $\mathcal{S}$—and the preferred context of the measuring apparatus $\mathcal{M}$ is defined by the basis $\{|p_i\rangle\}$ of $\mathcal{H}^M$—the Hilbert space that represents $\mathcal{M}$. More precisely, after the interaction:

- The measured observable A is a definite-valued observable of $\mathcal{S}$, and acquires one of its possible values (eigenvalues) a_i.
- The pointer observable P is a definite-valued observable of $\mathcal{M}$, and acquires one of its possible values (eigenvalues) p_i.

Although the BDMI is characterized by the central role played by the Biorthogonal Decomposition Theorem, two versions can be distinguished, which differ regarding ontological assumptions. According to one of them, proposed by Simon Kochen (1985), all quantum properties are relational since they are mediated by the relation of "witnessing": properties are not possessed by the system absolutely, but only when it is "witnessed" by another system. In the case of the measurement process described above, the pointer P of $\mathcal{M}$ "witnesses" the value acquired by the observable A of $\mathcal{S}$.

According to the other version, adopted by Dennis Dieks (1988, 1989) in his first works on the matter, the properties ascribed to the system do not have a relational nature. This view faces, however, a consistency problem with respect to the observables that acquire definite values when a composite system is split up into components in different ways. Let us consider, for example, a composite system with three components $\mathcal{S}^1$, $\mathcal{S}^2$, and $\mathcal{S}^3$. The biorthogonal decomposition can be applied to the two-component system (i) $\mathcal{S}^1 \cup (\mathcal{S}^2 \cup \mathcal{S}^3)$, or (ii) $\mathcal{S}^2 \cup (\mathcal{S}^1 \cup \mathcal{S}^3)$, or (iii) $\mathcal{S}^3 \cup (\mathcal{S}^1 \cup \mathcal{S}^2)$. Suppose that, as a result of this application, in case (i) the observable P of the component $\mathcal{S}^1$ is definite-valued, in case (ii) the observable Q of the component $\mathcal{S}^2$ is definite-valued, and in case (iii) the observable R of the component $\mathcal{S}^1 \cup \mathcal{S}^2$ is definite-valued. In this situation, the BDMI does not specify how the definite-valued observables of $\mathcal{S}^1$ and $\mathcal{S}^2$ relate to those of $\mathcal{S}^1 \cup \mathcal{S}^2$. For instance, it cannot be decided whether $P \otimes I^2$ and $I^1 \otimes Q$—where I^1 and I^2 are the identity observables corresponding to $\mathcal{S}^1$ and $\mathcal{S}^2$, respectively—, or R, or the three observables of the component $\mathcal{S}^1 \cup \mathcal{S}^2$

are definite-valued. This consistency problem was addressed by different authors during the 1990s (see Bacciagaluppi 1996, Vermaas 1999). Perhaps it was this problem that motivated Dieks to modify his original modal approach in favor of his Perspectival Modal Interpretation, which will be introduced later in this chapter.

3.4 The Spectral-Decomposition Modal Interpretation (SDMI)

The Spectral-Decomposition Modal Interpretation (SDMI), proposed by Pieter Vermaas and Dieks (1995), is a generalization of the BDMI to density operators. Instead of relying on the biorthogonal decomposition of the quantum state, the SDMI appeals to the spectral decomposition of the density operator representing the—pure or mixed—state of the system.

Spectral Decomposition Theorem: Any self-adjoint operator ρ on a Hilbert space $\mathcal{H}$ can be decomposed as

$$\rho = \sum_i \rho_i \Pi_i,$$

where the ρ_i are the eigenvalues and the Π_i are the eigenprojectors of ρ.

On the basis of this theorem, the actualization rule of the SDMI states that, when a system $\mathcal{S}$, represented by a Hilbert space $\mathcal{H}$, is in a state ρ, the definite-valued observables of $\mathcal{S}$ are the projectors Π_i defined by the Spectral Decomposition Theorem and those commuting with them.

Under the assumption that reduced states are the quantum states of open systems, the SDMI actualization rule can be applied to the measurement situation as described in the previous section. The interaction between the system $\mathcal{S}$ and the apparatus $\mathcal{M}$ leads the composite system to state $|\psi\rangle$ (see eq. (3-1)). Then, the reduced states of $\mathcal{S}$ and $\mathcal{M}$ are, respectively,

$$\rho_r^{\mathcal{S}} = Tr_{(\mathcal{M})} |\psi\rangle\langle\psi| = \sum_i |c_i|^2 |a_i\rangle\langle a_i| = \sum_i |c_i|^2 \Pi_i^a, \quad (3\text{-}2)$$

$$\rho_r^{\mathcal{M}} = Tr_{(\mathcal{S})} |\psi\rangle\langle\psi| = \sum_i |c_i|^2 |p_i\rangle\langle p_i| = \sum_i |c_i|^2 \Pi_i^p. \quad (3\text{-}3)$$

According to the SDMI, the definite-valued observables of $\mathcal{S}$ and $\mathcal{M}$—which define the corresponding preferred contexts—are those that commute with all the projectors $|a_i\rangle\langle a_i|$ and $|p_i\rangle\langle p_i|$, respectively. Therefore, also in the SDMI the observables A of $\mathcal{S}$ and P of $\mathcal{M}$ acquire definite values, and the probabilities of the corresponding eigenvalues are given by the coefficients of the diagonalized reduced states.

The SDMI agrees with the BDMI in cases in which this last one applies, and generalizes it by fixing the definite-valued observables in terms of multi-dimensional projectors when the biorthogonal decomposition is degenerate and the Biorthogonal Decomposition Theorem does not guarantee unicity.

3.5 Non-ideal measurements: A challenge for the BDMI and the SDMI

As shown in the previous sections, the BDMI and the SDMI solve the measurement problem in the case of the ideal von Neumann measurement, which introduces a perfect correlation between the eigenstates $|a_i\rangle$ of the observable A of the system $\mathcal{S}$ to be measured and the eigenstates $|p_i\rangle$ of the pointer P of the measuring apparatus $\mathcal{M}$. However, ideal measurements can never be achieved in practice: the interaction between $\mathcal{S}$ and $\mathcal{M}$ never produces a perfect correlation between A and P. Two kinds of non-ideal measurements are usually distinguished in the literature:

- *Imperfect measurement (first kind)*:

$$\sum_i c_i\,|a_i\rangle \otimes |p_0\rangle \quad \longrightarrow \quad \sum_{ij} d_{ij}\,|a_i\rangle \otimes |p_j\rangle\,,\ \text{where, in general,}\ d_{ij} \neq 0 \text{ with } i \neq j. \tag{3-4}$$

- *Disturbing measurement (second kind)*:

$$\sum_i c_i\,|a_i\rangle \otimes |p_0\rangle \quad \longrightarrow \quad \sum_i c_i\,\left|a_i^d\right\rangle \otimes |p_i\rangle\,,\ \text{where, in general,}\ \left\langle a_i^d\,\middle|a_j^d\right\rangle \neq \delta_{ij}. \tag{3-5}$$

However, the disturbing measurement can also be expressed as an imperfect measurement, and vice versa, by a change of basis:

$$\sum_i c_i\,\left|a_i^d\right\rangle \otimes |p_i\rangle = \sum_{ij} d_{ij}\,|a_i\rangle \otimes |p_j\rangle\,. \tag{3-6}$$

Non-ideal measurements pose a serious challenge to the BDMI and the SDMI, because their actualization rules for selecting the definite-valued observables do not pick out the right properties for the apparatus when the measurement is not ideal (see Albert and Loewer 1990, 1991, 1993; see also Ruetsche 1995). Let us consider that a non-perfectly correlated state is diagonalized so as to make it possible to apply the biorthogonal decomposition:

$$\sum_{ij} d_{ij}\,|a_i\rangle \otimes |p_j\rangle = \sum_i c'_i\,|a'_i\rangle \otimes |p'_i\rangle\,. \tag{3-7}$$

According to the BDMI and the SDMI, the diagonalized state does not select the pointer P as a definite-valued observable, but a different observable P' with eigenstates $|p'_i\rangle$. In this case, in which the selected definite-valued observable is different from that expected, the question arises as to what extent they are different from each other. In the case of an imperfect measurement, it may be assumed that the $d_{ij} \neq 0$, with

$i \neq j$, are small; then, the difference may also be small. But in the case of a disturbing measurement, the $d_{ij} \neq 0$, with $i \neq j$, need not be small and, as a consequence, the disagreement between the BDMI/SDMI selection and the experimental result might be unacceptable (see a full discussion in Bacciagaluppi and Hemmo 1996).

An example that clearly brings out the difficulties posed by non-ideal measurements to the BDMI and the SDMI was formulated in the context of Stern-Gerlach experiments (Elby 1993). This argument appeals to the fact that the wavefunctions in the vertical z-variable typically have infinite "tails" that introduce non-zero cross-terms; therefore, the "tail" of the wavefunction of the "down" beam may produce detection in the upper detector, and vice versa (see Dickson 1994 for a detailed discussion).

These challenges were considered as a "silver bullet" for killing Modal Interpretations of the BDMI and the SDMI types (Harvey Brown, cited in Bacciagaluppi and Hemmo 1996). Perhaps this opinion was one of the determining factors in the declining interest of philosophers of physics in Modal Interpretations from the 2000s onward.

Some authors have considered that the solution to the above difficulties rests on the appeal to quantum decoherence. According to the environment-induced approach to decoherence proposed by Wojciech Zurek (1981, 2003), any macroscopic measuring apparatus is an open system in continuous interaction with its environment; as a consequence of this interaction, the reduced state of the apparatus becomes, almost instantaneously, indistinguishable from a state that would represent a quantum mixed state, which can be interpreted in terms of ignorance over unknown but definite values of the apparatus' macroscopic pointer (see discussion in Schlosshauer 2007). The idea that decoherence might play an important role in Modal Interpretations, by making macroscopic quantities definite, was suggested by several authors early on (Dieks 1989, Healey 1995). In particular, considerations about decoherence acquired central relevance precisely in the discussion about non-ideal measurements, in which the BDMI and the SDMI do not generally pick out the right definite-valued observables for the measuring apparatus. In the context of this discussion, Guido Bacciagaluppi and Meir Hemmo (1996) showed that, when the apparatus is a finite-dimensional system in interaction with an environment of a huge number of degrees of freedom, decoherence guarantees that the spectral decomposition of the reduced state of the apparatus will lead to a result very close to the ideally expected result: as a consequence, an observable indistinguishable from the apparatus pointer will be selected as definite-valued.

This solution to the problem of non-ideal measurements depends on interpreting the reduced state of an open system as its quantum state. However, this view can be challenged by arguing that the reduced state of an open system is not a quantum state because reduced states may evolve non-unitarily and, therefore, do not obey the dynamical postulate of quantum mechanics. As Maximilian Schlosshauer emphasizes in his well-known book about decoherence, strictly speaking a reduced density operator is only "a calculational tool for computing the probability distribution of a set of possible outcomes of measurements, but they do not specify the *state* of the system." (Schlosshauer 2007, 48; emphasis in the original). In particular, it can be proved that reduced states can be conceived as a kind of coarse-grained states of a closed composite system, which disregard certain degrees of freedom considered as irrelevant (see Fortin and Lombardi 2014). From this perspective, it is not conceptually appropriate to appeal to a coarse-grained state to solve the problem concerning how the—fine-grained—quantum state defines the definite-valued observables of a system. These arguments are not further discussed at this point because they will be presented in detail in Sections 8.3 and 8.4 of Chapter 8. Nevertheless, here is the place to point out that this perspective on reduced states not only defies the solution to the problem of non-ideal measurements that appeals to decoherence, but also goes against the central idea of the SDMI, since it applies its actualization rule to the reduced states of open systems.

3.6 Consecutive measurements: A challenge for the BDMI and the SDMI

As seen in the two previous sections, the reduced states of interacting systems play a central role in the SDMI. This section will show why the SDMI actualization rule cannot explain the correlations between the outcomes of consecutive measurements.

Following a classification originally due to Pauli (1933), the literature on quantum mechanics usually distinguishes between *first kind* measurements—the interaction neither disturbs nor destroys the measured system—and *second kind* measurements—the interaction disturbs or destroys the measured system. Let us consider a first kind measurement under the standard von Neumann model, as described by eq. (3-1). The first question is to explain why the pointer P acquires a definite value: this is the traditional measurement problem. A further question is what result is obtained when a second measurement of the same observable is performed on the same system. As is well known in practice, first kind measurements yield the same result on immediate repetition. In Dirac's words: "When we measure a real dynamical variable ξ, [. . .] if we make a second measurement on the same dynamical variable ξ immediately after the first, the result of the second measurement must be the same as that of the first. Thus, after the first measurement has been made, there is no indeterminacy in the result of the second." (Dirac 1958, 36).

A paradigmatic example is given by consecutive Stern-Gerlach experiments: no matter what interpretation of the theory one uses to explain them, any physicist knows that, if the particle outcoming the first magnet is deflected in the $+z$-direction ($-z$-direction), a second magnet in the same direction will deflect it again in the $+z$-direction ($-z$-direction) with certainty. The non-trivial correlations between the outcomes of consecutive measurements of different observables are also well known: for instance, if the particle is deflected by the first magnet in the $+z$-direction

($-z$-direction), a second magnet in the x-direction will deflect it in the $+x$-direction or in the $-x$-direction with a probability of 0.5 for each case.

According to collapse interpretations, in each particular detection, the pure state $|\psi\rangle$ of eq. (3-1) indeterministically "collapses" to a component of the superposition. Let us suppose that, in the first measurement, that pure state collapses to $|a_k\rangle \otimes |p_k^1\rangle$, where $|p_k^1\rangle$ is an eigenstate of the pointer P^1 of the first measuring apparatus $\mathcal{M}^1$. This situation is interpreted as saying that the system $\mathcal{S}$ is in the state $|a_k\rangle$ and the apparatus $\mathcal{M}^1$ is in the state $|p_k^1\rangle$ with certainty; then, the observable A of $\mathcal{S}$ acquires the definite value a_k and the pointer P^1 of $\mathcal{M}^1$ acquires the definite value p_k^1 with certainty. The collapse hypothesis gives a straightforward account of the agreement between the outcomes of consecutive measurements of the same observable. In fact, a second measurement of the observable A will maintain the system $\mathcal{S}$ in the same state $|a_k\rangle$ resulting from the collapse in the first measurement. This means that the conditional probability of obtaining the value p_k^2 of the pointer P^2 of the apparatus $\mathcal{M}^2$ in the second measurement, given that the value p_k^1 of P^1 of $\mathcal{M}^1$ was obtained in the first measurement, is $pr(p_k^2 \, | p_k^1) = 1$.

The collapse hypothesis also offers a simple explanation to the non-trivial correlations between the outcomes of consecutive measurements of different observables (see, e.g., Messiah 1961, Cohen-Tannoudji et al. 1977). Let us consider, for example, a Stern-Gerlach experiment, where the observable A to be measured is the spin S^z in z-direction, with eigenvectors $|z_+\rangle$ and $|z_-\rangle$, and the role of the pointer P is played by the particle's momentum P^z in z-direction, with eigenvectors $|p_0^z\rangle$, $|p_+^z\rangle$, and $|p_-^z\rangle$. In general, the first interaction leads to a superposition $|\psi\rangle = c_+ \, |z_+\rangle \otimes |p_+^z\rangle + \; c_- \, |z_-\rangle \otimes |p_-^z\rangle$; so, the first measurement collapses it to, say, $|z_+\rangle \otimes |p_+^z\rangle$. In this case, the result of a second measurement of S^z will be the value p_+^z of P^z with certainty. But if the second measurement is performed on the spin S^x in x-direction, the state $|z_+\rangle$ must be expressed in the eigenbasis $\{|x_+\rangle, |x_-\rangle\}$ of the new observable,

$$|z_+\rangle = {}^1\!/\!\sqrt{2}\; (|x_+\rangle + |x_-\rangle), \tag{3-8}$$

and the interaction correlates $|x_+\rangle, |x_-\rangle$ with the eigenvectors $|p_+^x\rangle, |p_-^x\rangle$ of the momentum P^x in x-direction, respectively. Then, the second measurement collapses the second superposition to, say, $|x_+\rangle \otimes |p_+^x\rangle$ with probability 0.5. Therefore, the conditional probability of obtaining the value p_+^x of the momentum P^x in x-direction in the second measurement, given that the value p_+^z of the momentum P^z in z-direction was obtained in the first measurement, is 0.5, as experimentally expected.

In the SMDI (and, as a consequence, in the particular case of the BMDI), reduced states are conceived as the quantum mixed states of the measured system and of the measuring apparatus after the interaction. As Dieks asserts: "The projection operator $|\psi\rangle\langle\psi|$ is an observable of the total system [...]. But we are really interested in the individual properties of device and object taken by themselves. Therefore, we need substitutes for $|\psi\rangle\langle\psi|$ that represent the states of these individual systems. In the context of standard quantum mechanics such operators are readily available, namely the density operators for the partial systems" (Dieks 2007, 298); "we took [the reduced operator] W_1 for the state of system 1. This is standard practice in quantum mechanics; however, the usual justification relies on the probabilistic interpretation of the theory and the Born rule" (Dieks 2007, 299). Despite its apparent naturalness and the good results that it provides in the case of a single ideal measurement, the assumption that the reduced operator represents the quantum state of a system is not innocuous when consecutive measurements are considered (see Ardenghi, Lombardi, and Narvaja 2013). By using $|\psi^1\rangle$ to denote the state of the composite system $\mathcal{S} \cup \mathcal{M}^1$, let us take seriously that, in the case of consecutive measurements, the reduced operators

$$\rho_r^S = Tr_{(M^1)} |\psi^1\rangle\langle\psi^1| = \sum_i |c_i|^2 |a_i\rangle\langle a_i|, \tag{3-9}$$

$$\rho_r^{M^1} = Tr_{(S)} |\psi^1\rangle\langle\psi^1| = \sum_i |c_i|^2 |p_i^1\rangle\langle p_i^1| \tag{3-10}$$

really represent the quantum states of the measured system $\mathcal{S}$ and of the first measuring system $\mathcal{M}^1$, respectively. According to the SDMI, the pointer P^1 of $\mathcal{M}^1$ is a definite-valued observable which, as a consequence, indeterministically acquires one of its possible values, say p_k^1. But now the reduced state ρ_r^S is the state of the system $\mathcal{S}$ before the second measurement, when the second measuring apparatus $\mathcal{M}^2$ is in the ready-to-measure state $|p_0^2\rangle$, eigenstate of the pointer P^2 of $\mathcal{M}^2$. So, a new correlation is established when $\mathcal{S}$ interacts with $\mathcal{M}^2$:

$$\rho_0^2 = \rho_r^S \otimes |p_0^2\rangle\langle p_0^2| \quad \longrightarrow \quad \rho^2 = \sum_i |c_i|^2 |a_i\rangle\langle a_i| \otimes |p_i^2\rangle\langle p_i^2|. \tag{3-11}$$

Then, after the second interaction, the analogous account of measurement applies: $\rho_{r(2)}^S$ is the reduced operator corresponding to the system $\mathcal{S}$ and $\rho_r^{M^2}$ is the reduced operator corresponding to the second measuring apparatus $\mathcal{M}^2$:

$$\rho_{r(2)}^S = Tr_{(M^2)}\, \rho^2 = \sum_i |c_i|^2 |a_i\rangle\langle a_i|, \tag{3-12}$$

$$\rho_r^{M^2} = Tr_{(S)}\, \rho^2 = \sum_i |c_i|^2 |p_i^2\rangle\langle p_i^2|, \tag{3-13}$$

where the $|p_i^2\rangle$ are eigenstates of the pointer P^2 of $\mathcal{M}^2$. Therefore, the pointer P^2 indeterministically acquires one of its possible values, say p_l^2. The problem is that the value p_l^2 acquired by the pointer P^2 of $\mathcal{M}^2$ in the second measurement has no relation to the value p_k^1 acquired by the pointer P^1 of $\mathcal{M}^1$ in the first measurement because partial traces cancel the correlations between the interacting systems. As a consequence, correlations between consecutive measurements remain unexplained.

In summary, the SDMI cannot explain the agreement and the correlations between the readings of the pointers involved in consecutive measurements because reduced states are conceived as the legitimate quantum states of the interacting systems.

The issue of how to interpret reduced states will be discussed in detail in Sections 8.3 and 8.4 of Chapter 8.

3.7 The Atomic Modal Interpretation (AMI)

As explained in Sections 3.3, the BDMI (and its generalized version SDMI) leads to a consistency problem about the observables that acquire definite values when the total system is split into components in different ways. This problem can be generalized as follows. The Hilbert space of the universe $\mathcal{H}^{univ}$, like any Hilbert space, can be factorized in countless ways. Then, if each factorization is assumed to define a set of basic or atomic subsystems, the multiple factorizability of a Hilbert space implies that there is a multiplicity of ways to define the building blocks of nature. A theorem due to Bacciagaluppi (1995; see also Vermaas 1997) shows that if one applies the SDMI to every factorization and insists that the values of the observables corresponding to the different factorizations are factorization-independent, then one ends up with a mathematical contradiction of the Kochen-Specker type.

The Atomic Modal Interpretation (AMI), proposed by Bacciagaluppi and Dickson (1999), attempts to overcome this obstacle by assuming that, in nature, there is a set of mutually disjoint atomic quantum systems $\mathcal{S}^j$ that are the elementary subsystems of all the quantum systems of the universe. In other words, the set of the building blocks of the quantum reality is one and only one. From the mathematical point of view, this means that there is a "preferred" factorization of the Hilbert space $\mathcal{H}^{univ}$ of the entire universe. If each atomic quantum system $\mathcal{S}^j$ is represented by its corresponding Hilbert space $\mathcal{H}^j$, then the Hilbert space $\mathcal{H}^{univ}$ is expressed as

$$\mathcal{H}^{univ} = \mathcal{H}^1 \otimes \mathcal{H}^2 \otimes \ldots \otimes \mathcal{H}^N. \tag{3-14}$$

The definite-value ascription to the atomic quantum systems in the AMI follows the general idea of the traditional Modal Interpretations, in which the actualization rule depends on the state of the system. According to SQM, if ρ^j is the reduced state of the atomic quantum system $\mathcal{S}^j$, then the probability that some projector Π_a^j on $\mathcal{H}^j$ has value 1 is computed as $Tr\left(\rho^j \Pi_a^j\right)$. The AMI assigns the definite value 1 to the projector Π_a^j when that probability is $Tr\left(\rho^j \Pi_a^j\right) = 1$, that is, if and only if Π_a^j is an eigenprojector of ρ^j.

In turn, the definite-value ascription for a non-atomic quantum system $\mathcal{S}^\sigma$ is defined by means of the reduced states of its atomic subsystems. In fact, if the composite quantum system $\mathcal{S}^\sigma$ consists of the atomic components $\mathcal{S}^1, \mathcal{S}^2, \ldots, \mathcal{S}^n$, and ρ^σ is the state of $\mathcal{S}^\sigma$, then the reduced states of the atomic components can be computed by applying partial traces on ρ^σ, so that the eigenprojectors of those reduced states are the Π_i^1, the $\Pi_j^2, \ldots$, the Π_k^n, respectively. The AMI applied to $\mathcal{S}^\sigma$ assigns a definite value 1 to the projector $\Pi_\alpha^\sigma = \Pi_a^1 \otimes \Pi_b^2 \otimes \ldots \otimes \Pi_z^n$ if and only if the interpretation applied to the atomic quantum subsystems $\mathcal{S}^1, \mathcal{S}^2, \ldots, \mathcal{S}^n$ assigns simultaneously the definite value 1 to the projectors $\Pi^1_a, \Pi^2_b, \ldots, \Pi^n_z$. Since the atomic quantum systems are mutually disjoint, the probability that the projector Π^σ_α of $\mathcal{S}^\sigma$ has value 1, that is, the joint probability that all the projectors $\Pi^1_a, \Pi^2_b, \ldots, \Pi^n_z$ have value 1, can be computed as

$$Tr\left(\rho^\sigma \Pi^1_a \otimes \Pi^2_b \otimes \ldots \otimes \Pi^n_z\right). \tag{3-15}$$

This implies that the properties of composite non-atomic quantum systems can be deduced from the properties of their atomic components. In mathematical terms, this means that the algebra of properties for a composite system is just the Cartesian product of the algebras of properties for the component atomic subsystems (see Bacciagaluppi and Dickson 1999).

Despite its simplicity—or perhaps precisely due to this simplicity—, the AMI leads to a conceptual problem: a non-atomic quantum system $\mathcal{S}^\sigma$, defined as composite of atomic quantum systems, does not necessarily have the properties that measurement in quantum mechanics predicts. The reason is the following. Let us suppose that the system $\mathcal{S}^\sigma$ is in the quantum state ρ^σ with an eigenprojector Π^σ such that $Tr\left(\rho^\sigma \Pi^\sigma\right) = 1$. This implies that if the property represented by Π^σ is measured, a positive outcome with probability 1 is obtained. But it may be the case that the projector Π^σ is not composite of atomic properties and, therefore, according to the AMI, it is not a property possessed by the composite quantum system $\mathcal{S}^\sigma$.

Two answers to this conceptual problem have been proposed. The first one was given by Clifton (1996), who admits the existence of dispositional properties, that is, properties that may have probability 1 but, nevertheless, are not revealed by measurements. The second answer is that proposed by Dieks (1998), who claims that the value 1 of the projector Π^σ of the composite system $\mathcal{S}^\sigma$ shows that $\mathcal{S}^\sigma$ has a collective dynamical effect onto the measuring apparatus, that is, an effect that cannot be explained by the action of the atomic components.

Nevertheless, the main challenge for the AMI is to justify the assumption that there is a preferred partition of the universe, and to provide some idea about how such a factorization should look like. Perhaps the absence of a satisfactory answer to this question explains why the AIM has not aroused greater interest in the philosophy of physics community.

3.8 The Perspectival Modal Interpretation

Let us take a composite system $\mathcal{S}^{\alpha\beta}$, whose component subsystems $\mathcal{S}^\alpha$ and $\mathcal{S}^\beta$ are represented by the Hilbert spaces $\mathcal{H}^\alpha$ and $\mathcal{H}^\beta$, respectively, and consider a property represented by the projector Π^α defined on $\mathcal{H}^\alpha$. It is usual to assume that Π^α represents the same property as that represented by $\Pi^\alpha \otimes I^\beta$ defined on $\mathcal{H}^\alpha \otimes \mathcal{H}^\beta$, where I^β is the identity operator on $\mathcal{H}^\beta$. This assumption is based on the observational indistinguishability of the magnitudes represented by Π^α and by $\Pi^\alpha \otimes I^\beta$: if the Π^α-measurement on $\mathcal{S}^\alpha$ has a certain outcome, then the $\Pi^\alpha \otimes I^\beta$-measurement on $\mathcal{S}^{\alpha\beta}$ has exactly the same outcome. Now the question arises whether the BDMI and the SDMI, applied to the component system $\mathcal{S}^{\alpha}$, assign a value to Π^{α} that is the same as the value assigned to $\Pi^{\alpha} \otimes I^{\beta}$ when the interpretations are applied to the composite system $\mathcal{S}^{\alpha\beta}$ (Property Composition), and vice versa (Property Decomposition). The answer to this question is negative: the BDMI and the SDMI violate Property Composition and Property Decomposition (for a proof, see Vermaas 1998).

If the idea that the projectors Π^{α} and $\Pi^{\alpha} \otimes I^{\beta}$ represent the same property is retained (for a different position, see van Fraassen 1991), the violation of Property Composition and Property Decomposition is a serious problem. In fact, Frank Arntzenius (1990) finds the violation of Property Composition to be bizarre, since different truth values are assigned to propositions like "the left-hand side of the table is green" and "the table has a green left-hand side" (a similar argument is put forward by Clifton 1996; see also Clifton 1995c).

The above difficulties depend not only on the peculiarities of the BDMI and the SDMI, but also on assuming that quantum properties are monadic. Therefore, such difficulties can be avoided if an ontology of relational properties is adopted. An early relational modal interpretation was proposed by Kochen (1985); as already mentioned in Section 3.3 of the present chapter, according to him, in a bi-partite system one system possess properties "as witnessed" by the other system. Still in the modal framework, Gyula Bene and Dieks (2002) developed a more general and articulated relational view, the Perspectival Modal Interpretation (PMI), in which the properties of any physical system are defined with respect to other physical systems that serve as "reference systems" (Bene 1997). The PMI is closely related to the SDMI since similar rules are used to assign properties to quantum systems.

More precisely, according to the PMI, the state of $\mathcal{S}$ with respect to the reference system $\mathcal{R}$ is denoted by ρ_{R}^{S}. In the special case in which $\mathcal{R}$ is the same as $\mathcal{S}$, the state ρ_{S}^{S} is called "the state of $\mathcal{S}$ with respect to itself." If the system $\mathcal{S}$ is contained in a system $\mathcal{Q}$, the state ρ_{Q}^{S} is defined as the density operator that can be computed from ρ_{Q}^{Q} by taking the partial trace over the degrees of freedom in $\mathcal{Q}$ that do not pertain to $\mathcal{S}$:

$$\rho_{Q}^{S} = Tr_{(Q/S)}\, \rho_{Q}^{Q}. \tag{3-16}$$

The PMI begins with the dynamical state of the whole universe $\mathcal{U}$ with respect to itself, which it is assumed to be a pure state, $\rho_{U}^{U} = |\psi\rangle\langle\psi|$, evolving unitarily according to the Schrödinger equation. For any system $\mathcal{S}$ contained in the universe, its state with respect to itself, ρ_{S}^{S}, is postulated to be one of the projectors of the spectral resolution of ρ_{U}^{S}, where

$$\rho_{U}^{S} = Tr_{(U/S)}\, \rho_{U}^{U} = Tr_{(U/S)}\, |\psi\rangle\langle\psi|\,. \tag{3-17}$$

In particular, if there is no degeneracy in the eigenvalues of ρ_{U}^{S}, ρ_{S}^{S} is the one-dimensional projector $|\psi_{S}\rangle\langle\psi_{S}|$.

The ontological idea behind the PMI is that different relational descriptions, given from different perspectives, are to be considered as equally objective. Physical reality has a relational character and perspectival states cannot be reduced to more fundamental non-relational states. Therefore, it may happen that a system is localized from the perspective of one observer and, nevertheless, is delocalized from a different perspective. However, when different observers look at the same macroscopic object, this effect will be suppressed due to decoherence effects (for a criticism of the role of decoherence in explaining how the relativity of macroscopic facts can be ignored, see Muciño, Okon, and Sudarsky 2022).

According to the PMI, the course of events in experiments of the EPR-type may be described in a local manner. Indeed, the change in the relational state of particle 2 with respect to particle 1 can be understood as a consequence of the change in the reference system brought about by the local measurement interaction between particle 1 and the measuring device. This local measurement is responsible for the creation of a new perspective, and from this new perspective there is a new relational state of particle 2 (Dieks 2009, 2019). A conceptually similar argument can be applied to account for a non-collapse treatment of Wigner's friend paradox (Dieks 2019, 2021).

The PMI agrees with Bohr's statement that the definition of physical reality in the quantum realm should include the experimental context (see Dieks 2017 for a detailed discussion of Bohr's position and a comparison with the modal approach). However, the PMI is more general and technically precise, since it defines the state of a system with respect to arbitrary contexts, not necessarily related to measurements. This removes the threat of subjectivism, as the relational states follow unambiguously from the quantum formalism and the physics of the situation.

4

The Postulates of the Modal-Hamiltonian Interpretation

Traditional Modal Interpretations, such as the BDMI and the SDMI, were specifically designed to solve the quantum measurement problem, and they successfully achieved their purpose in the case of ideal measurements. However, as discussed in the previous chapter, a series of articles of the 1990s showed that those approaches did not pick out the right properties for the apparatus in non-ideal measurements. This might explain, at least partially, the decline of interest in Modal Interpretations since the end of the 1990s.

However, the difficulties of those original Modal Interpretations were due not to their modal nature, but to the fact that their actualization rule made the set of definite-valued observables to depend on the instantaneous state of the system. Jeffrey Bub (1992, 1997) noticed this fact, and stressed that the preferred context of definite-valued observables can be defined solely by a "privileged observable" representing a property that is always definite-valued irrespective of the state. From this perspective, Bohmian Mechanics turns out to be a modal interpretation, according to which the preferred context is a priori defined by the position observable. The fact that this becomes a natural alternative given the difficulties of the traditional Modal Interpretations maybe explains Bub's preference for Bohmian Mechanics during that decade. But as far as I know, nobody emphasized that this is not the only reasonable possibility for a modal interpretation with a state-independent and fixed "privileged" observable. In fact, as will be explained in the present chapter, the Modal-Hamiltonian Interpretation (MHI) endows the Hamiltonian with a determining role, both in the definition of systems and subsystems and in the selection of the preferred context.

In this chapter, the MHI will be introduced in a neutral way, without discussing its advantages over other proposals. The arguments supporting it will become clear in the following chapters, where its physical relevance will be explained (Chapter 5) and it will be applied to solve traditional interpretative challenges (Chapter 6).

4.1 Properties and systems

As stated in Chapter 1, from a realist perspective, providing an ontological interpretation of a theory amounts to saying how reality would be if the theory were true. Although, in general, physicists agree in their use of physical language, the relation between physical language and reality is not a self-evident matter: physical

The Modal-Hamiltonian Interpretation of Quantum Mechanics. Olimpia Lombardi, Oxford University Press.
© Olimpia Lombardi (2026). DOI: 10.1093/9780198951407.003.0004

theories do not provide their own ontological interpretations. Therefore, in order to endow quantum mechanics with an ontological interpretation, it is necessary to formulate *interpretative postulates* IP that establish the ontological reference of each of the theoretical elements introduced by the representative postulates RP and the quantum postulates QP presented in Section 2.2 of Chapter 2 (see Lombardi and Castagnino 2008, Castagnino and Lombardi 2008). Such interpretative postulates will specify which kind of items in the ontology (objects, properties, facts, etc.) is the reference of each physical term in the RP and the QP (systems, observables, states, etc.). In other words, they will describe what ontological categories populate the quantum-mechanical reality.

On the basis of how a quantum system was characterized by RP1 in Section 2.2 of Chapter 2, the first step in the presentation of the MHI is to identify the fundamental items of the ontology: *properties*.

> **IP1 (Properties):** Given a quantum system $\mathcal{S} : (\mathcal{O}, H)$, the observables $O \in \mathcal{O}$ ontologically represent *type-properties* $[O]$, and their corresponding eigenvalues o_i ontologically represent *case-properties* $[o_i]$ of the type-property $[O]$. In particular, the projectors $\Pi \in \mathcal{O}$ are observables that ontologically represent type-properties $[\Pi]$ with case-properties $[1]$ and $[0]$.

Here I advance that the difference between type-property and case-property corresponds, with a subtle difference, to the traditional philosophical distinction between determinable and determinate properties. The explanation of the reason for choosing the "type-case" terminology, as well as the detailed discussion of this distinction, are postponed until Chapter 10.

From the algebraic perspective of the MHI, a quantum system $\mathcal{S}$ is mathematically represented by a space $\mathcal{O}$ of observables, among which the Hamiltonian H plays a definitional role. Since the fundamental items of the ontology are type-properties with their case-properties, whose counterparts in the physical language are observables with their eigenvalues, the ontological counterpart of a space of observables and, as a consequence, of a quantum system is a bundle of properties.

> **IP2 (Systems):** Given a quantum system $\mathcal{S} : (\mathcal{O}, H)$, its ontological counterpart is a *bundle* of the type-properties $[O]$, represented by the observables $O \in \mathcal{O}$, with their corresponding case-properties $[o_i]$, represented by the eigenvalues o_i of each $O \in \mathcal{O}$.

Bundle theories emerged in the framework of the Humean rejection of the existence of substance: in this philosophical context, individuals are no longer conceived as a substrate in which properties inhere, but as a mere bundle of properties. Although the MHI proposal is framed in the general idea of bundles, the conception of quantum systems diverges from the traditional notion of individual as a bundle of properties: bundle-systems in the quantum domain do not belong to the ontological category of individual (this point will be discussed in detail in Chapter 10).

Any quantum system can be decomposed into parts in many ways; however, not any decomposition leads to parts which are, strictly speaking, quantum systems. This is the case only when the components' behaviors obey the Dynamical Postulate QP1, which states how quantum systems evolve. In this case, the components are dynamically independent of each other, and this dynamical independence is the consequence of the non-interaction between them. On this basis, the decomposition of quantum systems is characterized as follows.

IP3 (System Decomposition): A quantum system $\mathcal{S} : (\mathcal{O}, H)$ is *composite* when two quantum systems $\mathcal{S}^1 : (\mathcal{O}^1, H^1)$ and $\mathcal{S}^2 : (\mathcal{O}^2, H^2)$ can be defined, such that:

(i) $\mathcal{O} = \mathcal{O}^1 \otimes \mathcal{O}^2$ and
(ii) $H = H^1 \otimes I^2 + I^1 \otimes H^2 \in \mathcal{O}$, where I^1 and I^2 are the identity operators in $\mathcal{O}^1$ and $\mathcal{O}^2$ respectively.

In this case it can be said that $\mathcal{S}^1$ and $\mathcal{S}^2$ are *subsystems* of the *composite system* $\mathcal{S} = \mathcal{S}^1 \cup \mathcal{S}^2$. When a quantum system is not composite, it will be called *elemental system.*

The System Decomposition postulate IP3 provides a precise criterion for distinguishing between elemental and composite systems, and such a criterion is based on the systems' Hamiltonians.

The composition of quantum systems is given by the following postulate.

IP4 (System Composition): Given two quantum systems $\mathcal{S}^1 : (\mathcal{O}^1, H^1)$ and $\mathcal{S}^2 : (\mathcal{O}^2, H^2)$, a quantum system $\mathcal{S} : (\mathcal{O}, H)$ can always be defined, such that:

(i) $\mathcal{O} = \mathcal{O}^1 \otimes \mathcal{O}^2$ and
(ii) $H = H^1 \otimes I^2 + I^1 \otimes H^2 + H_{\text{int}} \in \mathcal{O}$, where H_{int} is usually called *interaction Hamiltonian.*

Although IP4 has been called "System Composition," strictly speaking two cases must be distinguished:

- If $H_{\text{int}} = 0$ ($\mathcal{S}^1$ and $\mathcal{S}^2$ do not interact), $\mathcal{S}^1$ and $\mathcal{S}^2$ are *subsystems* of the composite system $\mathcal{S} = \mathcal{S}^1 \cup \mathcal{S}^2$: they satisfy the System Decomposition postulate IP3.
- If $H_{\text{int}} \neq 0$ ($\mathcal{S}^1$ and $\mathcal{S}^2$ interact), from the time when the interaction begins, $\mathcal{S}^1$ and $\mathcal{S}^2$ are not subsystems of the system $\mathcal{S}$, because it does not satisfy the System Decomposition postulate IP3. In fact, strictly speaking, $\mathcal{S}^1$ and $\mathcal{S}^2$ are not quantum systems but mere *parts* of $\mathcal{S}$, to the extent that they do not evolve unitarily according to the Dynamical Postulate QP1.

Once the differences between composite systems, subsystems of a composite system, and parts of a system have been introduced, it is necessary to specify how the properties of a composite system are connected to the properties of its component subsystems.

IP5 (Properties of Systems and Subsystems): Given a composite quantum system $\mathcal{S} = \mathcal{S}^1 \cup \mathcal{S}^2 : (\mathcal{O}, H)$, with $\mathcal{S}^1 : (\mathcal{O}^1, H^1)$ and $\mathcal{S}^2 : (\mathcal{O}^2, H^2)$, and given the observables $A^1 \in \mathcal{O}^1$ of $\mathcal{S}^1$ and $A^2 \in \mathcal{O}^2$ of $\mathcal{S}^2$, and the observables $A = A^1 \otimes I^2 \in \mathcal{O}$ and $A^f = f(A^1 \otimes I^2, I^1 \otimes A^2) \in \mathcal{O}$ of $\mathcal{S}$, where f is an analytical function, then,

(i) the observables A and A^1 ontologically represent *the same type-property* $[A] = [A^1]$ with the same case-properties $[a_i] = [a_i^1]$, where the $a_i = a_i^1$ are the eigenvalues of both A and A^1, and
(ii) the observable A^f ontologically represents a type-property $[A^f]$ with case-properties $[f(a_i^1, a_j^2)]$, where the a_i^1, a_j^2 are the eigenvalues of A^1 and A^2 respectively.

The interpretative postulate IP5 expresses the usual assumption according to which the observable A^1 of a subsystem $\mathcal{S}^1$ and the observable $A = A^1 \otimes I^2$ of the composite system $\mathcal{S} = \mathcal{S}^1 \cup \mathcal{S}^2$ represent the same property; this assumption is based on the empirical indistinguishability of the two observables. Moreover, this postulate establishes the necessary connections between the properties of a composite system and the properties of its subsystems. These connections are not specific to quantum mechanics, but are also common in classical mechanics: for example, the energy of a two-particle composite system is a particular combination (expressed by the sum) of the energies of the component subsystems.

4.2 Facts and possibility

Up to this point, properties have been identified. But properties are not *facts*: properties and facts belong to different ontological categories. Here, facts are not conceived in terms of the assignment of a certain property to an object. In this case, possible facts are defined as the possible occurrence of certain case-properties, and probabilities will be applied to them.

IP6 (Possible Facts): Given a quantum system $\mathcal{S} : (\mathcal{O}, H)$:

(i) If $\Pi \in \mathcal{O}$ is a projector observable, the *possible fact* $[\Pi : 1]$ is the possible occurrence of the case-property $[1]$ of the type-property $[\Pi]$.
(ii) If $A = \sum_i a_i \Pi_i \in \mathcal{O}$, where $a_i \neq a_j$ for $i \neq j$ and the $\Pi_i \in \mathcal{O}$ are the eigenprojectors of A, the *possible fact* $[A : a_k]$ is the possible occurrence of the case-property $[a_k]$ corresponding to the type-property $[A]$.

Now the question is how the possible facts corresponding to a certain quantum system are related to each other. Let us recall how the concept of complete set of orthogonal projectors, CSOP, was defined in Section 2.2 of Chapter 2.

IP7 (Relations between Possible Facts): Given a quantum system $\mathcal{S} : (\mathcal{O}, H)$, and its CSOP $\{\Pi_\alpha\}$, with $\alpha = 1$ to M:

(i) If $\Pi_B = \sum_J \Pi_J \in \mathcal{O}$ is a projector, where $\Pi_J \in \{\Pi_\alpha\}$ and $J \in \mathcal{J} \subset \{1, \cdots, M\}$, then the possible fact $[\Pi_B : 1]$ is *equivalent* to the disjunction of the possible facts $[\Pi_J : 1]$, that is, $[\Pi_B : 1] \equiv V_J[\Pi_J : 1]$.

(i) If $A = \sum_i a_i \Pi_i \in \mathcal{O}$, where $a_i \neq a_j$ for $i \neq j$ and the $\Pi_i \in \mathcal{O}$ are the eigenprojectors of A, the possible fact $[A : a_k]$ is *equivalent* to the possible fact $[\Pi_k : 1]$.

It is clear that, given the CSOP $\{\Pi_\alpha\}$ of a quantum system $\mathcal{S} : (\mathcal{O}, H)$, the projector observables of the form $\Pi_B = \sum_J \Pi_J \in \mathcal{O}$ lead to a *set of possible facts* $\mathcal{F}^\alpha = \cup\{[\Pi_B : 1]\}$ that is isomorphic to the set $\mathcal{P}^\alpha = \cup\{\Pi_B\}$ of projectors (see Section 2.2 of Chapter 2); then, $\mathcal{F}^\alpha$ also defines a Boolean algebra. As a consequence, in each quantum system, the probability function f_ρ^α, defined by the state ρ over the set P^α through the Born Rule QP2, can ontologically be interpreted as the Kolmogorovian measure of an ontological *propensity*, determined by the state ρ over the set of possible facts $\mathcal{F}^\alpha$.

IP8 (Ontological Propensities): Given a quantum system $\mathcal{S} : (\mathcal{O}, H)$ in a state ρ, and the set of possible facts $\mathcal{F}^\alpha$ defined by the CSOP $\{\Pi_\alpha\}$ with $\alpha = 1$ to M, the probability function $f_\rho^\alpha : \mathcal{P}^\alpha \to [0,1]$ defined by the Born Rule in each context ontologically represents a measure p_ρ^α of the *ontological propensity to actualization* of the corresponding possible facts, $p_\rho^\alpha : \mathcal{F}^\alpha \to [0,1]$, a measure that satisfies the axioms of Kolmogorov expressed in terms of the Boolean connectives $\neg$, $\wedge$, and $\vee$.

This means that the quantum state does not belong to the ontological category of properties, as in the classical case, but rather encapsulates ontological propensities.

IP9 (Quantum State): Given a quantum system $\mathcal{S} : (\mathcal{O}, H)$, its state ρ *encodes the ontological propensities to actualization* of the possible facts corresponding to the case-properties of the type-properties of $\mathcal{S}$, and the time evolution of ρ, given by the Schrödinger equation, ontologically represents *the time evolution of those ontological propensities.*

Up to this point, the quantum system has been assumed to have its own state. However, this may not be the case when it is a subsystem of a composite system in an entangled state. Let us consider a composite system $\mathcal{S} = \mathcal{S}^1 \cup \mathcal{S}^2$, whose state ρ is entangled relative to the decomposition into the elemental systems $\mathcal{S}^1$ and $\mathcal{S}^2$ (the relativity of entanglement will be discussed in Section 8.2 of Chapter 8). Since the state ρ encodes the propensities to actualization of the possible facts corresponding to the system $\mathcal{S}$, strictly speaking the corresponding probabilities are ascribed to the possible facts involved in system $\mathcal{S}$. Therefore, if probabilities are to be assigned to the possible facts involved in, say, $\mathcal{S}^1$, they have to be inferred by taking into account the relation between the properties of systems and subsystems as introduced in the interpretative postulate IP5 (see the previous section). In particular, the probability corresponding to the possible fact $[A^1 : a_k]$, where A^1 is an observable of $\mathcal{S}^1$, should be obtained as the probability corresponding to the possible fact $[A^1 \otimes I^2 : a_k]$, where $A^1 \otimes I^2$ is an observable of $\mathcal{S}$. However, there is a calculation shortcut to obtain probabilities in the subsystems of a composite system, which is based on the concept of reduced state. In our case, the reduced state $\rho_r^{\mathcal{S}^1}$ of $\mathcal{S}^1$ is defined as the density operator by means of which the expectation values of all the observables belonging to $\mathcal{S}^1$ can be computed. Therefore, the probability assigned to the possible fact $[A^1 \otimes I^2 : a_k]$ obtained with the state ρ of $\mathcal{S}$ has the same value as the probability assigned to the possible fact $[A^1 : a_k]$ obtained with the reduced state $\rho_r^{\mathcal{S}^1}$ of $\mathcal{S}^1$. In general, this shortcut simplifies calculations. Nevertheless, this must not lead us to forget that the reduced state is a mere calculation tool to compute probabilities and expectation values and, therefore, must not be confused with the quantum state of the subsystem. The issue of the conceptual meaning of reduced states will be discussed in depth in Sections 8.3 and 8.4 of Chapter 8.

4.3 Actual facts

In the previous section, possible facts with their corresponding propensities to actualization were presented. Since some of those possible facts become actual, it is necessary to introduce the ontological category of *actual facts*.

> **IP10 (Actual Facts):** The *actual fact Act* $[A: a_k]$ is the result of the actualization of the possible fact $[A: a_k]$ and, therefore, it is the *actual occurrence* of the case-property $[a_k]$ corresponding to the type-property $[A]$.

With this terminology, *Act* $[A: a_k]$ is read as "*the occurrence of the case-property* $[a_k]$ *corresponding to the type-property* $[A]$ *is actual.*" Let us note the difference between the terminology proposed here and that of modal logic. Whereas in the language of modal logic possibility is introduced by means of a logical operator used to qualify the truth of a proposition, here propositions of the form "$[A: a_k]$" refer to possible facts, and *Act* is an operator that applies to possible facts to turn them into actual facts. In other words, the unqualified discourse refers to the realm of possibility, and actualization is denoted by a specific logical operator that represents the transition from the possible to the actual.

The Kochen-Specker theorem (1967) teaches us that, at a given time t, the actualization of the possible facts belonging to all the sets $\mathcal{F}^{\alpha}$, defined by all the CSOPs of the system, is not allowed by the theory. Therefore, actualization must be restricted in some way. Here, the restriction is established by the following interpretative postulate.

IP11 (Preferred Context): Given a quantum system $\mathcal{S} : (\mathcal{O}, H)$, there is a *preferred context*, that is, a *preferred CSOP* Π^{p}_{α} that determines a *preferred set of possible facts* $\mathcal{F}^{\text{p}} = \cup\{[\Pi_B : 1]\}$ (with $\Pi_B = \sum_J \Pi^{\text{p}}_J \in \mathcal{O}, \Pi^{\text{p}}_J \in \Pi^{\text{p}}_{\alpha}$, and $J \in \mathcal{J} \subset \{1, \cdots, M\}$) where actualization occurs, in the sense that:

(i) One and only one of the possible facts $[\Pi^{\text{p}}_{\alpha} : 1] \in \mathcal{F}^{\text{p}}$, say $[\Pi^{\text{p}}_k : 1]$, is also an actual fact, *Act* $[\Pi^{\text{p}}_k : 1]$.
(ii) Given the actual fact *Act* $[\Pi^{\text{p}}_k : 1]$, all the possible facts $[\Pi_B : 1] \equiv \vee_J [\Pi^{\text{p}}_J : 1] \in \mathcal{F}^{\text{p}}$ (see IP7 (i)), such that Π^{p}_k is one of the Π^{p}_J, are also actual facts, *Act* $[\Pi_B : 1]$.
(iii) Given the actual fact *Act* $[\Pi^{\text{p}}_k : 1]$, the possible fact $[A : a_k] \equiv [\Pi^{\text{p}}_k : 1] \in \mathcal{F}^{\text{p}}$ (see IP7 (ii)) is also an actual fact, *Act* $[A : a_k]$.

Since SQM is an essentially indeterministic theory, in general it does not determine which possible facts become actual. The actualization or non-actualization of a possible fact belonging to the preferred set $\mathcal{F}^{\text{p}}$ is determined only when the measure of the propensity to actualization of such a fact is 1 or 0, respectively.

IP12 (Determined Actualization): Given an elemental quantum system $\mathcal{S} : (\mathcal{O}, H)$ and the preferred set $\mathcal{F}^{\text{p}}$ of possible facts, if the state ρ ascribes a propensity to actualization with measure 1 (measure 0) to the possible fact $[A : a_k] \in \mathcal{F}^{\text{p}}$, then such a possible fact is (is not) actual:

$$\text{If } p^{\text{p}}_{\rho}([A : a_k]) = 1, \text{ then } Act\,[A : a_k]$$
$$\text{If } p^{\text{p}}_{\rho}([A : a_k]) = 0, \text{ then } \neg Act\,[A : a_k]$$

4.4 The Modal-Hamiltonian actualization rule

In the interpretative postulates IP11 and IP12, the preferred context was presupposed. Now it is necessary to provide a rule to identify it: this task is usually the most subtle step of any modal interpretation. In fact, since SQM does not account for actualization due to its indeterministic nature, the rule cannot be inferred from the formalism, but must be introduced as a purely interpretative assumption. Therefore, the adequacy of the actualization rule should be assessed in the light of its physical relevance and its ability to solve the interpretative problems of the theory (these matters will be discussed in detail in Chapters 5 and 6). However, before introducing the rule, I want to discuss the conceptual motivations underlying its adoption.

Since the 1960s, research on the mathematical properties of the formal structure of SQM has experienced enormous advances: many results, unknown by the founding fathers of the theory, were obtained, and this work greatly improved the understanding of the deep obstacles that any interpretation must face. However, in many cases this interest in the features of the formalism has led to blurring the physical content of the theory: the interpretative arguments are generally based on mathematical results and centered on formal models of the quantum measurement. The MHI moves away from this "formalist" trend by placing an element with a clear physical meaning, the Hamiltonian of the system, at the heart of the interpretation: the content of the actualization rule of the MHI will be based on the structure of the Hamiltonian.

Although in Chapter 2 only invariances under space-time transformations were commented, the Hamiltonian may have other symmetries. Precisely, the so-called Schrödinger group is the group of transformations that leave the Hamiltonian invariant (see Tinkham 1964). Since each symmetry of the Hamiltonian leads to an energy degeneracy (see, e.g., Meijer and Bauer 2004), much valuable information on the energy spectrum of the system can be obtained by applying the machinery of the group theory to the study of the symmetries of the Hamiltonian. In fact, if H is invariant under a symmetry transformation with generator K, then $[H, K] = 0$ (see eq. (2-17) in Chapter 2). Therefore, if $H|n\rangle = \omega_n |n\rangle$, then

$$KH|n\rangle = K\omega_n |n\rangle \Rightarrow HK|n\rangle = \omega_n K|n\rangle. \tag{4-1}$$

This means that any vector $K|n\rangle$ obtained by applying the operator K to the eigenvector $|n\rangle$ is also an eigenvector of H with the same eigenvalue ω_n. If H is expressed as $H = \sum_n \omega_n \Pi_n$, where Π_n is the eigenprojector corresponding to the eigenvalue ω_n, the index k_n corresponding to the degeneracy of ω_n can be included explicitly in such a way that

$$H|n, k_n\rangle = \omega_n |n, k_n\rangle \Rightarrow H = \sum_n \omega_n \sum_{k_n} |n, k_n\rangle \langle n, k_n|, \tag{4-2}$$

$$K|n, k_n\rangle = \kappa_{k_n} |n, k_n\rangle \Rightarrow K = \sum_n \sum_{k_n} \kappa_{k_n} |n, k_n\rangle \langle n, k_n|. \tag{4-3}$$

The degeneracies with origin in symmetries are called "normal" (Tinkham 1964) or "systematic" (Cohen-Tannoudji, Diu, and Lalöe 1977). On the contrary, degeneracies with no obvious origin in symmetries are called "accidental." However, deeper study usually shows that either the accidental degeneracy is not exact or a hidden symmetry can be found in the Hamiltonian that explains the degeneracy. A classic example is the degeneracy, in the hydrogen atom, of states of different angular momentum l but the same principal quantum number n, for instance, of $2s$ and $2p$ functions. In this case, Vladimir Fock (1935) showed that the degeneracy arises from a four-dimensional rotational symmetry of the Hamiltonian in momentum space. For this reason, it is assumed that, once all the symmetries of the Hamiltonian have been identified, a basis for the Hilbert space of the system is obtained and the "good quantum numbers" are well defined. This strategy is what underlies the group approach to quantum mechanics, where the physical features of the quantum system are studied by analyzing the symmetry properties of its Hamiltonian (Weyl 1950, Hamermesh 1962, Tinkham 1964, Tung 1985).

Now all the conceptual elements necessary to formulate the Modal-Hamiltonian actualization rule have been presented. The basic idea can be expressed by the classical Latin maxim "*Ubi lex non distinguit, nec nos distinguere debemus*": where the law does not distinguish, neither should we distinguish. The Hamiltonian of the system is the core of the dynamical law of the theory. Therefore, it is the Hamiltonian, with its symmetries, that rules actualization: no observable whose eigenvalues would distinguish among eigenvectors corresponding to a single degenerate eigenvalue of the Hamiltonian must take on a definite value, since the actualization of the possible facts corresponding to that observable would introduce in the system an asymmetry not contained in the Hamiltonian. Once this basic idea has been clearly understood, the Modal-Hamiltonian actualization rule can be formulated in a very simple way.

> **IP13 (Actualization Rule: AR):** Given an elemental quantum system $\mathcal{S} : (\mathcal{O}, H)$, where $H = \sum_n \omega_n \Pi_n$, the *preferred CSOP* is $\{\Pi_n\}$, where each Π_n is the eigenprojector of H corresponding to the eigenvalue ω_n.

In other words, the degeneracies of the Hamiltonian define a sort of "coarse-grained basis" of the Hilbert space $\mathcal{H}$, where $\mathcal{H}$ is such that $\mathcal{O} \subset \mathcal{B}(\mathcal{H})$ (see Section 2.1 of Chapter 2): actualization must not introduce a finer discrimination that would break the symmetries of the Hamiltonian.

From this formulation of the actualization rule, it is clear that the MHI follows the trend of traditional Modal Interpretations, such as the BDMI and the SDMI, in the sense of dropping the eigenstate-eigenvalue link. Indeed, even if the state ρ ascribes a propensity to actualization with measure 1 to a possible fact, if this fact does not belong to the preferred set $\mathcal{F}^{\mathrm{p}}$ of possible facts—defined by the preferred CSOP—, then it does not become actual.

In the literature on Modal Interpretations, the problem of identifying the preferred context is usually expressed in terms of deciding which observables are definite-valued. Let us see how the AR works in those terms for different cases:

(a) The Hamiltonian H does not have symmetries: it is non-degenerate. In this case,

$$H\,|n\rangle = \omega_n\,|n\rangle \qquad \text{with } \omega_n \neq \omega_{n'}, \tag{4-4}$$

where $\{|n\rangle\}$ is a basis of the Hilbert space $\mathcal{H}$. Here, the preferred CSOP is $\{|n\rangle\langle n|\}$; so, $\{|n\rangle\}$ can be called "preferred basis." Therefore, n is the only good quantum number: the definite-valued observables of the system are H and all the observables commuting with H.

(b) The Hamiltonian H has certain symmetries that lead to energy degeneracy. In this case, H can be written as

$$H\,|n,i_n\rangle = \omega_n\,|n,i_n\rangle \Rightarrow H = \sum_n \omega_n \sum_{i_n} |n,i_n\rangle\langle n,i_n| = \sum_n \omega_n \Pi_n, \qquad (4\text{-}5)$$

where $\omega_n \neq \omega_{n'}$ and the index i_n expresses the degeneracy of the energy eigenvalue ω_n. Let us consider an observable of the form

$$A = \sum_{n,i_n} a_n\,|n,i_n\rangle\langle n,i_n| = \sum_n a_n \sum_{i_n} |n,i_n\rangle\langle n,i_n| = \sum_n a_n \Pi_n, \qquad (4\text{-}6)$$

where $a_n \neq a_{n'}$. It is clear that $[H,A]=0$. Moreover, A has the same degeneracy as H since they both have the same eigenprojectors Π_n: the subspace spanned by the degenerate eigenvectors corresponding to a_n is the same as that spanned by the degenerate eigenvectors corresponding to ω_n. In other words, A has the same symmetries as H. Therefore, all the observables A commuting with H and having the form of eq. (4–6) are definite-valued. On the contrary, observables of the form

$$B = \sum_{n,i_n} b_{n,i_n}\,|n,i_n\rangle\langle n,i_n|, \qquad (4\text{-}7)$$

despite commuting with H, do not acquire definite values, since any of the actual facts $Act\,[B:\ b_{n,i_n}]$ would discriminate among the eigenvectors corresponding to a single degenerate eigenvalue ω_n of H: they would distinguish where the Hamiltonian does not distinguish.

On the basis of the above remarks, the MHI actualization rule AR can be expressed in a different but equivalent way.

IP13' (Actualization Rule: AR): Given an elemental quantum system $\mathcal{S} : (\mathcal{O}, H)$, the definite-valued observables of the system are the Hamiltonian H and the observables commuting with H and having, at least, the same symmetries as H—that is, those observables commuting with H and that do not break the symmetries of H.

As explicitly stated in the interpretative postulates IP13 and IP13', AR identifies the definite-valued observables of elemental quantum systems. However, when the elemental system is a subsystem of a composite system, the definite-valued observables of the composite system can also be identified. In fact, according to IP5, given a composite quantum system $\mathcal{S} = \mathcal{S}^1 \cup \mathcal{S}^2$, with $\mathcal{S}^1 : (\mathcal{O}^1, H^1)$ and $\mathcal{S}^2 : (\mathcal{O}^2, H^2)$, if the observable $A^1 \in \mathcal{O}^1$ of $\mathcal{S}^1$ acquires the definite value a^1_Ω, the same happens for the observable $A = A^1 \otimes I^2 \in \mathcal{O}$ of $\mathcal{S}$, since A^1 and A represent the same type-property $[A^1] = [A]$ with the same case-properties $[a^1_i] = [a_i]$, where the $a^1_i = a_i$ are the eigenvalues of both A^1 and A. Moreover, according to IP5, if the observable $A^1 \in \mathcal{O}^1$ of $\mathcal{S}^1$ acquires the definite value a^1_Ω and the observable $A^2 \in \mathcal{O}^2$ of $\mathcal{S}^2$ acquires the definite value a^2_Φ, any observable $A^f = f(A^1 \otimes I^2, I^1 \otimes A^2) \in \mathcal{O}$ of $\mathcal{S}$, representing the type-property $[A^f]$ with case-properties $[f(a^1_i, a^2_j)]$, is also definite-valued and acquires the definite value $f(a^1_\Omega, a^2_\Phi)$.

As explained in Section 3.3 of Chapter 3, the actualization rules of traditional Modal Interpretations, such as the BDMI and the SDMI, in general do not assign the same value to A^1 and $A = A^1 \otimes I^2$; this fact contradicts the common physical assumption—based on the observational indistinguishability of A^1 and A—that both observables represent the same property. In the MHI, no problems arise since that assumption is imposed as an interpretative postulate (IP5). This postulate can be formulated because the interpretation provides a precise criterion, given by IP2, to distinguish between elemental systems and composite systems. Once the elemental quantum systems have been univocally identified, there is no obstacle to applying the AR only to them and to identifying the definite-valued observables of the composite system by means of the interpretative postulate IP5.

Another difference between the MHI and traditional Modal Interpretations is worth noting. In the BDMI and the SDMI, the preferred context of definite-valued observables, where actualization occurs, depends on the instantaneous state of the system, which continuously changes over time. This means that actualization is a phenomenon that happens continuously at every instant. Such an interpretative stance leads to the need to account for the dynamics of actual properties (see Vermaas 1996). In the MHI, by contrast, this step is unnecessary because the dynamics of actual properties is trivial. Given that the preferred context depends on the Hamiltonian, the definite-valued observables always commute with the Hamiltonian and, therefore, are constants of motion of the system. In other words, the observables that receive definite values are the same during all the "life" of the quantum system as such—precisely, as a closed system—, since its "birth," when it arises as the result of an interaction between other systems, up to its final "death," when it disappears by interacting with another system. The same idea can be expressed in different terms. According to the MHI, the Hamiltonian of the system is a definite-valued observable in any case and, therefore, the energy is completely definite in all quantum systems. In turn, although it cannot be strictly said that energy "does not commute" with time—since time is not represented by an operator in quantum mechanics (Pauli 1933; see discussion, e.g., in Ballentine 1998, 343–347 and in Busch 2008)—, it is commonly accepted that if the energy of the quantum system is completely definite, time is completely indefinite. So, the question about the precise time when energy acquires a definite value—about the time when one of the possible facts corresponding to the value of the Hamiltonian becomes actual—does not seem to make much sense: the Hamiltonian and the rest of the definite-valued observables acquire their definite values when the system comes into being, and it is this event what defines time itself. More precisely, from the perspective of the MHI, in the domain of actuality time is discrete and defined by the interactions that give rise to new systems, whereas in the domain of possibility propensities evolve over the linear and continuous time represented by the variable t in the Schrödinger equation. This point, related to the nature of time in SQM, will be discussed in detail in Section 13.2 of Chapter 13.

4.5 Mathematics, physics, and ontology

This chapter was devoted to introduce the MHI in detail by means of the interpretative postulates IP1 to IP13. In accomplishing this task, the physical language, its mathematical representation, and its ontological counterpart have been used with the purpose of preserving the precision of the presentation. It is worthwhile, then, to summarize the main elements of the interpretation, which will be relevant in the following chapters:

- Observables, mathematically represented by self-adjoint operators of a von Neumann algebra $\mathcal{B}(\mathcal{H})$, ontologically represent *type-properties.*
- The values of an observable, mathematically represented by the eigenvalues of the corresponding self-adjoint operator, ontologically represent *case-properties.*
- A quantum system, mathematically represented by the space of self-adjoint operators of $\mathcal{B}(\mathcal{H})$, ontologically represents a *bundle* of type-properties with their corresponding case-properties.
- A quantum system is *elemental* when it cannot be decomposed into subsystems, that is, into unitary evolving quantum systems; otherwise, it is *composite.*
- Although possible facts have no mathematical representation, ontologically a *possible fact* is the possible occurrence of a case-property of a certain type-property.
- The state of a quantum system encodes the *propensities to actualization* of all the possible facts corresponding to the case-properties of all the type-properties of the system.
- A possible fact becomes an *actual fact* in a completely indeterministic way.
- The Kochen-Specker theorem proves that not all the type-properties of a system can simultaneously lead to actual facts. So, the *preferred context* where actualization occurs must be identified by an actualization rule.
- The *Modal-Hamiltonian actualization rule* states that the type-properties leading to actual facts (through the actual occurrence of one of their case-properties) are those represented by the Hamiltonian and by all the observables commuting with the Hamiltonian and having, at least, its same symmetries.

Although this chapter has dealt with the ontological reference of the physical language, many philosophical questions still remain unanswered. For example, the differences between the categories of individual and of property, the nature of properties, the conception of possibility, among other very relevant issues, have not been considered here. A detailed discussion of these ontological aspects will be offered in Chapters 9 and 10.

The content of the present chapter, although detailed, does not yet explain why anyone should pay attention to the MHI. The advantages of this proposal will become clear when its agreement with the practice of physics will be shown soon, in Chapter 5, and when it will be applied to the resolution of the quantum measurement problem, in Chapter 6.

5
Applications of the Modal-Hamiltonian Interpretation

Standard Quantum Mechanics is a theory that has been applied to many concrete physical situations and by means of which an impressive amount of experimental evidence has been accounted for. Therefore, a "good" interpretation must not only meet the traditional interpretative challenges of the theory, but must also show its agreement with the orthodox practice of physics. This chapter is devoted to argue for the physical relevance of the MHI by applying it to well-known models and experimental results, which will be presented in the usual physical language.

5.1 Free particle

The Hamiltonian of a free particle reads

$$H^P = \frac{P^2}{2m} = \frac{P_x^2 + P_y^2 + P_z^2}{2m}, \tag{5-1}$$

where P is the momentum observable, with components P_x, P_y, and P_z, and m is the mass of the particle. The particle is said to be "free" because there are no fields acting on it. Then, space is homogeneous and, as a consequence, H^P is invariant under space-displacements in any direction—an analogous argument could be given in terms of the isotropy of space and space-rotations in any direction. The components P_x, P_y, and P_z are the generators of the space-displacement symmetry of H^P and, for this reason, they are constants of motion of the system. Since the Hamiltonian has a symmetry, it is degenerate.

According to the Modal-Hamiltonian actualization rule AR, H^P acquires a definite value, and also P^2 since it is proportional to H^P and, then, has the same space-displacement symmetry—P^2 is the Casimir operator of the group generated by P_x, P_y, and P_z. Nevertheless, P_x, P_y, and P_z are not definite-valued observables because, being the generators of the symmetry, the actualization of any of the possible facts corresponding to their eigenvalues would break the symmetry of the free particle, in the sense of introducing an asymmetry not contained in the Hamiltonian.

Of course, the three components of P can be used for the theoretical description of the free particle. In fact, generally, any two of them are added to H^P to compose a complete set of commuting observables (CSCO), $\{H^P, P_x, P_y\}$, $\{H^P, P_y, P_z\}$, or $\{H^P, P_x, P_z\}$,

The Modal-Hamiltonian Interpretation of Quantum Mechanics. Olimpia Lombardi, Oxford University Press.
© Olimpia Lombardi (2026). DOI: 10.1093/9780198951407.003.0005

that defines a basis of the Hilbert space $\mathcal{H}^P$ of the free particle. Given the functional dependence among the four magnitudes, the CSCO $\{P_x, P_y, P_z\}$ can be used equivalently. But this fact does not mean that those observables have to be considered definite-valued. On the contrary, the application of AR to this system agrees with the empirical inaccessibility to the values of P_x, P_y, and P_z in the free particle. In fact, anyone who wants to know these values must perform a measurement on the particle. But a measurement always involves an interaction with the measured object, which breaks the symmetry of the original system by modifying its Hamiltonian. For example, consider a screen acting as a potential barrier that breaks the homogeneity of space in direction y. Under measurement, the particle is no longer free: the Hamiltonian of the new system is

$$H^{P'} = H^P + V(y) = \frac{P^2}{2m} + V(y), \tag{5-2}$$

where $V(y)$ is the potential energy representing the presence of the screen. This means that the screen breaks the space symmetry in a certain direction and this makes the value of the momentum in that direction empirically accessible. In other words, the symmetry breaking introduced by the interaction of the free system with the measuring apparatus is what allows the observer to have empirical access to an observable that was a generator of the symmetry of the original system's Hamiltonian and, as a consequence, was empirically inaccessible.

5.2 Free particle with spin

The spin S is an internal contribution to the total angular momentum and, therefore, adds further degrees of freedom to the particle: the Hilbert space is now $\mathcal{H}^{PS} = \mathcal{H}^P \otimes \mathcal{H}^S$, where $\mathcal{H}^P$ is the Hilbert space of the free particle and $\mathcal{H}^S$ is the Hilbert space of the spin. In this case, the Hamiltonian reads

$$H^{PS} = \frac{P^2}{2m} + E_0, \tag{5-3}$$

where E_0 can only be a multiple of S^2 and, then, may be conceived as an internal contribution to the energy (see Ballentine 1998).

According to the MHI, in this case the system is composite, because it can be decomposed into two non-interacting subsystems (see IP3, Section 4.1 of Chapter 4): a free particle without spin, corresponding to $\mathcal{H}^P$ and with Hamiltonian $H^P = P^2/2m$, and a spin system, corresponding to $\mathcal{H}^S$ and with Hamiltonian $H^S = k\,S^2$, with $k = const.$ Then, the actualization rule AR must be applied independently to each elemental subsystem.

The rule applies to the free particle subsystem as explained in the previous section. In turn, in the spin subsystem, the Hamiltonian H^S is invariant under space-rotation:

the generators of this symmetry are the three components J_x, J_y, and J_z of the total angular momentum J. But since in this case the orbital angular momentum L is zero, the total angular momentum $J = L + S$ turns out to be simply $J = S$, and the three components S_x, S_y, and S_z of the spin S are the generators of the space-rotation symmetry. Analogously to the case of the free particle, according to AR, in this case H^S acquires a definite value, and also S^2 since it is proportional to H^S—in fact, S^2 is the Casimir operator of the group generated by S_x, S_y, and S_z. Nevertheless, S_x, S_y, and S_z are not definite-valued since they are the generators of the symmetry of the Hamiltonian $H^S = k\,S^2$.

Again, this conclusion agrees with the fact that observers have no empirical access to the spin components of the free particle with spin. In order to know the value of these components, a measurement must be carried out on the system: it is necessary to introduce a magnetic field B of modulus $|B|$ in some direction, say z, which breaks the isotropy of space and, as a consequence, the original space-rotation symmetry. Under the action of B, the Hamiltonian $H^{S'}$ of the new system resulting from the interaction is no longer invariant under space-rotation, because now it includes the interaction $-\gamma\,|B|\,S_z$ that privileges a particular direction of space:

$$H^{S'} = k\,S^2 - \gamma\,|B|\,S_z, \tag{5-4}$$

where $\gamma = const.$ In other words, the experimental access to the spin component S_z is possible only by means of a measurement that breaks the space-rotation symmetry of the original Hamiltonian in direction z and, therefore, makes the system no longer free. This is the usual way in which a spin component is measured in a Stern-Gerlach experiment (the experiment will be considered in detail in the next chapter).

5.3 Harmonic oscillator

By definition, a harmonic oscillator is a system with a quadratic potential energy, which produces a restoring force against any displacement from equilibrium that is proportional to the displacement. In one dimension, the Hamiltonian of this system is

$$H^H = \frac{P^2}{2m} + \frac{m\,\Omega^2 Q^2}{2}, \tag{5-5}$$

where Q is the position observable and Ω is the frequency of oscillation. If the dimensionless position and momentum operators, $q = (m\Omega/\hbar)^{1/2}Q$ and $p = (1/m\Omega\hbar)^{1/2}P$, are introduced in eq. (5-5), the Hamiltonian becomes

$$H^H = \frac{1}{2}\,\hbar\,\Omega\,\left(q^2 + p^2\right). \tag{5-6}$$

In turn, if the number-of-modes observable $N = a^\dagger a$ is used,

$$N = a^\dagger a = \left(\frac{q - ip}{\sqrt{2}}\right)\left(\frac{q + ip}{\sqrt{2}}\right) \quad \Rightarrow \quad H^H = \hbar\,\Omega\left(N + \frac{1}{2}\right). \tag{5-7}$$

As is well known, in this case the spectra of H and N can be obtained algebraically:

$$H^H\,|n\rangle = \omega_n\,|n\rangle \qquad N\,|n\rangle = n\,|n\rangle\,. \tag{5-8}$$

The non-degeneracy of H^H expresses the fact that it has no symmetries: the CSCO $\{H^H\}$ defines the basis $\{|n\rangle\}$ of the Hilbert space $\mathcal{H}^H$ of the harmonic oscillator.

According to the actualization rule AR, since in this case the Hamiltonian H^H is non-degenerate, the preferred CSOP is $\{|n\rangle\langle n|\}$: the definite-valued observables of the system are H^H and all the observables commuting with H^H. In particular, the number of modes N acquires a definite value because $[H^H, N] = 0$.

The harmonic oscillator has a central relevance in quantum mechanics because it provides a model for many kinds of vibrating systems. In particular, the electromagnetic field can be decomposed in terms of linearly independent modes, each one of which behaves as a harmonic oscillator that is usually associated with a particle; in this case, N is conceived as the observable "number of particles." But it is worth emphasizing that the particle-picture of the system, with a definite number of particles, is not generic: it can only be metaphorically retained when the observable N meets the requirements imposed by the actualization rule AR and, as a consequence, is a definite-valued observable.

Another point to stress here is that, in all of those vibrating phenomena, energy—and its functions, as N—is the relevant physical magnitude of the system, whose value is assumed to be definite. The MHI agrees with this usual assumption, since it selects the non-degenerate Hamiltonian as the observable that defines the basis of actualization.

5.4 Free hydrogen atom

In the physical literature, the hydrogen atom is described as a two-particle system consisting of an electron and a proton interacting through a Coulombian interaction. In this case, the Hamiltonian reads

$$H^A = \frac{P_e^2}{2m_e} + \frac{P_p^2}{2m_p} + \frac{e^2}{|Q_e - Q_p|}, \tag{5-9}$$

where the subindexes e and p refer to the electron and the proton, respectively, and e is the electric charge of the electron. The usual strategy for solving the energy-eigenvalue equation in coordinate representation is to refer the Hamiltonian to the center of mass of the system by means of a canonical transformation, and to write the resulting equation in spherical coordinates (r, θ, ϕ). As is well known, with this strategy the solution of the equation can be expressed as the product of two functions, one only dependent on the radial coordinate and the other only dependent on the angular coordinates: $\Psi(r, \theta, \phi) = R(r)\,Y(\theta, \phi)$. By solving the radial and the angular equations, three "good" quantum numbers are obtained: the principal quantum number n, the orbital angular momentum quantum number l, and the magnetic quantum number m_l. These quantum numbers correspond to the eigenvalues of the observables H^A, L^2, and L_z respectively, where L is the orbital angular momentum, and L_x, L_y, and L_z are its components:

$$H^A\,|n, l, m_l\rangle = \omega_n\,|n, l, m_l\rangle, \tag{5-10}$$

$$L^2\,|n, l, m_l\rangle = l(l+1)\hbar^2\,|n, l, m_l\rangle, \tag{5-11}$$

$$L_z\,|n, l, m_l\rangle = m_l\hbar\,|n, l, m_l\rangle, \tag{5-12}$$

with $n = 0, 1, 2, \ldots$, $l < n$, and $-l \leq m_l \leq l$. In particular, the energy eigenvalues are computed as

$$\omega_n = -\frac{\mu\,e^4}{2\,\hbar^2\,n^2}, \tag{5-13}$$

where $\mu = \left(1/m_e + 1/m_p\right)^{-1}$ is the reduced mass of the atom. Therefore, the hydrogen atom is described in terms of the basis $\{|n, l, m_l\rangle\}$ defined by the CSCO $\{H^A, L^2, L_z\}$: the quantum numbers n, l, and m_l label the solutions Ψ_{nlm_l} of the energy-eigenvalue equation.

In this case, the Hamiltonian H^A is degenerate due to its space-rotation invariance. When the spin of the electron is not considered (for the effect of the spin, see below, Section 5.5 of the present chapter), the total angular momentum $J = L + S$ is simply $J = L$. Then, the three components L_x, L_y, and L_z of L are the generators of the symmetry group representing the space-rotation invariance, and L^2 is the Casimir operator of the group. As a consequence, although l and m_l are good quantum numbers in the sense of collaborating to the definition of a basis of the Hilbert space, the eigenvalues ω_n of the Hamiltonian do not depend on them: due to the symmetry of H^A, the values of L^2 and L_z have no manifestations in the energy spectrum. According to the actualization rule AR, as a result of the degeneracy of H^A, the observables L^2 and L_z do not acquire definite values: the only definite-valued observables of the system are H^A and the observables having, at least, the same space-rotation symmetry (at least, the same degeneracy) as H^A.

The fact that the MHI does not assign definite values to L^2 and L_z is in agreement with experimental evidence, in particular, with the data coming from spectroscopy. Let us consider each observable in detail:

i) In chemistry, the states Ψ_{nlm_l} of the atom, commonly called "orbitals," are labeled as $X\alpha$, where X is the principal quantum number n, and α is replaced with *s*, *p*, *d*, *f*, etc., that is, with letters corresponding to the value of the orbital angular momentum quantum number l. So, the orbitals of the free hydrogen atom are labeled as 1*s*, 2*s*, 2*p*, 3*s*, 3*p*, 3*d*, etc. As can be seen, the magnetic quantum number m_l is not included in those labels because, although Ψ_{nlm_l} depends on the three quantum numbers, the space-rotation symmetry of the Hamiltonian makes the selection of L_z completely arbitrary: since space is isotropic, L_x or L_y can be chosen to obtain an equally legitimate description of the free atom. The arbitrariness in the selection of the *z*-direction is manifested in spectroscopy by the fact that the spectral lines give no experimental evidence about the values of L_z: there is no empirical access to the number m_l of the free atom. The MHI, which does not assign a definite value to L_z, agrees with these experimental results. Analogously to the case of the free particle with spin, if the value of L_z is to be known, a magnetic field that breaks the isotropy of space must be applied (the detailed description of this case will be given in the next section).

ii) On the contrary, the value of the quantum number l is included in the traditional orbitals' labels as *s*, *p*, *d*, etc. Moreover, the value of l can be inferred from the observed energy spectrum of the hydrogen atom, and plays a role in the explanation of the well-known spectral series (Paschen, Balmer, Lyman, etc.). These facts might be interpreted as a symptom of the definite-valuedness of L^2 in the free hydrogen atom. However, the manifestation of the value of l requires the interaction between the atom and an electromagnetic field. The usual explanation runs as follows. Since energy transitions involve the absorption or emission of a photon (spin 1), conservation of the angular momentum forces the atom to experience a change of 1 in its orbital angular momentum L. For this reason, when a photon is absorbed by an atom in an *s* orbital, the atom acquires orbital momentum and makes a transition to a *p* orbital; when a photon is absorbed by an atom in a *p* orbital, the orbital momentum increases ($p \rightarrow d$ transition) or decreases ($p \rightarrow s$ transition), depending on the relative orientations of the photon's and the atom's angular momenta, but transitions $s \rightarrow d$ or $p \rightarrow f$ are forbidden. This traditional explanation is based on the fact that the manifestation of the value of l is the result of an interaction; but, then, the system is no longer the free hydrogen atom. The new system has a Hamiltonian of the form

$$H^{A'} = H^A + H^{EM} + H_{\text{int}}, \tag{5-14}$$

where H^A is the Hamiltonian of the free hydrogen atom (see eq. (5-9)), H^{EM} is the Hamiltonian of the electromagnetic field, which can be computed as the infinite sum of the Hamiltonians of the independent harmonic oscillators corresponding to the infinite modes of the field, and H_{int} is the interaction Hamiltonian, which depends on the dipole moment of the atom and on the electric field (see, e.g., Ballentine 1998, 548–549). The interaction breaks the

> original symmetry in L^2 and, as a consequence, removes the energy degeneracy in the quantum number l: now the energy eigenvalues ω_{nl} turn out to be functions of both the quantum numbers n and l. This fact is what leads to the manifestation of the value of l in the energy spectrum, and allows L^2 to become a definite-valued observable in the new non-free system.

The fact that L^2 is not a definite-valued observable in the free hydrogen atom does not imply that it never acquires a definite value in a free atom. The particular symmetry features of the hydrogen atom strongly depend on the Coulombian potential, generated by its one-proton nucleus. In more complex atoms, the potential may not be perfectly Coulombian, and this asymmetry removes the degeneracy in l of the Hamiltonian: the energy eigenvalues ω_{nl} are functions of both n and l with no need of interaction (see, e.g., Ballentine 1998, 280). This means that L^2 does no longer discriminates among the different eigenvectors corresponding to a single degenerate energy eigenvalue, but rather removes the degeneracy of the symmetric Coulombian case. According to AR, this implies that L^2 may be a definite-valued observable of free atoms with non-Coulombian potentials.

5.5 Zeeman effect

When an external magnetic field is applied to the hydrogen atom, the spectral lines split into multiple closely spaced lines. First observed by Pieter Zeeman in 1896, this phenomenon is known as *Zeeman effect.*

In the previous section, it was recalled that the Hamiltonian of the free hydrogen atom is endowed with space-rotation symmetry, which makes the energy eigenvalues to be independent of the magnetic quantum number m_l, that is, which makes the Hamiltonian to be degenerate in m_l. It is precisely due to this symmetry that the selection of L_z for completing the basis of the Hilbert space is completely arbitrary. The arbitrariness of the choice of the z-direction agrees with the fact that there is no experimental evidence about the value of m_l in the energy spectrum of the free hydrogen atom.

Analogously to the measurement on a free particle with spin (see above, Section 5.2 of the present chapter), a magnetic field B along the z-axis breaks the isotropy of space and, as a consequence, the space-rotation symmetry of the Hamiltonian. In this case, the breaking of the symmetry removes the energy degeneracy in m_l: now L_z is not arbitrarily chosen but is selected by the direction of the magnetic field. But this, in turn, implies that the atom is no longer free: the Hamiltonian of the new system is approximately (see, e.g., Ballentine 1998, 326)

$$H^{Zeeman} = H^A + \frac{e}{2m_e e} BL, \tag{5-15}$$

where, again, H^A is the Hamiltonian of the free atom. As a consequence, the original degeneracy of the $(2l + 1)$-fold multiplet of fixed n and l is now removed: the energy levels turn out to be displaced by an amount

$$\Delta\omega_{nlm_l} = \frac{e\hbar\,|B|}{2m_e c} m_l. \tag{5-16}$$

This means that the Hamiltonian H^{Zeeman}, with eigenvalues ω_{nlm_l}, is now non-degenerate: it constitutes by itself the CSCO $\{H^{Zeeman}\}$ that defines the preferred basis $\{|n, l, m_l\rangle\}$. According to AR, in this case H^{Zeeman} and all the observables commuting with it are definite-valued: since this is the case for L^2 and L_z, in the physical conditions leading to the Zeeman effect both observables acquire definite values, and this agrees with experimental evidence of the Zeeman effect.

5.6 Fine structure

When the spectral lines of the hydrogen atom corresponding to $n > 1$ are examined at very high resolution, they are found to be closely spaced doublets. This splitting was one of the first experimental evidence of the spin of the electron. This phenomenon, known as "spin-orbit interaction," is usually explained by saying that the energy levels of the atom are affected by the "coupling" between the electron spin S and the orbital angular momentum L. In this case, the Hamiltonian of the system reads

$$H^{A''} = H^A + H^S + H_{so}, \tag{5-17}$$

where H^A is again the Hamiltonian of the free atom, $H^S = k\,S^2$ is the Hamiltonian of the spin, and H_{so} is the Hamiltonian representing the spin-orbit interaction, a function of the product $L \cdot S$.

When the spin-orbit interaction is neglected ($H_{so} = 0$), the system is composite (see Section 5.2 of the present chapter) and can be described in terms of the basis $\{|n, l, m_l, s, m_s\rangle = |n, l, m_l\rangle \otimes |s, m_s\rangle\}$, where $s(s+1)\hbar^2$ are the eigenvalues of S^2, and the $m_s\hbar$ are the eigenvalues of S_z. But when the spin-orbit interaction is taken into account, the observables L_z and S_z no longer commute with the total Hamiltonian and, therefore, they are not constants of motion of the system: it is usually said that m_l and m_s are not good quantum numbers anymore. Nevertheless, the Hamiltonian is still invariant under space-rotation: the components J_x, J_y, and J_z of the total angular momentum J are the generators of the space-rotation symmetry group, and J^2 is the Casimir operator of the group, with eigenvalues $j(j+1)\hbar^2$. In turn, J is the sum of the orbital angular momentum L and the spin angular momentum S:

$$J = L + S \qquad m_j = m_l + m_s, \tag{5-18}$$

where m_j corresponds to the eigenvalue of J_z. So, now m_j is a good quantum number. But the following relations also hold:

$$J^2 = (L + S)^2 \Rightarrow L \cdot S = \frac{J^2 - L^2 - S^2}{2}. \qquad (5\text{-}19)$$

This means that H_{so} is a function of J^2, L^2, and S^2, and the corresponding quantum numbers j, l, and s are good quantum numbers. As a consequence, the eigenvalues of the total Hamiltonian $H^{A''}$ have the general form

$$\omega_{nljs} = \omega_{nl} + \xi(nl)\left[j(j+1) - l(l+1) - s(s+1)\right], \qquad (5\text{-}20)$$

where the ω_{nl} represent the energy eigenvalues with no spin-orbit coupling, and ξ is a function of nl (see, e.g., Tinkham 1964, 181–183). Then, the basis $\{|n, l, j, s, m_j\rangle\}$ of the Hilbert space of the system is defined by the CSCO $\{H, L^2, J^2, S^2, J_z\}$.

It is clear that the spin-orbit interaction removes the original degeneracy of the eigenvalues ω_{nl} of the atom with no interaction. Therefore, in this case, the actualization rule AR selects L^2, J^2, and S^2 as definite-valued observables, because all of them commute with $H^{A''}$ and have the same degeneracy in m_j as $H^{A''}$. But the space-rotation symmetry still present in the system leads to a degeneracy of $H^{A''}$, which is manifested by the fact that the energy eigenvalues ω_{nljs} do not depend on m_j. Then, according to AR, although in this case m_j is a good quantum number, J_z does not acquire a definite value, and this result agrees with the arbitrariness of the selection of the z-direction for J_z.

When a magnetic field is applied to the atom, the spectral lines split in different ways. The "normal" Zeeman effect, explained in the previous section, is observed in spin 0 states where, trivially, there is no spin-orbit coupling. In states where there is spin-orbit coupling, the action of the magnetic field produces a further splitting of the energy levels known as "anomalous" Zeeman effect. Nevertheless, the explanation of the anomalous effect is the same as that of the normal effect: the action of the magnetic field along the z-axis breaks the space-rotation symmetry of the Hamiltonian by privileging the z-direction, and this leads to the removal of the original degeneracy of the Hamiltonian in the quantum number m_j—instead of in the quantum number m_l as in the normal effect. In this case, AR prescribes that J_z is also a definite-valued observable.

5.7 Quantum decoherence

The key idea of the environment-induced decoherence approach (EID) is that macroscopic systems are never isolated but always interact with their environments. According to Wojciech Zurek, decoherence is what leads to the classical behavior of quantum systems: this process is the result of the interaction between an open quantum system and its environment, which selects the candidates to classical states (see also Zurek 1981, 1982, 1991, 1993, 1994; for a review, see Zurek 2003). In order to conform to the literature on decoherence, in this section, I will use the common terminology of those works, in which the term 'system' is also applied to the open case (the phenomenon of decoherence can also be approached from a closed-system perspective, as will be discussed in detail in Section 8.6 of Chapter 8).

Let us consider a system $\mathcal{S}$ corresponding to a Hilbert space $\mathcal{H}$, whose state $|\psi_S\rangle$ is initially in a superposition of the eigenstates $|a_i\rangle$ of the observable A of $\mathcal{S}$:

$$|\psi_S\rangle = \sum_i c_i \, |a_i\rangle \, . \tag{5-21}$$

If $\mathcal{S}$ is an open system in contact with its environment $\mathcal{E}$, the initial state of the whole system $\mathcal{S} \cup \mathcal{E}$ can be expressed as

$$|\psi_{SE}(0)\rangle = \left(\sum_i c_i \, |a_i\rangle\right) \otimes |\varepsilon_0\rangle , \tag{5-22}$$

where $|\varepsilon_0\rangle$ is the state of the environment before interacting with $\mathcal{S}$. It can be proved that, when the interaction Hamiltonian H_{int} satisfies certain conditions (see Zurek 1981), $|\psi_{SE}(0)\rangle$ evolves into

$$|\psi_{SE}(t)\rangle = \sum_i c_i \, |a_i\rangle \otimes |\varepsilon_i(t)\rangle , \tag{5-23}$$

where the $|\varepsilon_i(t)\rangle$ are the states of the environment correlated to the different states $|a_i\rangle$. In the context of the decoherence program, the state of the open system $\mathcal{S}$ is represented by the reduced density operator $\rho_r^S(t)$, resulting from tracing over the environmental degrees of freedom,

$$\rho_r^S(t) = Tr_{(E)}\left(|\psi_{SE}(t)\rangle \langle \psi_{SE}(t)|\right) = \sum_{ij} c_i c_j^* \, |a_i\rangle \langle a_i| \, \langle \varepsilon_i(t)| \, \varepsilon_j(t)\rangle , \tag{5-24}$$

where the factor $\langle \varepsilon_i(t)| \, \varepsilon_j(t)\rangle$ determines the size of the off-diagonal terms at each time. Many standard models for the interaction Hamiltonian H_{int} show that, when the environment has a very large number of degrees of freedom, the states $|\varepsilon_i(t)\rangle$ of the environment rapidly approach orthogonality:

$$\langle \varepsilon_i(t)| \, \varepsilon_j(t)\rangle \longrightarrow \delta_{ij}. \tag{5-25}$$

This means that the reduced density operator rapidly becomes approximately diagonal in the basis $\{|a_i\rangle\}$:

$$\rho_r^S(t) \longrightarrow \rho_r^S = \sum_i |c_i|^2 \, |a_i\rangle \langle a_i| . \tag{5-26}$$

This result has been interpreted as the proof that decoherence "explains" collapse, since quantum entanglement seems to become an effectively classical correlation as a consequence of the interaction between the system and the environment (Zurek 2003). This is the idea of Philip Anderson when he claims that "the word «decoherence» [...] describes the process that used to be called «collapse of the wave function»." (Anderson 2001, 492). From this perspective, decoherence would supply the definitive answer to the measurement problem (see, e.g., Auletta 2000).

Despite the optimism of some authors, many others have expressed serious warnings about the capability of decoherence to solve the measurement problem (see, e.g., Bub 1997, Zeh 2003, Bacciagaluppi 2020). In particular, it has been stressed that ρ_r^S, being a reduced state, refers to what Bernard d'Espagnat (1976) called "improper mixture," which cannot be interpreted in terms of ignorance: despite decoherence, the total system $\mathcal{S} \cup \mathcal{E}$ is still described by a superposition $|\psi_{SE}(t)\rangle$ at any time (the interpretation of the concept of reduced state will be discussed in detail in Sections 8.3 and 8.4 of Chapter 8). On this basis, in his response to Anderson, Stephen Adler concludes: "I do not believe that either detailed theoretical calculations or recent experimental results show that decoherence has resolved the difficulties associated with quantum measurement theory" (Adler 2003, 135). These and similar arguments have led even some contributors to the decoherence program to express their skepticism about the relevance of decoherence to the solution of the measurement problem; as Erich Joos (2000, 14) says: "Does decoherence solves the measurement problem? Clearly not."

Regarding the measurement problem, Maximilian Schlosshauer (2004) distinguishes between two questions:

- *The problem of definite outcomes*: Why do we seem to perceive the pointer to have a definite value but not a superposition of values? This is the traditional quantum measurement problem.
- *The problem of the preferred basis*: Since the expansion of the final composite state resulting from the interaction between the system and the measuring apparatus is in general not unique, the measured observable is not univocally defined.

Although currently there is a consensus that the decoherence theory alone does not solve the first problem, the solution to the preferred basis problem has been considered the main contribution of the decoherence program. For instance, Schlosshauer thinks that "based on the progress already achieved by the decoherence program, it is reasonable to anticipate that decoherence embedded in some additional interpretative structure could lead to a complete and consistent derivation of the classical world from quantum-mechanical principles." (Schlosshauer 2004, 1287; see also Elby 1994).

In his first papers on this subject, Zurek (1981, 1982) studied physical models of measurement in which the interaction between the system of interest, here the measuring apparatus, and the environment dominates the process. In those cases, the reduced state of the apparatus ends up being diagonal in the eigenvectors of an observable P that commutes with the Hamiltonian describing the apparatus-environment interaction. It is this property that makes P a dynamically stable observable that can thus play the role of the pointer observable of the apparatus: "One can then find a sufficient criterion for dynamically stable pointer states that preserve the system-apparatus correlations in spite of the interaction of the apparatus with the environment by requiring all pointer state projection operators to commute with the apparatus-environment Hamiltonian" (Schlosshauer 2004, 1278–1279; see also Elby 1994). Consequently, the commutation between P and the apparatus-environment interaction Hamiltonian has commonly been considered the criterion that defines the apparatus' preferred basis—sometimes called "*pointer basis*." In Zurek terms: "The form of the interaction Hamiltonian between the apparatus and its environment is sufficient to determine which observable of the measured quantum system can be considered «recorded» by the apparatus. The basis that contains that record—the pointer basis of the apparatus—consists of the eigenvectors of the operator which commutes with the apparatus-environment interaction Hamiltonian" (Zurek 1981, 1516).

During the years immediately following Zurek's 1981 paper, the idea that the apparatus-environment interaction picks out the preferred basis was widely accepted. However, almost a decade after his first works on the subject, Zurek realized that his original idea was a simplification: in more general situations, when the system's dynamics is relevant, the selection of the preferred basis is more complicated. On this basis, Juan Pablo Paz and Zurek (1999; see also Zurek 2003) introduced the distinction between three basically different regimes for the selection of the preferred basis. They differ in the relative strength of the self-Hamiltonian of the system of interest and the system-environment interaction Hamiltonian:

- The first regime is the quantum measurement situation, where the self-Hamiltonian of the system—here the measuring apparatus—can be neglected and the evolution is completely dominated by the interaction Hamiltonian. In such a case, the preferred states are directly the eigenstates of the interaction Hamiltonian (Zurek 1981).
- The second regime is the more realistic and complex situation, where neither the self-Hamiltonian of the system nor the interaction Hamiltonian are clearly dominant, but both induce a non-trivial evolution. In this case, the preferred basis arises from the interplay between self-evolution and interaction. The quantum Brownian motion belongs to this case (Paz 1994).
- The third regime corresponds to the situation where the dynamics is dominated by the self-Hamiltonian of the system. In this case, the preferred states are simply the eigenstates of this self-Hamiltonian (Paz and Zurek 1999).

According to Schlosshauer (2004, 1280), these three regimes explain why many systems, especially in the microscopic domain, are typically found in energy eigenstates although the interaction Hamiltonian depends on an observable different than the energy of the interacting systems.

At first glance, the open-system approach of the decoherence program seems to be at odds with the closed-system perspective of the MHI, which conceives closed systems as the only legitimate quantum systems. However, a careful analysis of the results obtained by decoherence theorists shows that this first impression is not correct. Let us consider again the open system $\mathcal{S}$ in interaction with its environment $\mathcal{E}$. If $\mathcal{H}^S$ and $\mathcal{H}^E$ are the Hilbert spaces of $\mathcal{S}$ and $\mathcal{E}$, respectively, $\mathcal{H}^{SE} = \mathcal{H}^S \otimes \mathcal{H}^E$ is the Hilbert space of the whole system $\mathcal{S} \cup \mathcal{E}$. According to the dynamical postulate of quantum mechanics, the entangled state $|\psi_{SE}(t)\rangle$ of $\mathcal{S} \cup \mathcal{E}$ (see eq. (5-23)) evolves according to the Schrödinger equation under the action of the total Hamiltonian H^{SE}:

$$H^{SE} = H^S \otimes I^E + I^S \otimes H^E + H_{\text{int}}, \tag{5-27}$$

where H^S and H^E are the self-Hamiltonians of $\mathcal{S}$ and $\mathcal{E}$, respectively, H_{int} is the interaction Hamiltonian, and I^S and I^E are the identity operators of $\mathcal{B}(\mathcal{H}^S)$ and $\mathcal{B}(\mathcal{H}^E)$, respectively. The pointer P is an observable of $\mathcal{S}$, that is, $P = P^S \in \mathcal{B}(\mathcal{H}^S)$. But the same property can be represented in $\mathcal{B}(\mathcal{H}^{SE})$ as

$$P^{SE} = P^S \otimes I^E \in \mathcal{B}(\mathcal{H}^{SE}). \tag{5-28}$$

The above-mentioned requirement of the dynamical stability of the pointer amounts to the demand that P^{SE} be a constant of motion of the whole system, that is,

$$\left[P^{SE},\ H^{SE}\right] = \left[P^S \otimes I^E,\ H^S \otimes I^E + I^S \otimes H^E + H_{\text{int}}\right] = 0. \tag{5-29}$$

Since always $[P^S \otimes I^E,\ I^S \otimes H^E] = 0$, then the stability requirement for the pointer observable becomes

$$\left[P^S \otimes I^E,\ H^S \otimes I^E + H_{\text{int}}\right] = 0. \tag{5-30}$$

Now, the three regimes distinguished by Paz and Zurek as the result of the application of the decoherence theory to a number of models turn out to be the three particular cases of the condition expressed by eq. (5-30), and can be redescribed in terms of that condition:

- First regime: When $H^S \otimes I^E \ll H_{\text{int}}$, the self-Hamiltonian of $\mathcal{S}$ can be neglected, and then the condition of eq. (5-30) becomes $[P^{SE},\ H_{\text{int}}] = 0$. Therefore, the preferred basis is defined by the interaction Hamiltonian H_{int}.
- Second regime: When $H^S \otimes I^E \simeq H_{\text{int}}$, neither the self-Hamiltonian of $\mathcal{S}$ nor the interaction with the environment are clearly dominant. In this case, the preferred basis is defined by the condition of eq. (5-30).
- Third regime: When $H^S \otimes I^E \gg H_{\text{int}}$, the dynamics is dominated by the self-Hamiltonian of $\mathcal{S}$ and, thus, the condition of eq. (5-30) becomes $[P^S \otimes I^E,\ H^S \otimes I^E] = [P^S,\ H^S] = 0$. Therefore, the preferred basis is simply given by the eigenstates of H^S.

These results show that the three regimes, identified by studying specific models case by case, become particular cases of a general definition: the preferred basis is defined by the total Hamiltonian of the whole system. It is for this reason that the preferred states depend on the Hamiltonian's component that dominates the whole evolution. Therefore, the fact, noted by Schlosshauer, that many systems are typically found in energy eigenstates, far from being surprising, necessarily results from the requirement of stability for the pointer and from the central role that the total Hamiltonian and its components play in meeting this requirement.

The overall conclusion of this analysis is that, contrary to what a first impression may suggest, the decoherence program and the MHI are not incompatible approaches (see Lombardi 2010). By contrast, once it is understood that the definition of the preferred basis must not be searched in the system-environment interaction alone, it is not difficult to show that the pointer basis selected by decoherence agrees with the basis corresponding to the preferred context selected by the actualization rule of the MHI.

5.8 The Born-Oppenheimer approximation

As already explained, the actualization rule of the MHI endows the Hamiltonian of the system with the role of selecting the preferred context and, as a consequence, the energy of the system always acquires an actual definite value. But this does not mean that the momentum P is a definite-valued observable in any case, since it does not always commute with the Hamiltonian. In fact, a general form of the Hamiltonian is

$$H = \frac{P^2}{2m} + V(Q), \tag{5-31}$$

where $V(Q)$ is a scalar field, function of the position Q. If the mass m of the system is small, the kinetic term prevails over the potential term, and the Hamiltonian approximately commutes with P^2. In turn, for very large masses, the kinetic term can be neglected and H approximately commutes with $V(Q)$. So, the MHI actualization rule supports the usual physical assertion that "small" systems have approximately definite momentum and "large" systems have approximately definite position. In particular, the MHI agrees with the common physical assumption that, in an atom, electrons have definite momentum but not definite position, and the nucleus has definite position but not definite momentum. In general, the rule explains the fact that macroscopic systems, with their huge masses,—approximately—possess a definite value of position.

This point has a particular relevance in chemistry, where the description of molecules is traditionally based on the Born-Oppenheimer approximation, whose first step is the so-called *clamped nuclei approximation*: the electron-nucleus interactions are conceived in terms of electrons in the Coulomb potential produced by nuclei fixed at definite positions. This amounts to describe the nuclei as classical-like particles, that is, as precisely localized objects. However, from a strictly quantum-mechanical viewpoint, this strategy contradicts the Heisenberg principle: "In this «clamping-down» approximation, the atomic nuclei are treated essentially as classical particles; [...] this picture is non-quantum in a very fundamental way as the simultaneous assignment of fixed positions and fixed momenta (namely, zero) to them violates the Heisenberg uncertainty principle." (Chang 2015, 198; this point was also noticed by other authors as, e.g., Woolley 1978, Primas 1983, Amann 1992). Of course, this does not call into question the use of the Born-Oppenheimer approximation, which is pervasive in quantum chemistry: its justification relies on its own success in the calculation of the energy levels of molecules.

This approximation strategy of holding the nucleus at rest in a definite position is commonly thought of as formally arising from making the masses of the nuclei infinite. However, from a strictly quantum-mechanical viewpoint, without interpretative postulates the assumption of infinite nuclear masses does not yet explain why nuclei can be treated as having a definite value of position (see discussion in Lombardi and Castagnino 2010, Sutcliffe and Woolley 2015).

Of course, the MHI does not overcome the conceptual difficulty involved in the clamped nuclei approximation. However, it provides a conceptual rationale for the strategy of fixing the nuclei. For large masses, the Hamiltonian is—approximately—invariant under boost transformations and, therefore, approximately commutes with position. As a consequence, according to the MHI actualization rule, the position observable acquires a definite value. If this idea is applied to the nuclei of a molecule, with their very large masses when compared to those of the electrons, the assumption that nuclei can be treated as having definite positions and forming a definite geometrical structure becomes conceptually plausible.

5.9 The problem of symmetry in quantum chemistry

One of the hottest topics in the philosophy of chemistry is that of the relation between chemistry and physics. The problem reveals its main manifestation in the debate about the nature of molecular structure, partially defined by the spatial arrangement of the nuclei in a molecule. The debate does not focus on an auxiliary or secondary notion, but on a central concept of chemistry: molecular structure is the main factor in the explanation of reactivity, it is "the central dogma of molecular science" (Woolley 1978, 1074). As Robin Hendry claims, "molecular structure is so central to chemical explanation that to explain molecular structure is pretty much to explain the whole of chemistry" (Hendry 2010, 183). The problem consists in the fact that the concept of molecular structure finds no comfortable place in the theoretical framework of quantum mechanics. Whereas "[t]he alpha and omega of molecular chemistry is the doctrine that molecules exist as individual objects and that every molecule has a shape, characterized by its molecular frame" (Primas 1994, 216), the classical idea of definite spatial positions for the atomic nuclei, conceived as individual objects, is, at least, strongly conflicting in the quantum context. Although this problem has been widely discussed, the debate is far from settled.

Some authors retain an ontologically reductionist standpoint by considering the problem of molecular structure as the result of deep epistemic limitations (Sutcliffe and Woolley 2011, 2012) or of an inadequate formulation of the issue (Hettema 2012). Others consider that quantum decoherence provides the reductive link between the domain of classical molecular chemistry and the quantum realm (Trost and Hornberger 2009, Scerri 2011, 2013). From an opposite viewpoint, it has been claimed that molecular structure cannot be strictly explained by quantum mechanics (Primas 1983, 1998, Amann 1992, Hendry 2010, Martínez González, Fortin, and Lombardi 2019, Fortin and Lombardi 2021), even when decoherence is taken into account (Fortin, Lombardi, and Martínez González 2016). Nevertheless, this limitation is not always understood in the same way. For some authors, the impossibility of strict explanation in quantum terms is due to the merely approximate nature of the concept of molecular structure (Woolley 1978, 1982, Bader 1990, 2011). Others, by contrast, take an ontologically non-reductionist stance and consider the case of molecular structure as an example of ontological emergence (Hendry 1998, 2004, 2010, Matta, Lombardi, and Jaimes Arriaga 2020).

Perhaps the most striking expression of the problem of molecular structure is what Hendry (2010), following Guy Woolley and Brian Sutcliffe (1977), calls "the symmetry problem." The total Hamiltonian of a molecule takes into account all the interactions between nuclei, between electrons, and between electrons and nuclei. Since such interactions are Coulombian, they endow the Hamiltonian with certain symmetries, which are mathematically inherited by its ground state. For example, the ground state of the Hamiltonian is rotationally symmetric. However, the rotational asymmetry of polyatomic molecules is essential in the explanation of their chemical behavior. As Hendry stresses, in the quantum domain no directional property can be assigned to an isolated molecule in the ground eigenstate of its total Hamiltonian. Consider, for instance, the hydrogen chloride molecule with its asymmetric charge distribution, which explains its acidic behavior and its boiling point; however, according to quantum mechanics, the expectation value of the electric dipole moment of the molecule in an arbitrary eigenstate of the total Hamiltonian is always zero: "the acidic behavior of the hydrogen chloride molecule is conferred by its asymmetry," but "the asymmetry is not conferred by the molecule's physical basis according to physical laws" (Hendry 2010, 186; see also Hendry 1998; for a precise account of the experimental distinction between polar and non-polar molecules, see Woolley 1976).

Instead of analyzing the case of hydrogen chloride, let us consider the more familiar case of the ammonia molecule, whose tetrahedral structure—with the negative nitrogen in one of the vertices and the positive hydrogens in the other three—accounts for its electric dipole moment. Philip Anderson recalls the classical chemical explanation of the phenomenon: since no stationary state of the molecule can have a non-zero value of the electric dipole moment, the molecule "oscillates" between two asymmetric states: "By means of quantum mechanical tunneling, the nitrogen can leak through the triangle of hydrogens to the other side." (Anderson 1972, 394). This explanation appeals to the so-called *nitrogen inversion* or *umbrella inversion*: a fluxional process by which the nitrogen "moves" through the plane formed by the hydrogens and the molecule "turns inside out" (nitrogen inversion is not exclusive of ammonia, but is also characteristic of other kinds of molecules). Although this is a traditional explanation in chemistry, it is a semi-classical description that takes the geometrical structure of the molecule for granted: at each time, the positions of the nuclei are definite and, as a consequence, the molecule has a precise shape. By contrast, in a quantum-mechanical description, the molecule is in a superposition of the two states, and a system in a quantum superposition state does not oscillate between the two components of the superposition: "it is tempting to consider that the nitrogen nucleus lies, at any given time, on a well-defined side of the plane of the three protons, and that it jumps from one side to the other at more or less regular intervals of time. Of course, this «naive» view is at variance with strict quantum theory, but it is nevertheless very commonly used, either for teaching purposes or in popularization papers (see e.g. Anderson 1972)." (Claverie and Diner 1980, 74). If the issue at stake is the problem of symmetry, the question is how to explain the non-zero value of the electric dipole moment exclusively in quantum terms and, then, semi-classical descriptions must be excluded.

Let us analyze the case of ammonia in more detail. Since the Coulombian interactions between the components of a molecule only depend on the observable corresponding to the "distance" between particles, the Hamiltonian H of the molecule has parity symmetry, that is, symmetry under spatial inversion. Therefore, H commutes with the parity operator P:

$$[P, H] = 0. \qquad (5\text{-}32)$$

This means that the eigenstates of the Hamiltonian are also parity-symmetric. Moreover, this feature is preserved during the time evolution of the system, because the parity operator commutes with the Hamiltonian and, as a consequence, is a constant of motion. Therefore, the ground state of the molecule—one of the eigenstates of H—must also be parity-symmetric. Let us now consider an observable D_z, "electric dipole moment in z-direction," whose eigenstates are $|d_+(z)\rangle$ and $|d_-(z)\rangle$ and whose eigenvalues $d_+(z)$ and $d_-(z)$ represent the moment pointing "up" and "down" in direction z, respectively. These eigenstates are parity-asymmetric since

$$\begin{aligned} P\,|d_+(z)\rangle &= |d_-(z)\rangle \\ P\,|d_-(z)\rangle &= |d_+(z)\rangle\,. \end{aligned} \qquad (5\text{-}33)$$

Therefore, the states $|d_+(z)\rangle$ and $|d_-(z)\rangle$ cannot be eigenstates of the Hamiltonian, which, as explained above, are parity-symmetric. Since the ground state $|\Psi_0\rangle$ is parity-symmetric, it can be expressed as a superposition of those asymmetric states with the same weight for the two components:

$$|\Psi_0\rangle = \frac{1}{\sqrt{2}}\left(|d_+(z)\rangle + |d_-(z)\rangle\right). \qquad (5\text{-}34)$$

Why, then, are ammonia molecules always observed to have a non-zero electric dipole moment, despite the fact that their ground states have zero electric dipole moment?

The actualization rule AR of the MHI applies to this situation analogously to the way it was applied to the cases analyzed in the first six sections of this chapter. According to AR, since the observable D_z does not commute with the Hamiltonian of the ammonia molecule, it is not definite-valued: the direction of the electric dipole moment in z of the isolated molecule is not definite. This agrees with the fact that observers have no empirical access to the electric dipole moment of an isolated molecule. In order to know the value of D_z, a measurement must be performed on the system: it is necessary to introduce an electric field E in direction z that breaks the original parity symmetry. But now the system is no longer the isolated ammonia molecule, but a new system with a Hamiltonian H' that includes the action of E and, as a consequence, is not parity-symmetric. In other words, the experimental access to the electric dipole moment of a molecule is possible only through an interaction that breaks the parity symmetry of the original Hamiltonian and, therefore, makes the molecule no longer isolated.

It is worth mentioning that, from the viewpoint of the MHI, symmetries are not generated by the spatial shapes of molecules: symmetries are internal features of the Hamiltonian of the whole closed system; they must not be understood in terms of spatial configurations. According to the MHI, it is incorrect to assume that the ammonia molecule has, before being observed, a definite value of the electric dipole moment resulting from its geometrical shape, and that such a property is discovered when the molecule interacts with some external environment. From the MHI perspective, isolated ammonia molecules have no definite value of the electric dipole moment; this property acquires a definite value when the molecule becomes a part of a whole system that is no longer parity-symmetric.

Finally, it is important not to extrapolate the above conclusions to the problem of molecular structure, which consists of explaining the spatial arrangement of the nuclei of a molecule in quantum terms. Solving the problem of molecular structure requires that the quantum state—approximately—accounts, at least, for the positions of all the nuclei. But this account is not given by the states $|d_+(z)\rangle$ and $|d_-(z)\rangle$. In fact, the ammonia molecule has many degrees of freedom represented in its Hilbert space $\mathcal{H}^M$, in particular, the quantum degrees of freedom related to the observables R_α corresponding to the positions of the nuclei. However, with its two eigenvalues, D_z is a highly degenerate "collective observable" in $\mathcal{H}^M$, whose definite value—explained by the interaction of the molecule with an external field—still says nothing about the definite—or almost definite—values of the nuclei's positions R_α. Broadly speaking, there are many different ways of being an ammonia molecule with a definite value of electric dipole moment: there are many different geometrical configurations of the nuclei that realize the values $d_+(z)$ and $d_-(z)$ of D_z. This means that solving the symmetry problem is still far from solving the problem of the molecular structure as a whole (for a detailed discussion, see Fortin and Lombardi 2021).

5.10 The problem of optical isomerism in quantum chemistry

In chemistry, the composition of a molecule is expressed by its chemical formula, which specifies in what stoichiometric proportion the component elements are present in the chemical compound. However, the formula furnishes no information about the geometrical arrangement of the component atoms. For example, $C_2H_4O_2$ corresponds to different chemical compounds: methyl formate, acetic acid, and glycolaldehyde. Compounds that contain the same number of atoms of each element (and, therefore, have the same chemical formula) but have different spatial arrangements of their atoms are called *isomers*. Isomerism is a phenomenon of great relevance in chemistry, since it explains the fact that substances with the same composition have different physical and chemical behaviors.

There are two main forms of isomerism: structural isomerism and stereoisomerism. In *structural isomers*, sometimes referred to as *constitutional isomers*, the component atoms—and the functional groups—are connected to each other in different ways. In *stereoisomers*, the component atoms are linked to each other in the same way, but the geometrical configurations of atoms and functional groups in space are different. The class of stereoisomers includes the subclass of *enantiomers*: the structures of the members of a pair of enantiomers are non-superimposable mirror-images of each other, that is, the molecules are *chiral*.

When immersed in a symmetric environment, the members of a pair of enantiomers have identical chemical and physical properties. However, they have different behaviors when interacting with polarized light: they rotate plane-polarized light by equal amounts but in opposite directions. It is for this reason that they are called *optical isomers* and it is usually said that they are *optically active*. When the plane of polarization is rotated clockwise (as seen by an observer receiving the light), the optical isomer is called *dextro-rotatory* (D); in the case of counterclockwise rotation, the optical isomer is called *levo-rotatory* (L).

Chiral molecules play a key role in the enzymatic reactions of biological systems: many pharmacological drugs are chiral and generally only one of the members of the pair exhibits biological activity. The action of drugs is usually explained by means of the idea of "receptor." Receptors are protein molecules in the human body that are exclusively built from L-amino acids. This means that receptors are chiral and, as a consequence, have different reactions with the two enantiomers of a chiral drug. For instance, aspartame is a sweetening agent that has two enantiomers: one of them is more than a hundred times sweeter than sucrose; the other, is tasteless or slightly bitter. The most dramatic example of the different biological activity of enantiomers is that of thalidomide, a drug of molecular formula $C_{13}H_{10}N_2O_4$ introduced to the market in 1957 by a West German pharmaceutical company. It was prescribed to pregnant women as a sedative and to prevent nausea; its use spread rapidly to forty-six countries. But the drug was made and marketed as a mixture of D-thalidomide and L-thalidomide: the D-thalidomide is an effective sedative, whereas the L-thalidomide is a teratogen, that is, an agent affecting the development of the fetus and causing structural or functional abnormality. The exact number of drug victims worldwide is not known, although estimates range from 10.000 to 100.000. These are only some examples of the fact that, at present, the control and production of enantiomers is a central activity in the pharmaceutical industry.

Chemists not only successfully explain the behavior of enantiomers in terms of the geometrical configuration of the molecules, but they have also developed many techniques for the preparation of compounds where the proportions of the enantiomers can be controlled. The problem of chirality arises not at the chemical level, but with the attempts to give an account of the phenomenon exclusively in quantum terms.

As explained in the previous section, when a molecule is described by quantum mechanics, the Coulombian Hamiltonian only depends on the observables corresponding to the "distances" between the particles composing the molecule. In particular, if only the atomic nuclei are considered, the Hamiltonian depends exclusively on the inter-nuclear distances. In the case of optical isomerism, all the inter-atomic distances are the same for the two members of the pair and, as a consequence, the Hamiltonian is exactly the same for both. This means that quantum mechanics offers the same description for two chemical structures that can effectively be distinguished in practice through their optical and biological activity.

It is very important to emphasize that the problem of optical isomerism is completely independent of the Born-Oppenheimer approximation. Let us suppose, for example, that we write down the full Coulombian Hamiltonian H (with no approximation) of a molecule of alanine $C_3H_7NO_2$. Even if we cannot solve the time-independent Schrödinger equation for this Hamiltonian due to its complexity, we know that it only depends on the distance of the component particles and, therefore, this full Hamiltonian cannot account for the difference between D-alanine and L-alanine. As Sutcliffe and Woolley claim: "Clearly then, an eigenstate of H does not correspond to a classical molecule with structure!" (Sutcliffe and Woolley 2012, 416). In short, the problem of the quantum-mechanical distinction between optical isomers of the same substance is beyond the Born-Oppenheimer approximation and its underlying assumptions.

The problem embodied in optical isomerism points to a deep difficulty in the attempts to account for molecular structure in quantum terms. Although, conceptually, it is a particular case of the problem of symmetry discussed in the previous section, historically it was treated as an independent problem. In fact, the problem of isomerism was already noticed in the 1920s by Friedrich Hund, a pioneer in the development of quantum chemistry. The so-called *Hund's paradox* can be formulated in two versions. The first one, due to Hund (1927) himself, reads as follows: since chiral states (states $|L\rangle$ and $|D\rangle$, corresponding to levo-rotation and dextro-rotation, respectively) are not eigenstates of the Hamiltonian (since it is invariant under spatial reflection), and none of them corresponds to the ground state, why do chiral molecules display an optical activity that is stable in time, associated to a well-defined chiral state?, why are they not in a superposition of the two possible chiral states? More recently, Hund's paradox was formulated in a slightly stronger version (Berlin, Burin, and Goldanskii 1996): why do certain molecules have the property of chirality?

As explained in the previous section, the total Hamiltonian H of any molecule only depends on the distances between the interacting particles. Therefore, H is parity-symmetric and commutes with the parity operator P. Let us consider a chiral molecule from the perspective of quantum mechanics. The eigenstates $|\omega_n\rangle$ of the molecule's Hamiltonian have parity symmetry:

$$P\,|\omega_n\rangle = \pm\,|\omega_n\rangle\,. \tag{5-35}$$

In particular, the even levels have even parity, and the odd levels have odd parity. For example, the ground state $|\omega_0\rangle$ is symmetric and the first excited state $|\omega_1\rangle$ is anti-symmetric:

$$P\,|\omega_0\rangle = +\,|\omega_0\rangle \qquad P\,|\omega_1\rangle = -\,|\omega_1\rangle\,. \tag{5-36}$$

However, experimental data show that the states of optical isomers do not have this symmetry. In fact, if the isomers are described by the chiral states $|L\rangle$ and $|D\rangle$, each description is the mirror image of the other:

$$\begin{aligned} P\,|L\rangle &= |D\rangle \\ P\,|D\rangle &= |L\rangle\,. \end{aligned} \tag{5-37}$$

Therefore, the states $|L\rangle$ and $|D\rangle$ cannot be eigenstates of the Hamiltonian, which, as explained above, are symmetric or anti-symmetric. On the contrary, $|L\rangle$ and $|D\rangle$ can be expressed as superpositions of $|\omega_0\rangle$ and $|\omega_1\rangle$:

$$\begin{aligned} |L\rangle &= \frac{1}{\sqrt{2}}\left(|\omega_0\rangle + |\omega_1\rangle\right) \\ |D\rangle &= \frac{1}{\sqrt{2}}\left(|\omega_0\rangle - |\omega_1\rangle\right). \end{aligned} \tag{5-38}$$

Consequently, the ground state is a superposition of the chiral states:

$$|\omega_0\rangle = \frac{1}{\sqrt{2}}\left(|L\rangle + |D\rangle\right). \tag{5-39}$$

Why, then, are chiral states always observed? Why are superpositions of them never observed? Hund's paradox might be expressed in the following terms: since the ground state of optical isomers is a superposition of chiral states, in that state isomers should not manifest the optical activity corresponding to states $|L\rangle$ and $|D\rangle$; but we have empirical access to the optical activity associated to those states; how to avoid this seeming contradiction?

Let us consider the observable C, whose eigenstates are $|D\rangle$ and $|L\rangle$, and call it *chiral-rotation*: the eigenvalues d and l of C represent the properties dextro-rotation and levo-rotation, respectively (for the detailed treatment of this case, see Fortin, Lombardi, and Martínez González 2018). The observable C should be definite-valued for the molecule to be dextro-rotatory or levo-rotatory. Moreover, it is easy to see that C does not commute with H. Analogously to several examples in this chapter, in this case the actualization of the observable C would determine the rotation property of the molecule in a completely arbitrary way: it would introduce in the molecule an asymmetry not contained in its Hamiltonian. As a consequence, from the viewpoint of the MHI, the observable C of the enantiomer molecule is not definite-valued, that is, it does not belong to the preferred context. In other words, strictly speaking, the isolated molecule is neither dextro-rotatory nor levo-rotatory: chiral rotation is a property that has no definite value.

The fact that chiral rotation does not have a definite value in the isolated molecule is compatible with experience: there is no experimental evidence of the handedness of an isolated molecule. In order to make chiral rotation manifest as a definite-valued property, it is necessary to interact with the molecule. But not just any interaction serves this purpose: the interaction must be such that it breaks the parity symmetry of the original Hamiltonian H. More precisely, the molecule must interact with another system in such a way that the Hamiltonian H' of the new whole system is no longer parity-symmetric. Additionally, the observable C (strictly speaking, $C \otimes I^{Asym}$, where I^{Asym} is the identity operator corresponding to the interacting asymmetric system) must commute with the total Hamiltonian H' in order to obtain a stable reading of chiral rotation. Under these conditions, according to the MHI, C acquires a definite actual value: dextro-rotation or levo-rotation can be measured. However, now the system is no longer the isolated molecule, but the molecule in interaction with a parity-asymmetric environment.

This is precisely what happens when the molecule rotates the plane of polarization of plane-polarized light. In this case, a good candidate for the asymmetric Hamiltonian H' is that commonly introduced in quantum chemistry to describe the interaction between molecules and polarized light (see, e.g., Shao and Hänggi 1997), which is a function of the electric field E and the magnetic field B of the light. In this way, the observable C turns out to be a definite-valued observable of the new whole system. In other words, optical activity is not an intrinsic property of the chiral molecule, but is a property of the "molecule plus light" system. Analogously, biological activity is not an intrinsic property of the molecules of pharmacological drugs, but is a property of the "drug molecule plus receptor" systems. In a certain sense, this answer to Hund's paradox agrees with the stance that finds the solution to the symmetry problem in the interaction of the molecule with its environment. However, the MHI approach does not appeal to decoherence, but relies on an interpretation of quantum mechanics that explicitly accounts for measurement from the perspective of the symmetries of the system.

As is well known by chemists, optical activity is stably correlated with other properties, generally conceived as structure-based as the property of optical activity itself: if, say, dextro-rotation is observed, the corresponding values of the correlated properties must also be observed. In theoretical terms, the definite value of chiral rotation must lead to the definite-valuedness of those properties, and the actual value acquired by chiral rotation must be correlated with the actual values acquired by the correlated properties. This can be easily explained in the context of the MHI. In fact, if a property is stably correlated with chiral rotation *C*, it must be represented by an observable that commutes with and is functionally related to the observable *C*. Therefore, according to the MHI, if the molecule interacts with another system in such a way that *C* acquires a definite value in the new whole, then the observable commuting with *C* will also acquire a definite value. In turn, the functional relation between *C* and that commuting observable guarantees the proper relations between their actual values. Summing up, from the viewpoint of the MHI, the same symmetry breaking that explains why a definite value of chiral rotation is observed also explains why the definite values of the properties correlated with chiral rotation are also observed.

I decided to present the problem of optical isomerism separately due to the chemical relevance of the phenomenon of chirality. However, as mentioned above, it is a particular case of the problem of symmetry. For this reason, solving the problem of optical isomerism does not yet solve the problem of molecular structure. Analogously to the case of the electric dipole moment treated in the previous section, there are many different ways of being a chiral molecule, many different geometrical configurations of the nuclei that realize, say, a levo-rotatory molecule. In other words, solving Hund's paradox amounts to explaining the chiral behavior of the molecule, but is still far from solving the problem of molecular structure as a whole.

6
The Quantum Measurement Problem

In rough terms, the quantum measurement problem is the problem of explaining how, when a system in a superposition state is measured, the pointer observable of the measuring apparatus is always found in a definite value. Although quantum mechanics involves several interpretative challenges, the measurement problem has been the main focus of the interest of physicists and philosophers working on the foundations of the theory: most interpretations were designed specifically to solve it. In this chapter, I will show that the MHI offers a solution to the measurement problem for both ideal and non-ideal measurements, and that, in the non-ideal case, the interpretation distinguishes between reliable and non-reliable measurements. I will also explain how the MHI accounts for consecutive measurements. Finally, I will emphasize that the MHI approach provides a new, more physical understanding of the concept of quantum measurement.

6.1 Single measurement, frequency measurement, and state measurement

According to the standard von Neumann model, a quantum measurement is an interaction between a system $\mathcal{S}$ and a measuring apparatus $\mathcal{M}$. Before the interaction, $\mathcal{M}$ is prepared in a ready-to-measure state $|p_0\rangle$, eigenvector of the pointer observable P of $\mathcal{M}$, and the state of $\mathcal{S}$ is a superposition of the eigenstates $|a_i\rangle$ of an observable A of $\mathcal{S}$. The interaction introduces a correlation between the eigenstates $|a_i\rangle$ of A and the eigenstates $|p_i\rangle$ of P. If the correlation is perfect, then

$$|\psi_0\rangle = \sum_i c_i\,|a_i\rangle \otimes |p_0\rangle \quad \longrightarrow \quad |\psi\rangle = \sum_i c_i\,|a_i\rangle \otimes |p_i\rangle . \qquad (6\text{-}1)$$

The problem consists in explaining why, the state $|\psi\rangle$ being a superposition of the $|a_i\rangle \otimes |p_i\rangle$, the pointer P acquires a definite value.

In the context of the collapse interpretations, the pure state $|\psi\rangle$ is assumed to "collapse" to a mixture ρ^c,

$$\rho^c = \sum_i |c_i|^2\,|a_i\rangle \otimes |p_i\rangle\langle a_i| \otimes \langle p_i| , \qquad (6\text{-}2)$$

where the probabilities $|c_i|^2$ are interpreted in terms of ignorance. Then, in this situation, the state of the apparatus is supposed to be one of the eigenvectors $|p_i\rangle$ of P,

The Modal-Hamiltonian Interpretation of Quantum Mechanics. Olimpia Lombardi, Oxford University Press.
© Olimpia Lombardi (2026). DOI: 10.1093/9780198951407.003.0006

say $|p_k\rangle$, and therefore P acquires the definite value p_k, that is, the eigenvalue corresponding to the eigenvector $|p_k\rangle$, with probability $|c_k|^2$. In the Modal Interpretations, the problem is to explain the definite reading of the pointer, with its associated probability, without the collapse hypothesis. In the case of the MHI, the actualization rule AR is what must perform this task.

The first step of the argumentation of this chapter consists in framing the von Neumann model in the context of measurement practices. In fact, due to the probabilistic nature of quantum mechanics, the maximum information about a quantum system is always obtained by means of repeated measurements on the same system or on—approximately—identical systems. Therefore, three concepts of measurement must be distinguished:

- *Single measurement*: It is a single process, in which the reading of the pointer is registered. A single measurement, considered in isolation, does not yet provide relevant information about the state of the system $\mathcal{S}$.
- *Frequency measurement*: It is a repetition of identical single measurements, whose purpose is to obtain the values $|c_i|^2$ on the basis of the frequencies of the pointer readings in single measurements. A frequency measurement provides relevant information about the state of $\mathcal{S}$, but is not yet sufficient to completely identify such a state.
- *State measurement*: It is a collection of frequency measurements, each one of them with its particular experimental arrangement. Each arrangement correlates the pointer P of the apparatus $\mathcal{M}$ with an observable A_i of the system, in such a way that the A_i are not only different, but also non-commuting with each other. The information obtained through such a collection of frequency measurements is sufficient to reconstruct the state of $\mathcal{S}$.

The von Neumann model addresses the quantum measurement problem in the framework of the single measurement. This is completely reasonable to the extent that, if an adequate explanation of the single case is not available, the results obtained by the repetition of single cases cannot be explained. Nevertheless, although I will mainly analyze the single-measurement case, I will not forget that a single measurement is always an element of a measurement procedure by which, in the end, frequencies are obtained. Frequency measurements and state measurements will be discussed below, in Section 6.10 of this chapter.

6.2 Ideal measurements

In the von Neumann model and, in general, in the discussions about the quantum measurement problem, the Hamiltonians involved in the process are not usually taken into account. In the context of the MHI, where the Hamiltonians of the systems play a key role, a more detailed description of the measurement process must be provided. Thus, a single measurement is characterized as a three-stage process:

- *Stage I* ($t \leq t_0$): The system S and a measuring apparatus $\mathcal{M}$ do not interact.
- *Stage II* ($t_0 < t < t_1$): During this stage, S and $\mathcal{M}$ interact, and the interaction establishes the measuring correlation.
- *Stage III* ($t \geq t_1$): The interaction ends at $t = t_1$.

Stage I: Let us suppose that the single measurement is an element of a state measurement whose purpose is to obtain the coefficients of the state of the elemental quantum system $\mathcal{S} : (\mathcal{O}^S, H^S)$, with $\mathcal{O}^S \subset \mathcal{B}(\mathcal{H}^S)$. Let us also consider a non-degenerate observable $A \in \mathcal{O}^S$:

$$A = \sum_i a_i \, |a_i\rangle \langle a_i|, \quad \text{where } \{|a_i\rangle\} \text{ is a basis of } \mathcal{H}^S. \tag{6-3}$$

At a time $t = t_0$, the state of $\mathcal{S}$ reads

$$|\psi_S(t_0)\rangle = \sum_i c_i \, |a_i\rangle \in \mathcal{H}^S. \tag{6-4}$$

For simplicity, let us assume that the Hamiltonian $H^S \in \mathcal{O}^S$ of $\mathcal{S}$ is non-degenerate (the degeneracy of H^S does not modify the core of the argument):

$$H^S = \sum_i \omega_i^S \, |\omega_i^S\rangle \langle \omega_i^S|, \quad \text{where } \{|\omega_i^S\rangle\} \text{ is a basis of } \mathcal{H}^S. \tag{6-5}$$

The measuring apparatus is an elemental quantum system $\mathcal{M} : (\mathcal{O}^M, H^M)$, with $\mathcal{O}^M \subset \mathcal{B}(\mathcal{H}^M)$. $\mathcal{M}$ has an observable $P \in \mathcal{O}^M$ that must possess different and macroscopically distinguishable eigenvalues in order to play the role of the pointer:

$$P = \sum_i p_i \, \Pi_i, \quad \text{where the } \Pi_i \in \mathcal{O}^M \text{ are the eigenprojectors of } P. \tag{6-6}$$

For simplicity, let us assume that the Hamiltonian $H^M \in \mathcal{O}^M$ of $\mathcal{M}$ is non-degenerate (the physical reasonability of this assumption will be discussed in Section 6.9 of the present chapter):

$$H^M = \sum_i \omega_i^M \, |\omega_i^M\rangle \langle \omega_i^M|, \quad \text{where } \{|\omega_i^M\rangle\} \text{ is a basis of } \mathcal{H}^M. \tag{6-7}$$

For the reading of the pointer to be possible, the eigenvectors $|p_i\rangle$ of P must be stationary. Thus, the apparatus $\mathcal{M}$ is constructed in such a way that P commutes with H^M:

$$\left[H^M, P\right] = 0. \tag{6-8}$$

At time $t = t_0$, the apparatus $\mathcal{M}$ is prepared in a ready-to-measure state $|p_0\rangle$, eigenvector of P:

$$|\psi_M(t_0)\rangle = |p_0\rangle \in \mathcal{H}^M. \tag{6-9}$$

Therefore, at time $t = t_0$ the state of the composite system $\mathcal{S} \cup \mathcal{M} = \mathcal{S}^I$ is

$$|\psi_{\mathrm{I}}(t_0)\rangle = |\psi_S(t_0)\rangle \otimes |\psi_M(t_0)\rangle = \sum_i c_i\,|a_i\rangle \otimes |p_0\rangle \in \mathcal{H}, \tag{6-10}$$

where $\mathcal{O} = \mathcal{O}^S \otimes \mathcal{O}^M$ and $\mathcal{H} = \mathcal{H}^S \otimes \mathcal{H}^M$. Since during Stage I there is no interaction between $\mathcal{S}$ and $\mathcal{M}$, the total Hamiltonian of $\mathcal{S} \cup \mathcal{M} = \mathcal{S}^I : (\mathcal{O},\ H^I)$ is

$$H^{\mathrm{I}} = H^S \otimes I^M + I^S \otimes H^M \in \mathcal{O}. \tag{6-11}$$

According to the System Decomposition postulate IP3, this means that $\mathcal{S}$ and $\mathcal{M}$ are subsystems of the composite system $\mathcal{S}^I$.

Stage II: In this second, interaction stage, the systems $\mathcal{S}$ and $\mathcal{M}$ interact through an interaction Hamiltonian H_{int}. This means that the composite system $\mathcal{S}^I = \mathcal{S} \cup \mathcal{M} : (\mathcal{O},\ H^{\mathrm{I}})$ becomes the system $\mathcal{S}^{II} : (\mathcal{O},\ H^{\mathrm{II}})$, whose Hamiltonian reads

$$H^{\mathrm{II}} = H^S \otimes I^M + I^S \otimes H^M + H_{\mathrm{int}} = H^I + H_{\mathrm{int}} \in \mathcal{O}. \tag{6-12}$$

In turn, the state $|\psi_{\mathrm{I}}(t_0)\rangle$ of $\mathcal{S}^I$ in Stage I becomes the initial state $|\psi_{\mathrm{II}}(t_0)\rangle$ of $\mathcal{S}^{II}$ in Stage II. According to the Dynamical Postulate QP1, $|\psi_{\mathrm{II}}(t_0)\rangle$ evolves to a state $|\psi_{\mathrm{II}}(t_1)\rangle$ after a $\Delta t = t_1 - t_0$:

$$|\psi_{\mathrm{II}}(t_1)\rangle = e^{-iH^{\mathrm{II}}\Delta t/\hbar}\,|\psi_{\mathrm{II}}(t_0)\rangle = e^{-iH^{\mathrm{II}}\Delta t/\hbar}\,|\psi_{\mathrm{I}}(t_0)\rangle \in \mathcal{H}. \tag{6-13}$$

It can be proved that, if the interaction Hamiltonian H_{int} is

$$H_{\mathrm{int}} = -\frac{\lambda\hbar}{\Delta t}\left(A \otimes R^P\right), \tag{6-14}$$

where λ is a constant and R^P is the observable conjugate to P, $[P, R^P] = i\hbar I^M$, where I^M is the identity in $\mathcal{O}^M$, then the final state of $\mathcal{S}^{II}$ in Stage II is (see Mittelstaedt 1998)

$$|\psi_{\mathrm{II}}(t_1)\rangle = \sum_i c_i\,|a_i\rangle \otimes |p_i\rangle \in \mathcal{H}. \tag{6-15}$$

Stage III: At time $t = t_1$ the interaction ends: the system $\mathcal{S}^{II}$ becomes the original composite system $\mathcal{S} \cup \mathcal{M}$, whose Hamiltonian is again $H^{III} = H^I = H^S \otimes I^M + I^S \otimes H^M \in$

$\mathcal{O}$. Since in this stage $H_{int} = 0$, according to the System Decomposition postulate IP3, $\mathcal{S}$ and $\mathcal{M}$ turn out to be again subsystems of the composite system $\mathcal{S}^{III} = \mathcal{S}^{I} = \mathcal{S} \cup \mathcal{M}$. In turn, the state $|\psi_{II}(t_1)\rangle$ of $\mathcal{S}^{II}$ in Stage II becomes the initial state $|\psi_{III}(t_1)\rangle$ of $\mathcal{S}^{III} = \mathcal{S} \cup \mathcal{M}$ in Stage III:

$$|\psi_{III}(t_1)\rangle = |\psi_{II}(t_1)\rangle \in \mathcal{H}. \tag{6-16}$$

Since in this stage $\mathcal{S}$ and $\mathcal{M}$ are elemental quantum systems, the actualization rule AR has to be applied to each one of them:

(a) In the system $\mathcal{S} : (\mathcal{O}^S, H^S)$, the preferred CSOP $\{|\omega_i^S\rangle\langle\omega_i^S|\}$ is defined by the eigenbasis of the—by assumption—non-degenerate H^S. Here, two cases must be distinguished:

(a.1) if $[H^S, A] = 0$, then $|\omega_i^S\rangle = |a_i\rangle$. This means that both H^S and A are definite-valued.

(a.2) if $[H^S, A] \neq 0$, then $|\omega_i^S\rangle \neq |a_i\rangle$. As a consequence, the observable A is not definite-valued.

Nevertheless, let us note that, if the initial state of $\mathcal{S}$ were prepared not exactly at t_0 but at an arbitrary time $t < t_0$—an experimentally more plausible situation since it does not require absolute precision—, it would evolve until t_0 under the action of the Hamiltonian H^S; then, the measured state would not be the prepared state but the state at t_0 resulting from the evolution. As a consequence, if the state of S is to be measured as originally prepared, the Hamiltonian H^S must be zero or a multiple of the identity—which are physically equivalent situations, since they differ in a time-displacement that only modifies the phase of the state vector. But, in this case, H^S is completely degenerate and commutes with all the observables of $\mathcal{S}$, in particular, with A. This means that the experimentally plausible situation is (a.1), in which not only H^S but also A are definite-valued.

(b) In the system $\mathcal{M} : (\mathcal{O}^M, H^M)$, the preferred CSOP $|\omega_i^M\rangle\langle\omega_i^M|$ is defined by the eigenbasis of the non-degenerate H^M. Moreover, $[H^M, P] = 0$ in order to guarantee the stationarity of the $|p_i\rangle$. In turn, P does not break the symmetries of H^M because H^M has no symmetries—it is non-degenerate. Therefore, both H^M and P are definite-valued.

Due to the perfect correlation between the $|p_i\rangle$ and the $|a_i\rangle$, the ideal quantum measurement is sometimes interpreted as if the reading of a particular value p_k of the apparatus' pointer P were the unambiguous indication of the corresponding value a_k of the system's observable A. However, according to the MHI, this might not be the case because A might be non-definite-valued. Although this fact sounds odd from a classical viewpoint, it turns out to be unproblematic when the difference between classical and quantum measurements is understood. The idea that the value of the system's observable can be inferred from the reading of the apparatus' pointer is modeled under the paradigm of classical measurements, which rely on the perfect correlation between the values of the apparatus' pointer and the values of the observable to be measured. But in quantum measurements, the ultimate goal is not to "discover" the actual value of a system's observable, as in the classical case, but *to reconstruct the state of the system* just before the beginning of the measurement process. Therefore, the relevant fact is the definite reading of the apparatus' pointer: the task consists in explaining how the repetition of single measurements in which the pointer is definite-valued makes it possible to reconstruct the state of the measured system. As just explained, according to the MHI, no matter whether the system's observable A acquires a definite value or not, in each single measurement *the apparatus' pointer is always definite-valued.* The next step is to show that the repetition of single measurements in a frequency measurement yields the correct coefficients of the system's state.

Propensities and actualization: The preferred CSOP $\{|\omega_i^M\rangle\langle\omega_i^M|\}$ of the apparatus $\mathcal{M}$ defines the set of the corresponding possible facts $\mathcal{F}^{\mathrm{P}}$. The measure of the propensity to actualization of each possible fact $[|p_i\rangle\langle p_i|:1] \in \mathcal{F}^{\mathrm{P}}$ can be computed as (see Section 4.2 of Chapter 4):

$$p^{\mathrm{P}}_{\rho_r^M(t_1)}\left([|p_i\rangle\langle p_i|:1]\right) = Tr\left(\rho_r^M(t_1)\,|p_i\rangle\langle p_i|\right) = \langle p_i|\,\rho_r^M(t_1)\,|p_i\rangle, \qquad (6\text{-}17)$$

where $\rho_r^M(t_1)$ is the reduced state of $\mathcal{M}$ in Stage III:

$$\rho_r^M(t_1) = Tr_{(S)}\left(|\psi_{\mathrm{III}}(t_1)\rangle\langle\psi_{\mathrm{III}}(t_1)|\right) = \sum_i |c_i|^2\,|p_i\rangle\langle p_i|\,. \qquad (6\text{-}18)$$

Therefore,

$$p^{\mathrm{P}}_{\rho_r^M(t_1)}\left([|p_i\rangle\langle p_i|:1]\right) = |c_i|^2. \qquad (6\text{-}19)$$

Since each possible fact $[|p_i\rangle\langle p_i|:1]$ is equivalent to the possible fact $[P:p_i]$, then

$$p^{\mathrm{P}}_{\rho_r^M(t_1)}\left([P:p_i]\right) = |c_i|^2. \qquad (6\text{-}20)$$

Let us emphasize that, as explained in Section 4.2 of Chapter 4, the reduced state of $\mathcal{M}$ is used here as a calculational tool for computing probabilities and must not be conceived as the quantum state of $\mathcal{M}$. As can be seen, the measures of the propensities so obtained agree with the probabilities assigned to the different readings of the pointer by the collapse interpretation. Moreover, those measures prove to be constant with time, a reasonable result to guarantee the repeatability of the single measurement.

Of course, if the $|c_i|^2$ are to be obtained experimentally, the single measurement must be repeated under the same conditions for propensities to manifest themselves as frequencies in the resulting frequency measurement. But, in each single measurement, one and only one possible fact $[|p_i\rangle\langle p_i|:1]$ becomes actual, say *Act* $[|p_k\rangle\langle p_k|:1]$. Therefore, the following actual facts obtain:

- *Act* $[H^M : \omega_k^M]$: the case-property $[\omega_k^M]$ corresponding to the type-property $[H^M]$ actually occurs (in the language of physics: the Hamiltonian H^M has the value ω_k^M).
- *Act* $[P : p_k]$: the case-property $[p_k]$ corresponding to the type-property $[P]$ actually occurs (in the language of physics: the pointer P has the value p_k).

Moreover, from the viewpoint of the composite system $\mathcal{S}^{III} = \mathcal{S} \cup \mathcal{M}$ (see the Properties of Systems and Subsystems postulate IP5 in Section 2.1 of Chapter 2):

- The case-property $[p_k]$ of the type-property $[I^S \otimes P]$ actually occurs in $\mathcal{S}^{III}$ when the same case-property of the type-property $[P]$ actually occurs in $\mathcal{M}$.
- The case-property $[\omega_l^S]$ of the type-property $[H^S \otimes I^M]$ actually occurs in $\mathcal{S}^{III}$ when the same case-property of the type-property $[H^S]$ actually occurs in $\mathcal{S}$.
- The case-property $[\omega_k^M]$ of the type-property $[I^S \otimes H^M]$ actually occurs in $\mathcal{S}^{III}$ when the same case-property of the type-property $[H^M]$ actually occurs in $\mathcal{M}$.
- The case-property $[\omega_l^S + \omega_k^M]$ of the type-property $[H] = [H^S \otimes I^M + I^S \otimes H^M]$ actually occurs in $\mathcal{S}^{III}$ when the case-property $[\omega_l^S]$ of the type-property $[H^S]$ actually occurs in $\mathcal{S}$ and the case-property $[\omega_k^M]$ of the type-property $[H^M]$ actually occurs in $\mathcal{M}$ (in the language of physics: the total energy of the composite system is the sum of the energies of the component subsystems).

6.3 Non-ideal measurements

Let us consider the most general formulation of a non-ideal measurement as presented in Section 3.5 of Chapter 3:

$$\sum_i c_i\,|a_i\rangle \otimes |p_0\rangle \quad\longrightarrow\quad \sum_{ij} d_{ij}\,|a_i\rangle \otimes |p_j\rangle, \tag{6-21}$$

where the d_{ij} embody both imperfection and disturbance. As explained in that section, the actualization rule of the BDMI (Kochen-Dieks) and the SDMI (Vermaas-Dieks), when applied to non-ideal measurements, leads to results that disagree with those obtained in the orthodox collapse interpretation (see Albert and Loewer 1990, 1991, 1993; see also Ruetsche 1995). If the properties ascribed by a modal interpretation are different from those ascribed by the collapse interpretation, the question is how different they are. As remarked, in the case of an imperfect measurement, it can be expected that the $d_{ij} \neq 0$, with $i \neq j$, are small; then, the difference may also be small. But in the case of a disturbing measurement, the $d_{ij} \neq 0$, with $i \neq j$, need not be small and, as a consequence, the disagreement between the definite-valued observables selected by the traditional Modal Interpretations and those selected by collapse might be unacceptable (see a full discussion in Bacciagaluppi and Hemmo 1996). Although attempts were made to address the disagreements with different strategies, this difficulty may explain, at least partially, the declining interest of philosophers of physics in Modal Interpretations from the 2000s onward.

The MHI is immune to the challenge posed by non-ideal measurements because the result of the application of the actualization rule AR does not depend on the coefficients d_{ij} of the off-diagonal terms. As a consequence, the observable P that plays the role of the apparatus' pointer acquires a definite value in any case. Let us justify this claim.

In a non-ideal measurement, Stage I is characterized in the same way as in the ideal case. The difference begins at Stage II, where the correlation introduced by the interaction Hamiltonian H_{int} is not perfect. Therefore, the final state $|\psi_{\text{II}}(t_1)\rangle$ of Stage II, which is the initial state $|\psi_{\text{III}}(t_1)\rangle$ of Stage III, reads

$$|\psi_{\text{III}}(t_1)\rangle = |\psi_{\text{II}}(t_1)\rangle = \sum_{ij} d_{ij}\,|a_i\rangle \otimes |p_j\rangle \,. \tag{6-22}$$

Since in Stage III $H_{\text{int}} = 0$, according to the System Decomposition postulate IP3, in this stage $\mathcal{S}$ and $\mathcal{M}$ are again subsystems of the composite system $\mathcal{S}^{III} = \mathcal{S} \cup \mathcal{M}$.

By focusing on the apparatus $\mathcal{M}$, in Stage III the reduced state of the system $\mathcal{S}^{III}$ is

$$\rho_r^M(t_1) = Tr_{(S)}\left(|\psi_{\text{III}}(t_1)\rangle\langle\psi_{\text{III}}(t_1)|\right) = \sum_{ij} \rho_{ij}^M\,|p_i\rangle\langle p_j|\,, \tag{6-23}$$

where

$$\rho_{ij}^M = \sum_n d_{ni}\,d_{nj}^{\star}. \tag{6-24}$$

Although $\mathcal{M}$ evolves unitarily under the action of H^M according to the Dynamical Postulate QP1, the preferred CSOP $\{|\omega_{Mi}\rangle\langle\omega_{Mi}|\}$ is time-invariant as it is defined by the eigenbasis of H^M. In turn, since H^M commutes with P, also in this non-ideal case both H^M and P are definite-valued observables as expected.

Summing up, David Albert and Barry Loewer (1990) are right in asserting that the ideal measurement is a situation that can never be achieved in practice: the interaction in Stage II never introduces a completely perfect correlation. In spite of this, physicists often perform successful measurements. The MHI explains this fact with its state-independent actualization rule.

6.4 Reliable and non-reliable measurements

Let us now consider propensities and actualization for non-ideal measurements. In this case, the measure of the propensity to actualization of each possible fact $[|p_i\rangle\langle p_i| : 1] \in \mathcal{F}^{\text{P}}$ is given by

$$p^{\mathrm{P}}_{\rho_r^M(t_1)}\left([|p_i\rangle\langle p_i|:1]\right) = Tr\left(\rho_r^M(t_1)\,|p_i\rangle\langle p_i|\right) = \rho_{ii}^M = |d_{ii}|^2 + \sum_{n\neq i}|d_{ni}|^2. \qquad (6\text{-}25)$$

If the coefficients d_{ni}, with $n \neq i$, of the off-diagonal terms of the reduced state $\rho_r^M(t_1)$ are zero, the ideal-measurement case is recovered, and $\rho_{ii}^M = |d_{ii}|^2 = |c_i|^2$. A non-ideal measurement is the case in which the coefficients d_{ni}, with $n \neq i$, are not zero. However, in this case two situations must be distinguished:

- If the d_{ni}, with $n \neq i$, are small in the sense that $\sum_{n\neq i}|d_{ni}|^2 \ll |d_{ii}|^2$, then $\rho_{ii}^M \simeq |d_{ii}|^2 \simeq |c_i|^2$. This means that the frequency measurement performed by repetition of this single measurement is *reliable*: the coefficients $|c_i|^2$ can be approximately obtained.
- If the d_{ni}, with $n \neq i$, are not small, then $\rho_{ii}^M \simeq |d_{ii}|^2$ does not hold. Therefore, the result of the frequency measurement is *non-reliable*.

Nevertheless, no matter whether the result of the frequency measurement is reliable or not, in each single measurement one and only one possible fact $[|p_i\rangle\langle p_i|:1]$ becomes actual, say, $Act\,[|p_k\rangle\langle p_k|:1]$. Therefore, as in the ideal case, it can be guaranteed that, in the measuring apparatus $\mathcal{M}$,

$$Act\,[H^M : \omega_k^M] \quad \text{and} \quad Act\,[P : p_k]. \qquad (6\text{-}26)$$

Moreover, since $Act\,[H^S : \omega_l^S]$ is an actual fact in the system $\mathcal{S}$ under measurement, then in the composite system $\mathcal{S}^{III} = \mathcal{S} \cup \mathcal{M}$, $Act\,[H : (\omega_l^S + \omega_k^M)]$ and $Act\,[I^S \otimes P : p_k]$ are also actual facts on the basis of the same arguments as in the ideal case.

This MHI account of the measurement process shows that perfect correlation is not a necessary condition for "good" measurements: the coefficients of the system's state at the beginning of the process can be approximately obtained even when the correlation is not perfect, to the extent that the reliability condition is satisfied. Nevertheless, in both reliable and non-reliable frequency measurements, each single measurement provides *a definite reading of the pointer P*: the MHI is immune to Harvey Brown's "silver bullet" (see Section 3.5 of Chapter 3).

6.5 Reliability from an informational perspective

As argued in the previous sections, the MHI proves to be able to adequately account for the measurement problem, both in its ideal and its non-ideal versions. Moreover, in the non-ideal case, it offers a precise criterion to distinguish between reliable and non-reliable measurements. However, that criterion depends on the particular state of the measured system and, as a consequence, must be applied case by case. Now that we are no longer burdened with the quantum measurement problem, we can spend some time asking for a criterion of reliability that does not depend on the features of what is being measured, but only on the properties of the measurement device. Such a criterion can be formulated from an informational perspective (see Lombardi, Fortin, and López 2015).

Let us recall that, from a very general viewpoint, an informational arrangement includes three elements: the source S, which generates the message to be received at the destination, the channel CH, which is the medium used to transmit the information from source to destination, and the destination D, which receives the message (Shannon 1948; for an updated presentation, see Cover and Thomas 1991). The source S is a system with a range of possible states $s_1, \ldots, s_n$, usually called "letters," whose respective probabilities of occurrence are $pr(s_1), \ldots, pr(s_n)$. Since S produces sequences of states, usually called "messages," the entropy of the source S is computed as a weighted average as

$$H(S) = -\sum_{i=1}^{n} pr(s_i) \log pr(s_i). \tag{6-27}$$

Analogously, the destination D is a system with a range of possible states $d_1, \ldots, d_m$, with respective probabilities $pr(d_1), \ldots, pr(d_m)$. The entropy of the destination D is computed as

$$H(D) = -\sum_{j=1}^{m} pr(d_j) \log pr(d_j). \tag{6-28}$$

Despite their formal similarity, the entropies of the source $H(S)$ and of the destination $H(D)$ are conceptually different. $H(S)$ is the average amount of information produced at the source and, then, only depends on its features (on the distribution of probabilities on its states). $H(D)$, by contrast, is the average amount of information that the destination receives and, then, depends not only on the information produced by the source, but also on the channel, that is, on the information lost during transmission and the spurious information not originated in the source. Therefore, the relation between $H(S)$ and $H(D)$ is mediated by the following magnitudes:

- The *mutual information* $H(S; D)$ is the average amount of information generated at the source S and received at the destination D.
- The *equivocity* E is the average amount of information generated at S but not received at D.
- The *noise* N is the average amount of information received at D but not generated at S.

The relations between these three magnitudes are given by

$$H(S; D) = H(S) - E = H(D) - N. \tag{6-29}$$

Equivocation E and noise N are measures of the dependence between the source S and the destination D:

- If S and D are completely independent, the values of E and N are maximum ($E = H(S)$ and $N = H(D)$), and the value of $H(S; D)$ is minimum ($H(S; D) = 0$).
- If the dependence between S and D is maximum, the values of E and N are minimum ($E = N = 0$), and the value of $H(S; D)$ is maximum ($H(S; D) = H(S) = H(D)$).

The values of equivocity E and noise N are functions not only of the source and the destination, but also of the communication channel CH, which incorporates the possibility of errors in the process of transmission: the channel CH is defined by the matrix $[pr(d_j \mid s_i)]$, where $pr(d_j \mid s_i)$ is the conditional probability of the occurrence of d_j in the destination D given that s_i occurred in the source S, and the elements in any row add up to 1. On this basis, E and N can be computed as

$$E = -\sum_{j=1}^{m} pr(d_j) \sum_{i=1}^{n} pr(s_i \mid d_j) \log pr(s_i \mid d_j), \tag{6-30}$$

$$N = -\sum_{i=1}^{n} pr(s_i) \sum_{j=1}^{m} pr(d_j \mid s_i) \log pr(d_j \mid s_i), \tag{6-31}$$

where $pr(s_i \mid d_j) = pr(d_j \mid s_i) pr(s_i) / pr(d_j)$. On this basis, the mutual information reads

$$H(S; D) = -\sum_{j=1}^{m} \sum_{i=1}^{n} pr(s_i, d_j) \log pr(s_i, d_j), \tag{6-32}$$

where $pr(s_i, d_j) = pr(s_i \mid d_j)\, pr(d_j) = pr(d_j \mid s_i)\, pr(s_i)$. When the value of $H(S; D)$ is maximum (the values of E and N are minimum, $E = N = 0$), the channel is *deterministic*: all the information produced at the source arrives at the destination. By contrast, in a *noisy* channel ($N \neq 0$), the mutual information is lower than the information at the destination since $H(S; D) = H(D) - N$: the destination receives spurious information that was not produced at the source. In an *equivocal* channel ($E \neq 0$), the mutual information is lower than the information of the source since $H(S; D) = H(S) - E$: part of the information generated at the source gets lost and does not arrive at the destination. With these elements, all the tools needed to reconstruct a quantum measurement as an informational process are available.

The first step is to describe the calibration of the apparatus, which can be performed by appealing to the eigenstates $|a_i\rangle$ of the observable A of the system $\mathcal{S}$. Let us consider a particular $|a_k\rangle$ obtained as the result of a filtering-type measurement (see, e.g., Ballentine 1998, 246). Here, a completely general case will be considered, in which the interaction not only introduces imperfection under the form of non-zero off-diagonal terms of the final state, but also disturbs the state of the system, leading to the following evolution:

$$|a_k\rangle \otimes |p_0\rangle \longrightarrow \sum_{ij} \beta_{ij} \left|a_i^d\right\rangle \otimes |p_j\rangle, \tag{6-33}$$

where the β_{ij} and the $|a_i^d\rangle$ are completely generic and, therefore, the $|a_i^d\rangle$ do not need to be orthogonal to each other and may be superpositions of the $|a_i\rangle$. When this state is expressed in the basis $|a_i\rangle \otimes |p_j\rangle$, the evolution reads

$$|a_k\rangle \otimes |p_0\rangle \quad \longrightarrow \quad \sum_{ij} \gamma_{ij}^k \, |a_i\rangle \otimes |p_j\rangle, \tag{6-34}$$

where the γ_{ij}^k embody both imperfection and disturbance.

Once the composite system is prepared in the initial state $|a_k\rangle \otimes |p_0\rangle$, the frequency of each value p_j of the pointer P can be measured. This frequency represents the probability of the fact that the pointer P has the value p_j given that the value of the observable A of the measured system was a_k: $pr(p_j \mid a_k)$. But since the measurement is not ideal, measuring p_j does not guarantee perfect correlation; on the contrary,

$$pr(p_j \mid a_k) = pr\big((|a_1\rangle \otimes |p_j\rangle \, or \, |a_2\rangle \otimes |p_j\rangle \, or \ldots) \mid a_k\big). \tag{6-35}$$

But since the states $|a_i\rangle \otimes |p_j\rangle$ are orthogonal and, thus, exclusive, this conditional probability can be computed as

$$pr(p_j \mid a_k) = pr\big(|a_1\rangle \otimes |p_j\rangle \mid a_k\big) + pr\big(|a_2\rangle \otimes |p_j\rangle \mid a_k\big) + \ldots, \tag{6-36}$$

and, according to eq. (6-34),

$$pr(p_j \mid a_k) = \left|\gamma_{1j}^k\right|^2 + \left|\gamma_{2j}^k\right|^2 + \ldots = \sum_i \left|\gamma_{ij}^k\right|^2. \tag{6-37}$$

The full calibration of the apparatus is achieved by following the same procedure for each of the eigenstates of the observable A, so that all the probabilities $pr(p_j \mid a_i)$ are obtained.

Once calibration has been performed, a generic measurement can be easily characterized. When the unknown state of the system S is a superposition of the eigenstates of the observable A, the evolution induced by the interaction is (see eq. (6-34))

$$\sum_k \alpha_k \, |a_k\rangle \otimes |p_0\rangle \quad \longrightarrow \quad \sum_k \alpha_k \sum_{ij} \gamma_{ij}^k \, |a_i\rangle \otimes |p_j\rangle = \sum_{ij} \sum_k \alpha_k \gamma_{ij}^k \, |a_i\rangle \otimes |p_j\rangle. \tag{6-38}$$

When this eq. (6-38) is compared with eq. (6-22), it is easy to recognize the dependence of the d_{ij} of that equation with respect to the coefficients α_k of the initial state:

$$d_{ij} = \sum_k \alpha_k \gamma_{ij}^k. \tag{6-39}$$

In this situation, the frequency of each value p_j of the pointer P can be measured, which represents the probability $pr(p_j)$ of the value p_j, given by

$$pr(p_j) = \sum_i |d_{ij}|^2 = \sum_i \left| \sum_k \alpha_k \gamma_{ij}^k \right|^2 . \qquad (6\text{-}40)$$

If the measurement is *ideal*, then $\gamma_{ij}^k = \delta_{ji}\, \delta_{ik}$ and, as a consequence,

$$pr(p_j) = \sum_i \left| \sum_k \alpha_k \delta_{ji}\, \delta_{ik} \right|^2 = |\alpha_j|^2 = pr(a_j). \qquad (6\text{-}41)$$

But, in the generic case considered here, there are both imperfections and disturbances and, thus, the measured frequencies do not give direct information about the coefficients of the measured state. Moreover, the original reliability condition $\sum_{n \neq i} |d_{ni}|^2 \ll |d_{ii}|^2$, applicable to frequency measurements, cannot be used to know if the measuring apparatus in itself is reliable or not, because the coefficients d_{ij} are functions of the coefficients α_k of the unknown initial state. In this case, the informational conception of the quantum measurement proves to be fruitful.

If the just described generic measurement is conceived as a process of transference of information, the measured quantum state is the source S of information, with a range of possible "states" $a_1, a_2, \ldots$, that is, the eigenvalues of the observable A, whose respective probabilities of occurrence are $pr(a_i) = |\alpha_i|^2$. In turn, the apparatus' pointer is the destination D of information, with a range of possible "states" $p_1, p_2, \ldots$, that is, the eigenvalues of P, with probabilities of occurrence $pr(p_j)$, respectively. The measuring apparatus is the channel, which can be mathematically characterized independently of the peculiar features of source and destination. In particular, the channel is defined by the matrix $[pr(p_j \mid a_i)]$ that, as explained above, can be empirically determined through the process of calibration, prior to any particular measurement. From this informational perspective, it can be said that:

- When the channel is *deterministic*, then the measurement is *ideal*. In this case, all and only the information of the measured quantum state is recovered by the pointer. The perfect correlation introduced by the measuring apparatus amounts to the fact that the conditional probabilities that define the channel are trivial: $pr(p_j \mid a_i) = \delta_{ji}$.
- When the channel is *equivocal or/and noisy*, then the measurement is *non-ideal*. In this case, there is some loss of relevant information E or/and addition of spurious information N through the process: this is expressed by the fact that the conditional probabilities that define the channel are not trivial. But since those probabilities characterize the measuring device through calibration, they can be used to give a criterion of reliability independent of the particular measurement carried out:

- If the conditional probabilities are approximately trivial, $pr\left(p_j \mid a_i\right) \simeq \delta_{ji}$, then $pr(p_j) \simeq pr(a_j) = |\alpha_j|^2$. This means that, in the frequency measurements performed by the considered apparatus, the measured frequencies approximately yield the value of the coefficients $|\alpha_j|^2$ and, therefore, those frequency measurements are *reliable.*
- If the conditional probabilities are not approximately trivial, then $pr(p_j) \simeq pr(a_j) = |\alpha_j|^2$ does not hold. Therefore, in the considered apparatus, the results of frequency measurements are *non-reliable.*

Besides providing a reliability criterion that depends on the previously calibrated measuring apparatus and not on the state to be measured, the informational view of measurement supplies the tools to quantify the degree of non-reliability of a particular device. In fact, given the conditional probabilities $pr(p_j \mid a_i)$ of the channel-apparatus, the amounts of mutual information, equivocity, and noise can be computed. Of course, these magnitudes will depend on the particular quantum state to be measured (see eqs. (6-30) and (6-31)). Nevertheless, the reliability of different measuring apparatuses can be compared on the basis of some standard initial state. A reasonable standard is the situation in which the information generated at the source is maximum, that is, all its states are equiprobable. In this case, all the values of the coefficients α_k of the initial quantum state of the system $\mathcal{S}$ are $\sqrt{1/n}$; so $pr(s_i) = 1/n$. If the frequencies of detection of the eigenvalues p_j of P, which represent the probabilities $pr(p_j)$ of the values p_j, are measured in this situation, then E, N, and $H(S; D)$ can be computed with eqs. (6-30), (6-31), and (6-32), respectively. On this basis, the degree of reliability R of the measuring apparatus can be defined as

$$R = \frac{H(S;D)}{H(S;D)_{\max}} = \frac{H(S;D)}{H(S)} = -\frac{H(S;D)}{\log n}, \tag{6-42}$$

where $H(S;D)_{\max} = H(S) = -\log n$ corresponds to the ideal case. Therefore, if the measurement apparatus leads to ideal measurements, then its degree of reliability is maximum, $R = 1$.

Moreover, in the case of non-reliability, the values of equivocity E and noise N give clues about the source of this undesirable feature. Let us recall that in a noisy channel, $N \neq 0$, the destination receives spurious information that was not produced at the source, and in an equivocal channel, $E \neq 0$, part of the information generated at the source does not arrive at the destination. If each state of the destination is intended to indicate unambiguously which state of the source occurred, it is necessary that the backward probabilities $pr\left(s_i \mid d_j\right)$ have value 0 or 1, and this holds in an equivocation-free channel, $E = 0$, no matter the value of noise. This explains why noise does not prevent observation: indeed, in practice, informational situations commonly include noisy channels, and much technological effort is devoted to designing appropriate filters and/or shielding the channel to block spurious noise-carrying signals. On the contrary, an equivocal channel leads to an information loss that cannot be remediated by means of filters or shields, but requires the addition of redundant information, which is a common strategy in the processes of transmission of information. This means that the different sources of imperfection are not equally serious and demand different strategies for their solution. In summary, when the performances of different measuring apparatuses are compared on the basis of a standard state, knowing whether the source of non-reliability is equivocity or noise (or both) may give hints about what kind of actions on the apparatus may be more efficient to solve the problem.

6.6 The Stern-Gerlach experiment

Since the Stern-Gerlach experiment is the paradigm of quantum measurement, it is interesting to see how the MHI account of measurement is applied to this case.

The experiment is usually described as follows. A neutral free spin 1/2 particle, with constant velocity in the y-direction, passes between the poles of a magnet that generates an inhomogeneous magnetic field B, with components $B_x = B_y = 0$ and $B_z = zB'$, where B' is the field gradient. The particle is described in the plane zy and in a reference frame moving uniformly in the y-direction, such that the momentum in y-direction is $P_y = 0$. The gradient of the magnetic field produces a force that deflects the particle in the z-direction: the deflection depends on the component of spin in that direction.

As explained in Section 5.2 of the previous chapter, the free particle with spin is a composite system $\mathcal{S}^P \cup \mathcal{S}^S$. In this measurement situation:

- The spin subsystem $\mathcal{S}^S : (\mathcal{O}^S, H^S)$, with $\mathcal{O}^S \subset \mathcal{B}(\mathcal{H}^S)$ and $H^S = k\,S^2$, is the system $\mathcal{S}$ to be measured.
- The free particle without spin $\mathcal{S}^P : (\mathcal{O}^P, H^P)$, with $\mathcal{O}^P \subset \mathcal{B}(\mathcal{H}^P)$ and $H^P = P_z^2/2m$, is part of a measuring apparatus $\mathcal{M}$; so, in this case, $\mathcal{H}^M = \mathcal{H}^P$ and $H^M = H^P$. Moreover, condition $[H^M, P_z] = 0$ guarantees that the eigenvectors of P_z are stationary and, then, P_z can play the role of the pointer.

On this basis, at Stage I, we find that:

- The observable A is the spin in z-direction, $S_z \in \mathcal{O}^S$:

$$S_z\,|\uparrow\rangle = s_\uparrow\,|\uparrow\rangle, \quad S_z\,|\downarrow\rangle = s_\downarrow\,|\downarrow\rangle, \tag{6-43}$$

  where $s_\uparrow = -s_\downarrow = (1/2)\,\hbar$.
- The momentum in z-direction plays the role of the pointer, $P_z \in \mathcal{O}^M$:

$$P_z\,|+\rangle = p_+\,|+\rangle, \quad P_z\,|-\rangle = p_-\,|-\rangle, \quad P_z\,|0\rangle = p_0\,|0\rangle, \tag{6-44}$$

where $\{|+\rangle, |-\rangle, |0\rangle\}$ is a basis of $\mathcal{H}^M$. Again, this assumption is adopted for simplicity, and will be discussed in Section 6.9 of the present chapter.

- The states of $\mathcal{S}$ and $\mathcal{M}$ at $t_0 = 0$ are, respectively, $|\psi_S\rangle = c_1 |\uparrow\rangle + c_2 |\downarrow\rangle$ and $\psi_M = |0\rangle$. Then,

$$|\psi_{\mathrm{I}}(t_0 = 0)\rangle = c_1 |\uparrow\rangle \otimes |0\rangle + c_2 |\downarrow\rangle \otimes |0\rangle . \tag{6-45}$$

- As said above, the Hamiltonian of $\mathcal{S}$ is $H^S = k\, S^2$, and the Hamiltonian of $\mathcal{M}$ is such that $[H^M, P_z] = 0$. Therefore,

$$H^M |+\rangle = \omega_+ |+\rangle , \quad H^M |-\rangle = \omega_- |-\rangle , \quad H^M |0\rangle = \omega_0 |0\rangle . \tag{6-46}$$

Ideal measurement: At Stage II, the total Hamiltonian $H^{\mathrm{II}} = H^S \otimes I^M + I^S \otimes H^M + H_{\mathrm{int}}$ introduces a perfect correlation. Then, the initial state of $\mathcal{S} \cup \mathcal{M}$ in Stage III is

$$|\psi_{\mathrm{III}}(t_1)\rangle = c_1 |\uparrow\rangle \otimes |+\rangle + c_2 |\downarrow\rangle \otimes |-\rangle . \tag{6-47}$$

The preferred CSOP of $\mathcal{M}$ is given by the eigenbasis of the Hamiltonian H^M and, then, it is $\{|+\rangle\langle+| , |-\rangle\langle-| , |0\rangle\langle 0|\}$. Since $[H^M, P_z] = 0$, both the Hamiltonian $H^M = H^P$ and the particle's momentum P_z in z-direction are definite-valued.

The reduced state of subsystem $\mathcal{M}$ in Stage III reads

$$\rho_r^M(t_1) = Tr_{(S)}\left(|\psi_{\mathrm{III}}(t_1)\rangle\langle\psi_{\mathrm{III}}(t_1)|\right) = |c_1|^2 |+\rangle\langle+| + |c_2|^2 |-\rangle\langle-| . \tag{6-48}$$

Therefore, the measure of the propensity to actualization of the possible facts corresponding to P_z can be computed as

$$p^{\mathrm{P}}_{\rho_r^M(t_1)}\left([|+\rangle\langle+| : 1]\right) = |c_1|^2, \tag{6-49}$$

$$p^{\mathrm{P}}_{\rho_r^M(t_1)}\left([|-\rangle\langle-| : 1]\right) = |c_2|^2, \tag{6-50}$$

$$p^{\mathrm{P}}_{\rho_r^M(t_1)}\left([|0\rangle\langle 0| : 1]\right) = 0. \tag{6-51}$$

In turn, since the possible facts with zero propensity measure do not become actual (see the interpretative postulate IP12), only one of the following two situations will become actual:

$$Act\,[H^M : \omega_+] \text{ and } Act\,[P_z : p_+], \tag{6-52}$$

$$Act\,[H^M : \omega_-] \text{ and } Act\,[P_z : p_-]. \tag{6-53}$$

Non-ideal measurement: In this case, H_{int} does not introduce a perfect correlation. The initial state of $\mathcal{S} \cup \mathcal{M}$ in Stage III is, then,

$$|\psi_{\mathrm{III}}(t_1)\rangle = d_{11} |\uparrow\rangle \otimes |+\rangle + d_{12} |\uparrow\rangle \otimes |-\rangle + d_{21} |\downarrow\rangle \otimes |+\rangle + d_{22} |\downarrow\rangle \otimes |-\rangle . \tag{6-54}$$

The reduced state of subsystem $\mathcal{M}$ in Stage III reads

$$\rho_r^M(t_1) = \rho_{11}^M \, |+\rangle \langle +| + \rho_{12}^M \, |+\rangle \langle -| + \rho_{21}^M \, |-\rangle \langle +| + \rho_{22}^M \, |-\rangle \langle -|, \tag{6-55}$$

where

$$\rho_{ij}^M = \sum_{n=1}^{2} d_{ni} \, d_{nj}^{\star}. \tag{6-56}$$

In other words,

$$\rho_r^M(t_1) = \begin{pmatrix} |d_{11}|^2 + |d_{21}|^2 & d_{11} \, d_{12}^{\star} + d_{21} \, d_{22}^{\star} \\ d_{12} \, d_{11}^{\star} + d_{22} \, d_{21}^{\star} & |d_{12}|^2 + |d_{22}|^2 \end{pmatrix}. \tag{6-57}$$

The preferred CSOP of $\mathcal{M}$ is again given by the eigenbasis of the Hamiltonian H^M and, since $[H^M, P_z] = 0$, both the Hamiltonian $H^M = H^P$ and the particle's momentum P_z in z-direction are definite-valued. But now the measures of the propensities to actualization of the corresponding possible facts are computed as

$$p^{\mathrm{P}}_{\rho_r^M(t_1)} \left([|+\rangle \langle +| \, : \, 1] \right) = |d_{11}|^2 + |d_{21}|^2. \tag{6-58}$$

$$p^{\mathrm{P}}_{\rho_r^M(t_1)} \left([|-\rangle \langle -| \, : \, 1] \right) = |d_{22}|^2 + |d_{12}|^2. \tag{6-59}$$

$$p^{\mathrm{P}}_{\rho_r^M(t_1)} \left([|0\rangle \langle 0| \, : \, 1] \right) = 0. \tag{6-60}$$

In this non-ideal case, the frequency measurement resulting from the repetition of this single measurement is reliable if $|d_{21}|^2 \ll |d_{11}|^2$ and $|d_{12}|^2 \ll |d_{22}|^2$; if not, the frequency measurement does not supply the necessary information for the reconstruction of the original state of the measured system. Nevertheless, the observable P_z acquires a definite value in any case: this is the prediction that can be directly tested in each single measurement.

6.7 The problem of infinite tails

An argument that underlines the difficulties posed by non-ideal measurements is that raised by Andrew Elby (1993) in the context of the Stern-Gerlach experiment. This argument points to the fact that the wavefunctions in z-variable typically have infinite "tails" that introduce non-zero cross-terms; therefore, the "tail" of the wavefunction of the "down" beam may produce detection in the upper detector, prepared to detect p_+, and vice versa. Of course, this is a case of non-ideal measurement as discussed in the previous section. But since it manifests non-ideality in a clear way, it is worth treating it specifically.

Let us suppose that, in Elby's argument, the imperfection is due to a non-perfect collimation of the incoming beam. In this case, with the magnetic field still turned off, a fuzzy spot instead of a definite point is obtained on the screen. Therefore, the perfect ready-to-measure state $|p_0\rangle = |0\rangle$ must be replaced by a narrow Gaussian $|\varphi_0(z)\rangle$. As a consequence, the measurement process turns out to be expressed as

$$|\psi_{\mathrm{I}}\rangle = (c_1 |\uparrow\rangle + c_2 |\downarrow\rangle) \otimes |\varphi_0(z)\rangle \longrightarrow |\psi_{\mathrm{III}}\rangle = c_1 |\uparrow\rangle \otimes |\varphi_+(z)\rangle + c_2 |\downarrow\rangle \otimes |\varphi_-(z)\rangle, \tag{6-60}$$

where now $|\varphi_+(z)\rangle$ and $|\varphi_-(z)\rangle$ are Gaussians that do not need to be as narrow as the initial one. Let us call the widths of the upper and the lower detectors Δz_+ and Δz_- respectively. Thus, the long tail of the Gaussian $|\varphi_+(z)\rangle$ arrives to Δz_- and the long tail of the Gaussian $|\varphi_-(z)\rangle$ arrives to Δz_+. The probabilities corresponding to the four possible cases can be computed as

$$p(\uparrow,+) = |(\langle\uparrow| \otimes \langle\varphi_+(z)|)\,|\psi_{\mathrm{III}}\rangle|^2 = |c_1|^2 \int_{\Delta z_+} ||\varphi_+(z)\rangle|^2 dz = |c_{11}|^2, \tag{6-61}$$

$$p(\uparrow,-) = |(\langle\uparrow| \otimes \langle\varphi_-(z)|)\,|\psi_{\mathrm{III}}\rangle|^2 = |c_1|^2 \int_{\Delta z_-} |\langle\varphi_-(z)\,|\varphi_+(z)\rangle|^2 dz = |c_{12}|^2, \tag{6-62}$$

$$p(\downarrow,+) = |(\langle\downarrow| \otimes \langle\varphi_+(z)|)\,|\psi_{\mathrm{III}}\rangle|^2 = |c_2|^2 \int_{\Delta z_+} |\langle\varphi_+(z)\,|\varphi_-(z)\rangle|^2 dz = |c_{21}|^2, \tag{6-63}$$

$$p(\downarrow,-) = |(\langle\downarrow| \otimes \langle\varphi_-(z)|)\,|\psi_{\mathrm{III}}\rangle|^2 = |c_2|^2 \int_{\Delta z_-} ||\varphi_-(z)\rangle|^2 dz = |c_{22}|^2. \tag{6-64}$$

Now the measure of the propensity to actualization of the possible facts corresponding to P_z are computed as

$$p^{\mathrm{P}}_{\rho_r^M(t_1)}([|+\rangle\langle+| : 1]) = p(\uparrow,+) + p(\downarrow,+) = |c_{11}|^2 + |c_{21}|^2, \tag{6-65}$$

$$p^{\mathrm{P}}_{\rho_r^M(t_1)}([|-\rangle\langle-| : 1]) = p(\downarrow,-) + p(\uparrow,-) = |c_{22}|^2 + |c_{12}|^2. \tag{6-66}$$

According to Elby's argument, these cases can be read as follows:

- $|c_{11}|^2$ is the probability that $|\uparrow\rangle$ be detected by Δz_+
- $|c_{21}|^2$ is the probability that $|\downarrow\rangle$ be detected by Δz_+ (tail)
- $|c_{22}|^2$ is the probability that $|\downarrow\rangle$ be detected by Δz_-
- $|c_{12}|^2$ is the probability that $|\uparrow\rangle$ be detected by Δz_- (tail)

The MHI shows that, if the reliability condition, $|c_{21}|^2 \ll |c_{11}|^2$ and $|c_{12}|^2 \ll |c_{22}|^2$, holds, then the collimation, even if not perfect, is good enough for measurement, because in this case $|c_{11}|^2 \simeq |c_1|^2$ and $|c_{22}|^2 \simeq |c_2|^2$. But if the original Gaussian is not very narrow or the screen is placed too far from the magnet, the measurement will be non-reliable because the c_{ij}, with $i \neq j$, are not small enough. Nevertheless, according to the MHI actualization rule AR, since the preferred CSOP is defined by the eigenbasis of H^M and the pointer P_z commutes with H^M, a definite reading of the pointer, that is, a definite detection on Δz_+ or on Δz_- is obtained.

6.8 Consecutive measurements

In Section 3.6 of Chapter 3, it was explained why traditional Modal Interpretations face a serious difficulty when the task is to account for the agreement between the outcomes of consecutive measurements of the same observable: since partial traces cancel the correlations between the interacting systems, that agreement remains unexplained. A professional physicist, trained in the framework of the collapse version of SQM, may believe that the difficulty results from the fact that, according to Modal Interpretations, the quantum state always evolves unitarily according to the Schrödinger equation. From this perspective, the challenge that consecutive measurements poses to Modal Interpretations would be reason enough to discard them. Furthermore, our physicist friend might even argue that this challenge is a clear manifestation of the fact that quantum measurements cannot be correctly explained without the collapse hypothesis. However, this is a too hasty conclusion, because the correlations between the outcomes of consecutive measurements can be easily explained without collapse when partial traces are not introduced but the measuring apparatuses are taken into account (see Ardenghi, Lombardi, and Narvaja 2013).

Let us consider the case of consecutive Stern-Gerlach measurements: in the first one, spin S_z in z-direction is measured by the pointer P_z, and in the second one, spin S_x in x-direction is measured by the pointer P_x. After the first measurement, the correlated state of the system and the first apparatus is $|\psi\rangle$, and the second S_x-measuring apparatus is in its ready-to-measure state $|0_x\rangle$:

$$|\psi_1\rangle = |\psi\rangle \otimes |0_x\rangle = (c_+ |\uparrow\rangle \otimes |+_z\rangle + \ c_- |\downarrow\rangle \otimes |-_z\rangle) \otimes |0_x\rangle . \qquad (6\text{-}67)$$

In order to show the correlations to be introduced by the second measurement, the states $|\uparrow\rangle$ and $|\downarrow\rangle$ must be expressed in the eigenbasis $\{|\rightarrow\rangle, |\leftarrow\rangle\}$ of S_x:

$$|\psi_1\rangle = \left[\frac{1}{\sqrt{2}}(c_+ |+_z\rangle + c_- |-_z\rangle) \ \otimes |\rightarrow\rangle + \frac{1}{\sqrt{2}}(c_+ |+_z\rangle - c_- |-_z\rangle) \ \otimes |\leftarrow\rangle\right] \otimes \ |0_x\rangle . \qquad (6\text{-}68)$$

Since there is no collapse, the second interaction with a magnetic field in x-direction establishes the correlation between $|\psi_1\rangle$ and the eigenstates $|+_x\rangle$ and $|-_x\rangle$ of the second pointer P_x:

$$|\psi_2\rangle = \frac{1}{\sqrt{2}}(c_+ |+_z\rangle + c_- |-_z\rangle) \ \otimes |\rightarrow\rangle \otimes |+_x\rangle + \frac{1}{\sqrt{2}}(c_+ |+_z\rangle - c_- |-_z\rangle) \ \otimes |\leftarrow\rangle \otimes |-_x\rangle , \qquad (6\text{-}69)$$

or equivalently,

$$
\begin{aligned}
|\psi_2\rangle = {} & \frac{c_+}{\sqrt{2}} \; |{\rightarrow}\rangle \otimes |+_z\rangle \otimes |+_x\rangle + \frac{c_-}{\sqrt{2}} \; |{\rightarrow}\rangle \otimes |-_z\rangle \otimes |+_x\rangle + \\
& + \frac{c_+}{\sqrt{2}} \; |{\leftarrow}\rangle \otimes |+_z\rangle \otimes |-_x\rangle - \frac{c_-}{\sqrt{2}} \; |{\leftarrow}\rangle \otimes |-_z\rangle \otimes |-_x\rangle .
\end{aligned} \tag{6-70}
$$

Now we can compute any conditional probability by combining the outcomes of the first and the second measurement. For instance, we can compute the probability that the second pointer—here the momentum P_x in x-direction—acquires the value $+_x$, given that the first pointer—here the momentum P_z in z-direction—acquired the value $+_z$, as follows:

$$pr(+_x \,|+_z) = \frac{pr(+_x \wedge +_z)}{pr(+_z)}. \tag{6-71}$$

The crucial point to stress here is that $pr(+_x \wedge +_z)$ is a legitimate probability for two reasons. First, the pointers P_x and P_z are commuting observables (see Laura and Vanni 2008). In this particular case $[P_x, P_z] = 0$ because they are orthogonal components of the momentum of the particle. But, in general, the pointers of consecutive measurements commute because they belong to different measuring apparatuses which, as a consequence, are represented by different Hilbert spaces. Second, given that, by hypothesis, the measurements are consecutive, they are not simultaneous. Thus, from the first term of eq. (6-70), $pr(+_x \wedge +_z)$ can be obtained:

$$pr(+_x \wedge +_z) = \frac{|c_+|^2}{2}. \tag{6-72}$$

In turn, $pr(+_z)$ can be computed from the first and the third terms of eq. (6-70):

$$pr(+_z) = |c_+|^2. \tag{6-73}$$

Therefore,

$$pr(+_x \,|+_z) = \frac{pr(+_x \wedge +_z)}{pr(+_z)} = \frac{|c_+|^2/2}{|c_+|^2} = \frac{1}{2} \tag{6-74}$$

as expected.

Of course, in the second measurement the same observable as in the first measurement can be measured. In this case, the basis rotation introduced in eq. (6-68) is not necessary, and the second interaction leads the system to the state

$$|\psi_2\rangle = \; c_+ \,|{\uparrow}\rangle \otimes |+_z\rangle \otimes |+_x\rangle + c_- \,|{\downarrow}\rangle \otimes |-_z\rangle \otimes |-_x\rangle . \tag{6-75}$$

Now the conditional probabilities are

$$pr(+_x|+_z) = \frac{pr(+_x \wedge +_z)}{pr(+_z)} = \frac{|c_+|^2}{|c_+|^2} = 1, \tag{6-76}$$

$$pr(-_x|+_z) = \frac{pr(-_x \wedge +_z)}{pr(+_z)} = \frac{0}{|c_+|^2} = 0, \tag{6-77}$$

as expected. These results can be generalized for any different observables and for any number of consecutive measurements.

As a consequence, and by contrast to what our physicist friend may believe, the experimentally observed correlations between the outcomes of consecutive measurements can be explained without the need of the collapse hypothesis. The MHI, from its non-collapse perspective, can account for consecutive measurements to the extent that reduced states are not conceived as the quantum states of the components of a composite system. The discussion about the meaning of reduced states will continue in Sections 8.3 and 8.4 of Chapter 8.

6.9 Defining quantum measurement

On the basis of the analysis of the quantum measurement introduced in the present chapter, the conditions that must be satisfied for a quantum process to be considered as a single measurement can now be formulated explicitly:

(i) There must be two quantum systems: a system to be measured, $\mathcal{S} : (\mathcal{O}^S, H^S)$, and a measuring apparatus, $\mathcal{M} : (\mathcal{O}^M, H^M)$, which do not interact at $t < 0$.
(ii) The apparatus $\mathcal{M}$ must be so constructed as to have a pointer observable $P \in \mathcal{O}^M$ such that $[P, H^M] = 0$, P has, at least, the same degeneracy as H^M, and its eigenvalues are different and macroscopically distinguishable.
(iii) Since $t = 0$ and during a period $\Delta t = t_1$, $\mathcal{S}$ and $\mathcal{M}$ must interact through an interaction Hamiltonian $H_{\text{int}} \in \mathcal{O}^S \otimes \mathcal{O}^M$, $H_{\text{int}} \neq 0$. The interaction is intended to introduce a correlation between an observable $A \in \mathcal{O}^S$ of $\mathcal{S}$ and the observable $P \in \mathcal{O}^M$ of $\mathcal{M}$.

Let us note that the requirement of perfect correlation was not included as a defining condition of single measurement. In fact, even if the correlation is not perfect, in Stage III the energy of the system $\mathcal{M}$ is always definite-valued and, as a consequence, the pointer P also acquires a definite value. However, although a definite reading of the pointer is always obtained in each single measurement, the frequency measurement resulting from the repetition of single measurements is not always reliable. If the interaction Hamiltonian introduces an imperfect correlation in such a way that

$$|\psi_{\text{I}}\rangle = \sum_i c_i \, |a_i\rangle \otimes |p_0\rangle \quad \longrightarrow \quad |\psi_{\text{III}}\rangle = \sum_{ij} d_{ij} \, |a_i\rangle \otimes |p_j\rangle, \tag{6-78}$$

then

- the frequency measurement will be *reliable* only when the d_{ij}, with $i \neq j$, are small enough to make $|d_{ii}|^2 \simeq |c_i|^2$,
- when the d_{ij}, with $i \neq j$, are not small enough, the frequency measurement will be *non-reliable*, because the obtained frequencies are not close to the values $|c_i|^2$.

Nonetheless, in both cases each element of the frequency measurement is legitimately a *single measurement*, in which *a definite reading of the pointer is obtained.*

Although in the previous sections the Hamiltonian H^M of the apparatus was supposed to be non-degenerate, this assumption can be relaxed in such a way that

$$H^M = \sum_i \omega_i^M \, \Pi_i^M, \tag{6-79}$$

where the eigenprojectors Π_i^M of H^M do not need to be one-dimensional. In this case, the preferred CSOP is $\{\Pi_i^M\}$. Then, according to the MHI actualization rule, the definite-valuedness of the pointer requires that P has, at least, the same degeneracy—the same symmetries—as H^M:

$$P = \sum_i p_i \, \Pi_i^M. \tag{6-80}$$

However, in the effective practice of physics, the apparatus is a macroscopic system, whose Hamiltonian is the result of the interaction between a huge number of degrees of freedom. Since, in general, symmetries are broken by interactions, the symmetry of the Hamiltonian decreases with the complexity of the system. Then, a macroscopic system having a Hamiltonian with symmetries is a highly exceptional situation: in the generic case, the energy is its only constant of motion. As a consequence, in practice H^M is non-degenerate, and $\{|\omega_i^M\rangle\}$ is a basis of the Hilbert space $\mathcal{H}^M$ of the apparatus. In turn, the pointer P cannot have such a huge number of eigenvalues as H^M, because experimental physicists must be able to discriminate between them (see, for example, P_z in the Stern-Gerlach experiment, with its three eigenvalues). This means that, in general, P is a "collective" observable (see Omnès 1994, 1999), that is, a highly degenerate observable whose eigenprojectors introduce a sort of "coarse graining" in $\mathcal{H}^M$. Nevertheless, given that $[P, H^M] = 0$, and since P has more degeneracies than H^M (which in general is non-degenerate), according to the MHI actualization rule, the pointer P is a definite-valued observable.

The discussion in the preceding paragraph, in turn, also points to an issue that still remains rather obscure in the current literature on quantum measurement. During the last decades, and under the influence of the works of Wojciech Zurek and his collaborators (see, e.g., Paz and Zurek 2002, Zurek 2003), the claim that the measuring device necessarily interacts with its environment has become a commonplace. However, such a claim faces the conceptual challenge of supplying a precise definition of quantum system, a challenge that has not been solved in the context of that theoretical framework: "In particular, one issue which has been often taken for granted is looming big, as a foundation of the whole decoherence program. It is the question of what are the 'systems' which play such a crucial role in all the discussions of the emergent classicality. This issue was raised earlier, but the progress to date has been slow at best" (Zurek 1998, 1818). In the light of this shortcoming, it has been argued that this account of decoherence is misleading, since it hides the fact that a single, unitary evolving, quantum system can be partitioned in many different ways depending on which observables are considered "relevant" and "irrelevant" in each case (Castagnino, Laura, and Lombardi 2007, Castagnino, Fortin, and Lombardi 2010a, 2010b, Lombardi, Fortin, and Castagnino 2012; see also Harshman and Wickramasekara 2007a, 2007b, Omnés 1999): each partition leads to non-unitarily evolving parts which may or may not decohere. Therefore, decoherence is a phenomenon relative to the space of the relevant observables selected in each particular situation. The relativity of decoherence will be discussed in detail in Sections 8.6 and 8.7 of Chapter 8.

In contrast to Zurek's view, the MHI offers a definition of quantum system that is precise and, at the same time, does not contradict the Dynamical Postulate of quantum mechanics. On the basis of this definition, the measuring apparatus cannot be conceived as a material, macroscopic device surrounded by a "bath" of particles interacting with it. The measuring apparatus is the entire quantum system $\mathcal{M}$ that interacts with the measured system $\mathcal{S}$: it is such a system that has a pointer P commuting with its, in general, non-degenerate Hamiltonian H^M. As pointed out above, in the generic case $\mathcal{M}$ is a macroscopic system with a huge number of degrees of freedom, and P must be a "collective" and empirically accessible observable to play the role of the pointer. For this reason, the many degrees of freedom corresponding to the degeneracies of P can be conceived as an "internal environment" (see Omnés 1994, 1999).

Finally, let us emphasize again that the requirement $[P, H^M] = 0$ has a clear physical meaning: it is essential to guarantee the stationary behavior of P. If that requirement were not met because of the uncontrollable interaction between the macroscopic device and an external "bath," the reading of P would constantly change and measurement would be impossible. It is precisely at this point that the skills of experimental physicists play a central role: they have to be capable of designing an experimental setup such that the uncontrollable degrees of freedom of the complete system interacting with the measured system do not significantly affect the stationarity of the pointer. This goal may be achieved by different means: by shielding enough the macroscopic device, by making the Hamiltonian of the apparatus much greater than the sum of the Hamiltonian corresponding to the external environment and the interaction Hamiltonian, etc. But, in any case, the measurement must be a controlled situation, in which the behavior of the observable P can be used to obtain meaningful information about the measured system $\mathcal{S}$.

6.10 Frequency and state measurements

According to the ensemble interpretation of quantum mechanics (Ballentine 1970), the theory does not describe single systems but ensembles of systems: probability assignments only make sense on ensembles. The MHI does not endorse this view: the quantum state, either pure or mixed, refers to a single system, and probabilities measure the ontological propensity to actualization of the possible facts involving that single system.

However, even though probability is not *defined* in terms of frequency, probabilistic predictions can only be *tested* by means of frequencies on collection of identical systems, that is, on ensembles: a frequency measurement, whose purpose is to estimate the coefficients of the measured state in a particular basis, is a repetition of single measurements on an ensemble. In general, the statistical predictions of the theory, as the expectation value of an observable in a given state, can be empirically tested only on ensembles. When the concept of ensemble comes into play, not only can frequency measurement be correctly understood, but also state measurement.

Up to this point, single and frequency measurements were analyzed in the case in which the measured system $\mathcal{S}$ is initially in a pure state $|\psi_S\rangle$:

$$|\psi_S\rangle = \sum_i c_i \, |a_i\rangle \, . \tag{6-81}$$

In this case, the frequency measurement, if reliable, will provide—at least, approximately—the values $|c_i|^2$. But the same analysis can be made in the general case, where the initial state of $\mathcal{S}$ is expressed as ρ_S (pure or mixed):

$$\rho_S = \sum_{ij} \rho_{ij} \, |a_i\rangle \langle a_j| \, . \tag{6-82}$$

In this general case, the frequency measurement, if reliable, approximately yields the values of the coefficients ρ_{ii} of the main diagonal of ρ_S in the basis $\{|a_i\rangle\}$ (where $\rho_{ii} = |c_i|^2$ if $\rho_S = |\psi_S\rangle \langle \psi_S|$). For this purpose, the observable A of S is used,

$$A = \sum_i a_i \, |a_i\rangle \langle a_i| \, , \tag{6-83}$$

and the apparatus $\mathcal{M}$, with a Hamiltonian H^M and a pointer P such that $[P, H^M] = 0$, interacts with the system S through an interaction Hamiltonian H_{int} (see eq. (6-14)):

$$H_{\text{int}} = -\frac{\lambda \hbar}{t_1} \left(A \otimes R^P \right) . \tag{6-84}$$

However, one can also want to know—at least, approximately—the remaining coefficients ρ_{ij} with $i \neq j$. For this purpose, further frequency measurements must be performed, in which the observables B_{ij} and C_{ij} of $\mathcal{S}$ are used (see Ballentine 1998),

$$B_{ij} = 1/2\left(|a_i\rangle\langle a_j| + |a_j\rangle\langle a_i|\right), \quad C_{ij} = 1/2\left(|a_i\rangle\langle a_j| - |a_j\rangle\langle a_i|\right), \tag{6-85}$$

since

$$\langle B_{ij}\rangle_{\rho_S} = Tr\left(\rho_S\, B_{ij}\right) = 1/2\left(\rho_{ij} + \rho_{ji}\right) = \mathrm{Re}\left(\rho_{ij}\right), \tag{6-86}$$

$$\langle C_{ij}\rangle_{\rho_S} = Tr\left(\rho_S\, C_{ij}\right) = 1/2\left(\rho_{ij} - \rho_{ji}\right) = \mathrm{Im}\left(\rho_{ij}\right). \tag{6-87}$$

In this state measurement, all the frequency measurements are performed on the same ensemble, that is, on a collection of identical systems in the same state ρ_S. However, each frequency measurement requires its own experimental arrangement. In particular, the apparatus $\mathcal{M}$ must be designed so that it interacts with the system $\mathcal{S}$ through interaction Hamiltonians H_{int}^{Bij} or H_{int}^{Cij} such that

$$H_{\mathrm{int}}^{Bij} = -\frac{\lambda\hbar}{t_1}\left(B_{ij} \otimes R^P\right), \quad H_{\mathrm{int}}^{Cij} = -\frac{\lambda\hbar}{t_1}\left(C_{ij} \otimes R^P\right). \tag{6-88}$$

These interaction Hamiltonians introduce the desired correlation between the eigenstates of the observables B_{ij} and C_{ij} of $\mathcal{S}$, respectively, and the eigenstates of the pointer observable P of $\mathcal{M}$.

In summary, if the state ρ_S of a system $\mathcal{S}$ is to be reconstructed, a state measurement must be performed on an ensemble of systems identical to $\mathcal{S}$. Such a state measurement consists of a collection of frequency measurements, each one relying on the correlation between the pointer observable of the apparatus $\mathcal{M}$ and a particular observable of $\mathcal{S}$. In particular, the observables of $\mathcal{S}$ used for this purpose must be not only different, but also non-commuting with each other. The information obtained by means of this state measurement is sufficient to reconstruct the state ρ_S of $\mathcal{S}$.

6.11 A physical model of measurement

The discussions in this and the preceding chapters reveal a feature of the quantum measurement that cannot be noticed in the purely formal treatments of the process. In fact, in the von Neumann model, the observable A of the system $\mathcal{S}$ to be measured is considered in formal terms and deprived of its physical content. Then, the interaction between $\mathcal{S}$ and the measuring apparatus $\mathcal{M}$ has the sole function of introducing the correlation between A and the pointer P of $\mathcal{M}$.

However, the several and varied physical situations described in Chapter 5 show that observers have no empirical access to the observables that are generators of the symmetries of the system's Hamiltonian. In the context of measurement, A may be one of those observables. This is precisely the case in the Stern-Gerlach experiment, where S_z is a generator of the space-rotation symmetry of $H^S = k\,S^2$. It is the interaction with the magnetic field $B = B_z$ that breaks the isotropy of space by privileging the z-direction and, as a consequence, also breaks the space-rotation symmetry of H^S. In this way, the originally inaccessible observable S_z becomes empirically accessible.

This *physical account of quantum measurement* can be formulated from a general viewpoint as follows. When the observable A of the measured system $\mathcal{S}$ is a generator of a symmetry of the Hamiltonian H^S of $\mathcal{S}$, the interaction between $\mathcal{S}$ and the apparatus $\mathcal{M}$ not only establishes the correlation between A and the pointer P of $\mathcal{M}$, but also breaks that symmetry. Therefore, from a physical perspective, a quantum measurement is a process that breaks the symmetries of the system to be measured and, in this way, makes it possible to reconstruct its state in terms of some symmetry-generator observables that would otherwise be empirically inaccessible.

In summary, the MHI introduces a novel idea about the very nature of quantum measurement. The idea is that the formal von Neumann model of quantum measurement must be complemented with a physical model, according to which *a quantum measurement is a symmetry-breaking process* that renders a generator of a symmetry of the system's Hamiltonian empirically accessible.

7
A Galilei-invariant Version of the Modal-Hamiltonian Interpretation

As stressed by Jean-Marc Lévi-Leblond (1974), although it is usual to read that non-relativistic quantum mechanics is covariant, and even invariant, under the Galilei transformations, this topic has scarcely been addressed in the standard literature on the theory. This fact has, at least, two undesirable consequences. On the one hand, the meanings of the concepts of covariance and invariance are not precisely elucidated. On the other hand, one is left without a clear idea about in what sense the covariance and/or invariance of quantum mechanics can be asserted.

This situation has its counterpart in the field of the interpretation of quantum mechanics: the relevance of the Galilei group is rarely discussed in the impressive amount of literature on the subject. The general premise underlying the present chapter is that the relation between interpretation and Galilei transformations deserves to be seriously analyzed: the fact that the theory is covariant under the Galilei group does not guarantee the same property for the interpretation because, in general, interpretations add interpretative postulates to the formal structure of the theory.

On this basis, and after clarifying the concepts of covariance and invariance, in this chapter I will consider in what sense quantum mechanics is covariant and under what conditions it is invariant under the transformations of the Galilei group. This conceptual framework will provide the tools to study the result of the application of the Galilei transformations to the modal-Hamiltonian interpretative postulates.

7.1 Covariance and invariance

The covariance and the invariance of quantum mechanics under the Galilei transformations are commonly assumed as a well-known fact. However, only in very few cases this assumption is grounded on a conceptual elucidation of the involved notions: in general, the meanings of the words "invariance" and "covariance" are taken for granted. Therefore, it is worth beginning with a clarification of those concepts.

The meaning of the term "symmetry" is rooted in ordinary language: symmetry is a geometrical property of a body whose parts are equal in a certain sense. In mathematics, the term acquires a precise meaning in terms of *invariance*: a generic item is symmetric with respect to a certain transformation when it is invariant, that is,

The Modal-Hamiltonian Interpretation of Quantum Mechanics. Olimpia Lombardi, Oxford University Press.
© Olimpia Lombardi (2026). DOI: 10.1093/9780198951407.003.0007

remains unchanged under such transformation. The concept of group, originally proposed by Galois in the first half of the nineteenth century, comes to supplement the notion of symmetry: a group clusters different transformations into a specific structure.

However, this does not yet explain what kind of items can be endowed with the property of invariance. As Katherine Brading and Elena Castellani (2007) emphasize, the first step is to distinguish between symmetries of objects and symmetries of laws: although related to each other, both cases should not be confused. In fact, as will be stressed later, the symmetry of a law does not imply the symmetry of the objects (states and operators) contained in the law. Then, the philosophical implications of the symmetries under a particular group of transformations both of a law and of the objects involved in it must be analyzed.

Despite its mathematical precision, in physics the concept of symmetry has given rise to some disagreements on the meaning and the scope of the concept of invariance and of the closely related concept of covariance. Sometimes, the property of invariance is only applied to mathematical objects and, derivatively, to the physical items they refer to, and the property of covariance is reserved for equations and, derivatively, for the physical laws they express. However, some authors claim that the difference between invariance and covariance not only makes sense but is also relevant when applied to laws (Ohanian and Ruffini 1994, Suppes 2000, Brading and Castellani 2007). Hans Ohanian and Remo Ruffini (1994), for instance, assert that an equation is said to be covariant under a certain transformation when its form remains unaltered under the transformation, and it is said to be invariant when it is covariant and its content, that is, its absolute objects (constants and non-dynamical quantities) are also unchanged by the transformation. Although inspired by this idea, I will not follow it in every detail. Here, I will consider that an equation is invariant under a certain transformation when it does not change under the application of such a transformation, and it is covariant under that transformation when its form remains unchanged with such a transformation (Suppes 2000). From this perspective, the invariance of a law does not imply the invariance of the objects contained in its representing equation. On the basis of these conceptual clarifications, some formal definitions can now be introduced (see Lombardi, Castagnino, and Ardenghi 2010).

Let us consider a set $\mathcal{X}$ of objects $X_i \in \mathcal{X}$, and a group G of transformations $T_\alpha \in G$, where the $T_\alpha : \mathcal{X} \to \mathcal{X}$ act on the X_i as $X_i \to \tilde{X}_i$. An *object* $X_k \in \mathcal{X}$ is *invariant* under the transformation T_α if, for that transformation, $\tilde{X}_k = X_k$; in turn, $X_k \in \mathcal{X}$ is *invariant* under the group G if it is invariant under all the transformations $T_\alpha \in G$. In physical theories, the objects to which the transformations apply are usually states s, observables $\mathcal{O}$, and differential operators $\mathcal{D}$, and each transformation acts on them in a particular way. For example, in the Hamilton equations of classical mechanics, the state is $s = (\boldsymbol{q}, \ \boldsymbol{p})$, where $\boldsymbol{q}$ is position and $\boldsymbol{p}$ is momentum, the relevant observable $\mathcal{O}$ is the Hamiltonian $\mathcal{H}$, and the differential operators are $D_1 = d/dt$, $D_2 = \partial/\partial \boldsymbol{p}$, and $D_3 = \partial/\partial \boldsymbol{q}$. The time-reversal transformation, which acts on the variable t as $t \to -t$, reverses all the objects whose definitions in function of t are non-invariant under the transformation:

$$s = (\boldsymbol{q}, \boldsymbol{p}) \rightarrow \tilde{s} = (\tilde{\boldsymbol{q}}, \tilde{\boldsymbol{p}}) = (\boldsymbol{q}, -\boldsymbol{p}), \quad O = H \rightarrow \tilde{O} = \tilde{H},$$

$$D_1 = d/dt \rightarrow \tilde{\boldsymbol{D}}_1 = \tilde{d}/dt = -d/dt, \quad D_2 = \partial/\partial\boldsymbol{p} \rightarrow \tilde{\boldsymbol{D}}_2 = \tilde{\partial}/\partial\boldsymbol{p} = -\partial/\partial\boldsymbol{p},$$

$$D_3 = \partial/\partial\boldsymbol{q} \rightarrow \tilde{\boldsymbol{D}}_3 = \tilde{\partial}/\partial\boldsymbol{q} = \partial/\partial\boldsymbol{q}.$$

In Newton's second law, the state is $s = \boldsymbol{x}$, where $\boldsymbol{x}$ is position, the relevant observables are $O_1 = \boldsymbol{F}$ and $O_2 = m$, and the differential operator is $D = d^2/dt^2$. They transform under time-reversal as

$$s = \boldsymbol{x} \rightarrow \tilde{s} = \tilde{\boldsymbol{x}} = \boldsymbol{x} \qquad D = d^2/dt^2 \rightarrow \tilde{\boldsymbol{D}} = \tilde{d}^2/dt^2 = d^2/dt^2$$

$$O_1 = \boldsymbol{F} \rightarrow \tilde{O}_1 = \tilde{\boldsymbol{F}} \qquad O_2 = m \rightarrow \tilde{O}_2 = \tilde{m} = m$$

In physics, these objects are combined in equations representing the laws of a theory. A dynamical law is represented by a differential equation $E\left(s, O_i, D_j\right) = 0$ including the state s, certain observables O_i, and certain differential operators D_j. When a transformation is applied to all these objects, the law may remain exactly the same, that is, its form may be left invariant by the transformation. This means that the nomological relations between the transformed objects are the same as those linking the original objects. But it may also be the case that the equation still holds when only the state is transformed, and this implies that the evolution of the state is not affected by the transformation. Precisely, let L be a *law* represented by an equation $E\left(s, O_i, D_j\right) = 0$, and let G be a group of transformations $T_\alpha \in G$ acting on the objects involved in the equation as $s \rightarrow \tilde{s}$, $O_i \rightarrow \tilde{O}_i$, and $D_j \rightarrow \widetilde{D}_j$, L is *covariant* under the transformation T_α if $E\left(\tilde{s}, \tilde{O}_i, \widetilde{D}_j\right) = 0$, and L is *invariant* under the transformation T_α if $E\left(\tilde{s}, O_i, D_j\right) = 0$. Moreover, L is covariant/invariant under the group G if it is covariant/invariant under all the transformations $T_\alpha \in G$. A group G of transformations is said to be the *symmetry group* of a theory if the laws of the theory are covariant under the group G; this means that the laws preserve their validity even when the transformations of the group are applied to the involved objects. It is easy to see that the Hamilton equations of classical mechanics, $d\boldsymbol{q}/dt = \partial H/\partial\boldsymbol{p}$, and $d\boldsymbol{p}/dt = -\partial H/\partial\boldsymbol{q}$, are covariant under time-reversal when $\tilde{H} = H$, a condition satisfied when H is time-independent; nevertheless, they are not invariant under time-reversal because $\tilde{d}\boldsymbol{p}/dt \neq -\partial H/\partial\boldsymbol{q}$. In turn, Newton's second law is covariant under time-reversal when $\widetilde{\boldsymbol{F}} = \boldsymbol{F}$, a condition that is satisfied when $\boldsymbol{F}$ is time-independent, and it is also invariant since $\tilde{d}^2\boldsymbol{x}/dt^2 = \boldsymbol{F}/m$.

Once it is accepted that the concept of invariance makes sense in its application to laws, the relation between the invariance/covariance of a law and the invariance of the objects included in it deserves attention. When a law is covariant under a transformation, and all the objects it contains are also invariant under the same transformation, the law is invariant under the transformation as well. However, this is not the only way for a law to be invariant: if a law is covariant under a certain transformation, it can turn out to be invariant under the transformation even in the case that some of the objects it contains are not invariant under the same transformation. I will come back to this point in Section 7.3 of the present chapter, when discussing the invariance of the Schrödinger equation.

Some authors prefer to speak of the symmetry of a law instead of the covariance of a law. For example, John Earman (2004a) defines symmetry in the language of the model-view of theories as follows. Let $\mathcal{M}$ be the set of the models of a certain mathematical structure, and let $\mathcal{M}_L \subset \mathcal{M}$ be the subset of the models satisfying a law L. A symmetry of the law L is a map $S: \mathcal{M} \rightarrow \mathcal{M}$ that preserves $\mathcal{M}_L$, that is, for any $m \in \mathcal{M}_L$, $\tilde{m} = S(m) \in \mathcal{M}_L$. In our case, where L is represented by a differential equation $E\left(s, O_i, D_j\right) = 0$, each model $m \in \mathcal{M}_L$ corresponds to a solution $s = F(O_i, s_0)$ of the equation, representing a nomologically possible evolution of the system. Then, the covariance of L under a transformation T—that is, $E\left(s, \tilde{O}_i, \tilde{d}_j\right) = 0$—implies that $\tilde{s} = \tilde{F}\left(\tilde{O}_i, s_0\right)$ is also a solution of the equation and, then, it corresponds to a model $\tilde{m} \in \mathcal{M}_L$. This means that the above definition of covariance and the definition of symmetry in the model-view language are equivalent.

It is interesting to notice that the covariance of a dynamical law—represented by a differential equation—does not imply the invariance of the nomologically possible evolutions—represented by the solutions of the equation (see Castagnino, Lara, and Lombardi 2003, Earman 2004b). In fact, the covariance of L, represented by $E\left(s, O_i, D_j\right) = 0$, implies that $s = F(O_i, s_0)$ and $\tilde{s} = \tilde{F}\left(\tilde{O}_i, s_0\right)$ are both solutions of the equation, but it does not imply that $s = \tilde{s}$. In the model-view language, the symmetry of L does not imply that $m = \tilde{m}$. Nevertheless, since the invariance of L means that $E\left(\tilde{s}, O_i, D_j\right) = 0$, in this case $s = \tilde{s} = F(O_i, s_0)$ or, in the model-view language, $m = \tilde{m}$.

The above general definitions can be applied to the Schrödinger equation to explicitly state the conditions of covariance and invariance for quantum mechanics. Here, I will only consider the evolution equation of the theory, leaving aside the projection postulate, since it formalizes the collapse hypothesis, which is an interpretative postulate of SQM. Given a transformation T acting as $|\varphi\rangle \rightarrow |\tilde{\varphi}\rangle$, $O \rightarrow \tilde{O}$, $d/dt \rightarrow \tilde{d}/dt$, and $i \rightarrow \tilde{i}$ (considering i as the shorthand for the operator iI), by making $\hbar = 1$, the Schrödinger equation is *covariant* under T when

$$\frac{\tilde{d}\,|\tilde{\varphi}\rangle}{dt} = -\tilde{i}\tilde{H}\,|\tilde{\varphi}\rangle, \tag{7-1}$$

and it is *invariant* under T when

$$\frac{d\,|\tilde{\varphi}\rangle}{dt} = -iH\,|\tilde{\varphi}\rangle. \tag{7-2}$$

The above conditions can be specified for the case of the Galilei transformations. As explained in Section 2.3 of Chapter 2, the Galilei group $\mathcal{G}$ is a Lie group, so the Galilei transformations T_α can be represented by unitary operators U_α over the Hilbert space, with the exponential parametrization $U_\alpha = e^{iK_\alpha s_\alpha}$, where s_α is a continuous parameter and K_α is a self-adjoint operator independent of s_α called "generator" of the transformation T_α. Then, the Galilei group is defined by ten group generators which, when the system is free from external fields, represent the basic magnitudes of the theory: the energy H, the three momentum components P_i, the three angular-momentum components J_i, and the three boost components G_i, satisfying the commutation relations of eqs. (2-11). Finally, let us recall that, in this quantum context, the invariance of an observable O under a Galilei transformation T_α amounts to the commutation between O and the corresponding generator K_α (see eq. (2-14)).

Now all the theoretical elements necessary to analyze how the Galilei transformations affect the Schrödinger equation,

$$\frac{d\,|\varphi\rangle}{dt} = -i\,H\,|\varphi\rangle, \tag{7-3}$$

are available. Let us premultiply both members of this equation by $U = e^{iKs}$. By using the property $UU^{-1} = I$ and, then, adding $(dU/dt)\,|\varphi\rangle$ to both members, the following equation is obtained:

$$U\frac{d\,|\varphi\rangle}{dt} + \frac{dU}{dt}\,|\varphi\rangle = -i\,UHU^{-1}U\,|\varphi\rangle + \frac{dU}{dt}\,|\varphi\rangle. \tag{7-4}$$

Therefore,

$$\frac{d\left(U\,|\varphi\rangle\right)}{dt} = -i\left[UHU^{-1} + i\frac{dU}{dt}U^{-1}\right]U\,|\varphi\rangle. \tag{7-5}$$

If the action of the Galilei transformations on states and observables is recalled (see eqs. (2-12) and (2-13)), the above equation becomes

$$\frac{d\,|\tilde{\varphi}\rangle}{dt} = -i\left[\tilde{H} + i\frac{dU}{dt}U^{-1}\right]\,|\tilde{\varphi}\rangle. \tag{7-6}$$

On the basis of this equation, the covariance and the invariance of the Schrödinger equation will be analyzed in the next two sections.

7.2 The covariance of the Schrödinger equation

In order to understand the covariance of the Schrödinger equation, let us rewrite eq. (7-6) as

$$\frac{d\,|\tilde{\varphi}\rangle}{dt} - \frac{dU}{dt}U^{-1}\,|\tilde{\varphi}\rangle = -i\,\tilde{H}\,|\tilde{\varphi}\rangle. \tag{7-7}$$

It is quite clear that covariance obtains when the differential operator transforms as

$$\frac{d}{dt} \to \frac{\tilde{d}}{dt} = \frac{d}{dt} - \frac{dU}{dt}U^{-1}. \tag{7-8}$$

This means that the transformed differential operator $\tilde{d}/dt$ is a covariant time-derivative D/Dt, which makes the Schrödinger equation Galilei-covariant in the following sense:

$$\frac{\tilde{d}\,|\tilde{\varphi}\rangle}{dt} = \frac{D\,|\tilde{\varphi}\rangle}{Dt} = -i\tilde{H}\,|\tilde{\varphi}\rangle. \tag{7-9}$$

In a closed, constant-energy system free from external fields, H is time-independent and the P_i and the J_i are constants of motion (see eqs. (2-11)). Then, for time-displacements, space-displacements, and space-rotations, $dU/dt = de^{iKs}/dt = 0$, where K and s stand for H and τ, P_i and ρ_i, and J_i and θ_i, respectively. As a consequence, the time-derivative is invariant under time-displacements, space-displacements, and space-rotations (see eq. (7-8)): $d/dt \to \tilde{d}/dt = d/dt$. But for boost-transformations this is not the case: the covariance of the Schrödinger equation implies the transformation of the differential operator as $d/dt \to D/Dt$. This means that covariance under boosts amounts to a sort of "non-homogeneity" of time, which requires the covariant adjustment of the time-derivative. This conclusion should not be surprising since, when the system is described in a reference frame $\tilde{F}$ at a uniform motion corresponding to a velocity u_x with respect to the original frame F, the boost-transformed state depends on a generator that is a linear function of time:

$$G_x = mQ_x = m\,(Q_{x0} + V_x t) = mQ_{x0} + P_x t. \tag{7-10}$$

Then, if the Schrödinger equation is to be valid in $\tilde{F}$, where the state is $|\tilde{\varphi}\rangle$, the transformed time-derivative has to be adjusted to compensate the time-dependent transformation of the state.

7.3 The invariance of the Schrödinger equation

As seen in the previous section, with no external fields applied on the system, H, the P_i, and the J_i are time-independent and constants of motion. Then, for time-displacements, space-displacements, and space-rotations, eq. (7-6) yields

$$\frac{d\,|\tilde{\varphi}\rangle}{dt} = -i\tilde{H}\,|\tilde{\varphi}\rangle. \tag{7-11}$$

Moreover, for those transformations, $\tilde{H} = H$ because (see eqs. (2-11)):

- Time-displacements: $\tilde{H} = e^{iH\tau}H\,e^{-iH\tau} = H$ because $[H, H] = 0$.

- Space-displacements: $\tilde{H} = e^{iP_i\rho_i} H e^{-iP_i\rho_i} = H$ because $[P_i, H] = 0$.
- Space-rotations: $\tilde{H} = e^{iJ_i\theta_i} H e^{-iJ_i\theta_i} = H$ because $[J_i, H] = 0$.

By applying these results to eq. (7-11), the invariance of the Schrödinger equation under time-displacements, space-displacements, and space-rotations can be proved in the absence of external fields:

$$\frac{d\,|\tilde{\varphi}\rangle}{dt} = -iH\,|\tilde{\varphi}\rangle. \tag{7-12}$$

The case of boost-transformations is different from the previous cases because the Hamiltonian is not boost-invariant even when the system is free from external fields (for the same claim in classical mechanics, see Butterfield 2007, 6). In fact, under a boost-transformation corresponding to a velocity u_x, H changes as (see eqs. (2-11))

$$\tilde{H} = e^{iG_xu_x} H e^{-iG_xu_x} \neq H \quad \text{because } [G_x, H] = iP_x \neq 0, \tag{7-13}$$

and the generator G_x is given by eq. (7-10). Since G_x is not time-independent, $dU/dt = de^{iG_xu_x}/dt \neq 0$, and eq. (7-6) yields

$$\frac{d\,|\tilde{\varphi}\rangle}{dt} = -i\left[\tilde{H} + i\frac{de^{iG_xu_x}}{dt}e^{-iG_xu_x}\right]|\tilde{\varphi}\rangle. \tag{7-14}$$

In order to know the value of the bracket in the right-hand side of eq. (7-14), both terms in the bracket must be computed (see details in Lombardi, Castagnino, and Ardenghi 2010). By using the Hadamard lemma applied to the Baker-Campbell-Hausdorff formula,

$$e^{B}Ae^{-B} = A + [B,A] + (1/2!)[B,[B,A]] + (1/3!)[B,[B,[B,A]]] + \cdots, \tag{7-15}$$

and by applying the commutation relations of eqs. (2-11i) and (2-11f), $\tilde{H}$ becomes

$$\tilde{H} = e^{iG_xu_x} H e^{-iG_xu_x} = H - u_xP_x + \frac{1}{2}Mu_x^2 = H + T_B, \tag{7-16}$$

where

$$T_B = -u_xP_x + \frac{1}{2}Mu_x^2 \tag{7-17}$$

is the boost contribution to the energy. In turn, by means of the lemma and the commutation relation (2-11f), $\tilde{P}_x$ results

$$\tilde{P}_x = e^{iG_xu_x} P_x e^{-iG_xu_x} = P_x - Mu_x \;\Rightarrow\; \tilde{P} = P + P_B, \tag{7-18}$$

where $P_B = (-Mu_x, 0, 0)$ is the boost contribution to the momentum. Let us recall that, when there are no external fields, the internal energy W is a Casimir operator;

therefore, the Hamiltonian can be written as

$$H = \frac{P^2}{2m} + W. \tag{7-19}$$

By means of eqs. (7-18) and (7-19), it is easy to show that the transformed Hamiltonian can be expressed as

$$\tilde{H} = \frac{(P + P_B)^2}{2m} + W. \tag{7-20}$$

In turn, the time-derivative $de^{iG_x u_x}/dt$ of eq. (7-14) must be computed. By using the identity $e^{A+B} = e^A e^B e^{-[A,B]/2}$, which holds when $[A, [A, B]] = [B, [A, B]] = 0$, and by applying the commutation relation $[P_i, F(Q_j)] = -i\,\partial F/\partial Q_j$ valid on the Galilei algebra, it can be proved that (for details, see Lombardi, Castagnino, and Ardenghi 2010)

$$\frac{de^{iG_x u_x}}{dt} = -i\left(u_x P_x - \frac{1}{2} M u_x^2\right) e^{iG_x u_x}. \tag{7-21}$$

When the results eq. (7-16) and (7-21) are introduced into eq. (7-14), the terms added to H in $\tilde{H}$ cancel with those coming from the term containing the time-derivative. Therefore, in absence of external fields, the invariance of the Schrödinger equation can also be proved for boost-transformations:

$$\frac{d\,|\tilde{\varphi}\rangle}{dt} = -iH\,|\tilde{\varphi}\rangle\,. \tag{7-22}$$

The case of boost-transformations illustrates a claim previously mentioned in Section 7.1 of the present chapter: even though a law is invariant under a transformation when it is covariant and all the involved objects are invariant, this is not the only way to obtain invariance. When the quantum system is free from external fields, the Schrödinger equation is invariant under boost-transformations, despite the fact that the Hamiltonian and the differential operator d/dt are not boost-invariant objects.

Let us summarize the results obtained up to this point. When there are no external fields acting on the system, the Hamiltonian is invariant under time-displacements, space-displacements, and space-rotations, but not under boost-transformations. In spite of this, the Schrödinger equation is invariant under the full Galilei group, and this conceptually means that the state vector $|\varphi\rangle$ does not "see" the effect of the transformations: the evolutions of $|\varphi\rangle$ and $|\tilde{\varphi}\rangle$ are identical. In other words, the time-behavior of the system is independent of the Galilei-transformed reference frame used for the description.

7.4 Galilei group and external fields

When the system is under the action of external fields, they modify the evolution of the system. But, in non-relativistic SQM, fields are not quantized: they do not play the role of quantum systems that interact with other systems. For this reason, the effect of the fields on a system must be included in its Hamiltonian, because it is the only observable involved in the dynamical law. It can be proved that the most general form of the Hamiltonian in the presence of external fields is (see, e.g., Ballentine 1998)

$$H = \frac{(P - A(Q))^2}{2M} + V(Q), \tag{7-24}$$

where $A(Q)$ is a vector potential and $V(Q)$ is a scalar potential. The covariance of the Schrödinger equation, as expressed in eq. (7-7), fixes the way in which the potentials $A(Q)$ and $V(Q)$ must transform under the Galilei group. The electromagnetic field may be derived from a vector potential and a scalar potential; thus, a fully Galilei-covariant quantum theory of systems interacting with an external electromagnetic field is possible. However, the electric and magnetic fields that are Galilei-transformed as required to preserve Galilei-covariance, although related to the scalar and vector potentials in the usual way, are ruled by one of two sets of electromagnetic "field" equations, which can be considered the non-relativistic limits of the Maxwell equations in cases where either (i) magnetic effects predominate over electric ones ("magnetic limit": $c|B| >> |E|$) or (ii) electric effects predominate over magnetic ones ("electric limit": $c|B| << |E|$) (see Brown and Holland 1999, Colussi and Wickramasekara 2008). In turn, $A(Q)$ and $V(Q)$ should not be necessarily identified with the electromagnetic potentials, because they are arbitrary functions that do not need to satisfy the Maxwell equations; for example, the Newtonian gravitational potential can also be included in the scalar $V(Q)$ (Ballentine 1998).

At this point, a relevant conceptual issue must be stressed. Space-displacements and space-rotations are purely geometric operations of displacement and rotation of the system self-congruently to another place and to another direction, respectively. Analogously, time-displacements are purely geometric operations of displacement of the system self-congruently to another time, and they may agree or not with dynamical evolutions. The commutation of two transformation generators means that the corresponding geometric operations can be performed in either order with the same result; for instance, the commutation $[K_{p_i}, K_{p_j}] = 0$ (see eq. (2-10a)) means that the order in which space-displacements in different directions are performed does not modify the result. In particular, the validity of the Galilei group implies that time-displacements commute both with space-displacements and with space-rotations (see eqs. (2-10g) and (2-10h)). When the system is free from of external fields, this feature is given by the commutation relations involving the Hamiltonian, the three momentum components P_i, and the three angular momentum components J_i, $[P_i, H] = 0$

and $[J_i, H] = 0$ (eqs. (2-11g) and (2-11h)), because here the Hamiltonian *is* the time-displacement generator. But, in the presence of external fields, as the action of the fields is incorporated into the Hamiltonian of the system, *the Hamiltonian is no longer the generator of time-displacements*: it only retains the role of generator of the dynamical evolution (see Laue 1996, Ballentine 1998). For this reason, the commutation relations with the momentum components and with the angular momentum components break down: $[P_i, H] \neq 0$ and $[J_i, H] \neq 0$. However, to the extent that the covariance of the Schrödinger equation is retained, the commutation of time-displacements with both space-displacements and space-rotations still holds, and is still represented by $[K_{\rho_i}, K_\tau] = 0$ and $[K_{\theta_i}, K_\tau] = 0$, respectively (see eqs. (2-10g) and (2-10h)), where the momentum components are still the generators of space-displacements, $P_i = \hbar K_{\rho_i}$, and the angular momentum components are still the generators of space-rotations, $J_i = \hbar K_{\theta_i}$, but the Hamiltonian is no longer the generator of time-displacements, $H \neq \hbar K_\tau$.

7.5 The relevance of invariance for interpretation

As is well known, in principle a space-time transformation admits two interpretations: active and passive. Under the active interpretation, the transformation corresponds to the change from a system to the transformed system; for example, a system translated in space with respect to the original one. Under the passive interpretation, the transformation consists of a change of the viewpoint—the reference frame—from which the system is described; for example, the space-translation of the observer that describes the system. In the case of continuous space-time transformations, both active and passive interpretations are equally allowed. But such a situation is not so clear for discrete transformations: in general, it is accepted that only the active interpretation makes sense in this case (Sklar 1974, 363). Nevertheless, no matter which interpretation is adopted, the covariance of the fundamental law of a theory under its continuous symmetry group implies that the law still holds when the transformations of the group are applied: in the active-interpretation language, the original and the transformed systems are equivalent; in the passive-interpretation language, the original and the transformed reference frames are equivalent.

As typically accepted, the Galilei group is the symmetry group of continuous space-time transformations of classical and quantum mechanics. In the language of the passive interpretation, the covariance of the dynamical laws amounts to the equivalence among inertial reference frames—time-translated, space-translated, space-rotated, or moving uniformly with respect to each other. In other words, Galilei transformations do not introduce any modification in the physical situation, but only express *a change in the perspective from which the system is described.*

The above remarks are related to the fact that certain quantities are physically irrelevant in the light of the symmetries of the theory. For instance, the space-translation symmetry of a dynamical law means that the specific place where the system is located in space is irrelevant to its evolution, governed by such a law: "A global symmetry reflects the irrelevance of absolute values of a certain quantity: only relative values are relevant" (see Brading and Castellani 2007, 1360). This is the case in classical mechanics: space-translation invariance implies that absolute position is irrelevant to the system's behavior, that is, the equations of motion do not depend on absolute positions, only relative positions matter. The physical irrelevance of certain quantities is closely related to the issue of *objectivity*.

The intuition about a strong link between invariance and objectivity is rooted in a natural idea: what is objective should not depend on the particular perspective used for the description. When this intuition is translated into the group-theoretical language, it amounts to the assumption that what is objective according to a theory is what is invariant under the symmetry group of the theory. This idea appeared in the domain of formal sciences in Felix Klein's "Erlangen Program" of 1872, with the attempt to characterize all known geometries by their invariants (see Kramer 1970). The idea passed to physics with the advent of relativity, regarding the ontological status of space and time (Minkowski 1923). The claim that objectivity means invariance becomes a main thesis of Hermann Weyl's book *Symmetry* (1952). Max Born also expressed very clearly his conviction about the strong link between invariance and objectivity: "I think the idea of invariance is the clue to a rational concept of reality" (Born 1953, 144). In recent times, the idea has reappeared in several works. For instance, in her deep analysis of quantum field theory, Sunny Auyang (1995) grounds her general concept of "object" on its invariance under transformations between all representations. The assumption of invariance as the root of objectivity is also the central theme of Robert Nozick's book *Invariances: The Structure of the Objective World* (2001). In the same vein, David Baker (2010) argues that symmetries are a guide to finding out which quantities represent fundamental natural properties in a physical theory.

If the ontological import of symmetry is accepted, it is easy to realize that symmetry must play an active role in the understanding of a physical theory. In the particular case of quantum mechanics, the consideration of its Galilei-covariance cannot be overlooked in the discussions about interpretation.

As is well known, the Kochen-Specker theorem (Kochen and Specker 1967) establishes a barrier to any realist classical-like interpretation of quantum mechanics: it proves the impossibility of ascribing definite values to all the physical quantities (observables) of a quantum system simultaneously, while preserving the functional relations between commuting observables. This result amounts to the contextuality of quantum mechanics: the ascription of definite values to the observables of a quantum system is always contextual. As already stated in Chapter 3, as a consequence of the Kochen-Specker theorem, any realist interpretation of quantum mechanics is committed to selecting a subset of definite-valued observables from the set of all the observables of the system—or a preferred basis from all the formally equivalent bases of the Hilbert space. The observables of that subset will be those that acquire definite values without violating quantum contextuality. It is at this point that the symmetry group of the theory becomes a leading character: as noticed by Harvey Brown, Mauricio Suárez, and Guido Bacciagaluppi (1998), any interpretation that selects the set of the definite-valued observables of a quantum system must consider how that set is transformed under the Galilei group.

But now the link between invariance and objectivity comes into play. The study of the role of symmetry is particularly compelling for realist interpretations of quantum mechanics, which conceive a definite-valued observable as a physical property that *objectively* acquires an actual definite value among all its possible values: the fact that a certain observable acquires a definite value must be an objective fact that should not depend on the descriptive perspective adopted by the observer. Therefore, the set of the definite-valued observables of a system, which is picked out by the interpretation, should be left invariant by the Galilei transformations: from a realist viewpoint, it would be unacceptable for that set to change as a result of a mere change in the perspective from which the system is described.

In his article "Aspects of objectivity in quantum mechanics," Harvey Brown (1999) explicitly tackles the problem when discussing the objectivity of "sharp values." In particular, he focuses on interpretations that specify state-dependent rules for assigning sharp values to some of the self-adjoint operators representing quantum magnitudes: for example, the interpretations whose value-assignment rule is given by the eigenstate-eigenvalue link, or the traditional Modal Interpretations that make the set of definite-valued observables to depend on the instantaneous state of the system. Brown clearly explains the difference between the classical and the quantum case. In classical mechanics, Galilei non-invariant magnitudes modify their values with the change of reference frame; for this reason, if their objectivity is to be retained, they must be regarded not as intrinsic properties but as relational properties. For example, the values of classical position and momentum can be conceived as relational properties that link the system and the reference frame. In quantum mechanics, by contrast, the relational nature of properties acquires a further degree: whereas in the classical case the sharp value of a magnitude depends on the reference frame, in the quantum case the *very sharpness* of an observable's value should be relational in order to preserve its objectivity. For instance, the fact that the position of a system has a sharp (definite) value in a certain reference frame—as a consequence, that the system can be conceived as a localized particle—is itself relational: in a different reference frame, the system may have an unsharp value of position and, then, may behave as a delocalized particle. Brown also correctly stresses that it is not just boosts that produce this type of situation: passive spatial translations can cause some sharp-valued observables to become unsharp. On the basis of this analysis, he concludes that "[i]f, in the hope of providing an ontological interpretation of quantum mechanics, we introduce *state-dependent* rules for assigning sharp values to magnitudes associated with a specific quantum system, we should recognise that the objective status of such sharp values is relational, not absolute." (Brown 1999, 66–67; emphasis in the original).

In the same article, Brown considers an interpretation in which the rule of definite-value ascription is not state-dependent: he analyzes the problem of covariance in the de Broglie-Bohm pilot-wave interpretation, but he discards it because, although no privileged frame is picked out by the hidden dynamics of the corpuscles, the forces acting upon the corpuscles and generated by the guiding wave are Aristotelian, not Newtonian, because they produce velocities, not accelerations. This seems to lead him to interpretations that introduce state-dependent rules of definite-value ascription as the only alternative. This stance is opposite to the one that led Jeffrey Bub (1997) to advocate for Bohmian Mechanics, conceived as a modal interpretation whose rule of definite-value ascription picks out the position observable: the difficulties of the original Modal Interpretations in coping with non-ideal measurements due to their state-dependent rules turned Bohmian Mechanics into a natural alternative. But what both Bub and Brown seem to overlook is that other interpretative strategies can be conceived beyond Bohmian Mechanics, which postulates position as the definite-valued observable, and traditional Modal Interpretations, which make the definite-valued observables depend on the state of the system. One of those strategies is even more natural than the Bohmian proposal when the aim is to preserve the objectivity of definite-valuedness (or of sharpness, in Brown's terms) in the light of Galilei symmetry. In fact, the natural way to reach this goal, without making the objective status of definite-valuedness relational, is to appeal to the Casimir operators of the Galilei group: if the interpretation has to select a Galilei-invariant set of definite-valued observables, such a set must depend on those Casimir operators, insofar as they are invariant under all the transformations of the Galilei group. As will be discussed in the next sections, the MHI naturally leads to that conclusion.

7.6 Transformation of observables

Some authors adopt a different strategy to address the question of the Galilei-covariance of the Schrödinger equation. By assuming the Galilei-invariance of the differential operator d/dt, the Galilei-covariance of the Schrödinger equation is preserved by redefining the action of the boost-transformation on certain dynamical magnitudes, in particular, on the Hamiltonian. According to this strategy, a boost-transformation given by $U = e^{iG_x u_x}$ does not act on H as $\tilde{H} = UHU^{-1}$, but as (see Brown, Suárez, and Bacciagaluppi 1998, eq. (16), Brown and Holland 1999, eq. (33))

$$H \to \widehat{H} = UHU^{-1} + i\frac{dU}{dt}U^{-1}, \tag{7-25}$$

whereas the state and the differential operator transform as

$$|\varphi\rangle \to |\widehat{\varphi}\rangle = U|\varphi\rangle \qquad d/dt \to \widehat{d}/dt = d/dt. \tag{7-26}$$

Since these transformations were deliberately devised to preserve the Galilei-covariance of the Schrödinger equation, by introducing eqs. (7-25) and (7-26) into eq. (7-6), such a covariance is immediately obtained:

$$\frac{\widehat{d}\,|\widehat{\varphi}\rangle}{dt} = -i\,\widehat{H}\,|\widehat{\varphi}\rangle. \tag{7-27}$$

If $\tilde{H}$ is taken to be the boost-transformed H, when there are no external fields acting on the system the Hamiltonian turns out to be invariant also under boost-transformations, since $\widehat{H} = H$ (introduce eqs. (7-18) and (7-21) into eq. (7-25)). Therefore, the invariance of the Schrödinger equation under boosts follows from its covariance, given by eq. (7-27), and the boost-invariance of the objects involved in the equation, $\widehat{d}/dt = d/dt$ and $\widehat{H} = H$. In this case, the boost-transformation of H is still unitary—it is the identity—, and the choice between $\tilde{H}$ and $\widehat{H}$ as the boost-transformed Hamiltonian seems to be a matter of convention. However, the preference of $\widehat{H}$ over $\tilde{H}$ leads to undesirable consequences, both from a mathematical and from a physical point of view.

First, this strategy has an unpalatable ad hoc flavor. In fact, the Galilei-covariance of the Schrödinger equation, rather than a result, turns out to be an a priori truth: no matter the particular form the equation takes, the transformations are specifically defined to preserve its covariance. But the price to pay for the so-obtained covariance is to admit that a given transformation acts in different ways on different observables. Precisely, the transformation $H \to \widehat{H}$ given by eq. (7-25), although unitary when there are no external fields, becomes non-unitary in the presence of those fields. This means that some observables, in particular the Hamiltonian, transform with what Brown, Suárez, and Bacciagaluppi (1998, 297) call a "*sui generis*" non-unitary transformation. But non-unitarity breaks the basic features of the Galilei group. On the one hand, non-unitary transformations cannot be combined in a single operation as that given by eq. (2-9) (see Section 2.3 of Chapter 2), which expresses the action of the ten elemental transformations. On the other hand, non-unitary transformations do not preserve the commutation relations among transformed observables. Precisely, given two observables A and B such that $[A, B] = C$, it is easy to see that the application of a transformation represented by a unitary operator U yields

$$[A, B] = C \quad \Rightarrow \quad [\tilde{A}, \tilde{B}] = [UAU^{-1}, UBU^{-1}] = UCU^{-1} = \tilde{C}. \tag{7-28}$$

This property is what preserves the commutation relations that define the Galilei group also for the transformed observables; for instance, $[\tilde{P}_i, \tilde{P}_j] = 0$ or $[\tilde{J}_i, \tilde{G}_j] = i\varepsilon_{ijk}\tilde{G}_k$. But if the "sui generis" transformation (7-25) is used, this property gets lost. In particular, if $A \to \widehat{A}$ and $B \to \widehat{B}$, with some algebra it can be obtained

$$[\widehat{A}, \widehat{B}] = UCU^{-1} + i\left[U(B-A)\,\frac{dU^{-1}}{dt} + \frac{dU}{dt}\,(B-A)\,U^{-1}\right]. \tag{7-29}$$

The right-hand side of this equation can be identified neither with $\widehat{C}$—C transformed by eq. (7-25)—nor with $\tilde{C}$—C unitarily transformed. Therefore, the adoption of non-unitary transformations seems too high a price to pay for preserving the Galilei-covariance of the Schrödinger equation, because this strategy leaves us with no clear idea about what the Galilei group means when deprived of its basic features (see Lombardi, Castagnino, and Ardenghi 2010).

From the physical viewpoint, a transformation as that of eq. (7-25) also leads to undesirable consequences. According to that equation, with no external fields, the boost-transformed Hamiltonian is $\widehat{H} = H$. This means that, if H is the Hamiltonian of the system when described in the reference frame F, then $\widehat{H} = H$ is the Hamiltonian of the same system when described in the uniform-motion reference frame $\widehat{F} = \tilde{F}$. But this conclusion disagrees with the physical fact that the total energy of a system changes in an additive kinetic value when the descriptive perspective is changed from F to $\widehat{F} = \tilde{F}$. And this change in the total energy has its empirical manifestation as a Doppler shift in the energy spectrum (see Cohen-Tannoudji, Diu, and Lalöe 1977), which cannot be accounted for by the boost-transformation under the form of eq. (7-25).

In summary, if the formal structure of the Galilei group and the physical meaning of its transformations are to be preserved, there is no freedom to decide the form of the transformations. The argument of this section shows that any claim about the Galilei-covariance of the Schrödinger equation must be based on the adequate transformations of the observables, in particular, of the Hamiltonian. This conclusion not only contributes to the understanding of the role of the Galilei group in quantum mechanics, but also has relevant implications for interpretation, as will be seen in the following sections.

7.7 Transformation of systems

According to the first representative postulate of quantum mechanics RP1, introduced in Section 2.2 of Chapter 2, a quantum system $\mathcal{S}$ is represented by a pair $(\mathcal{O}, H)$, where $\mathcal{O}$ is the space of the self-adjoint elements of the von Neumann algebra $\mathcal{B}(\mathcal{H})$ of linear bounded operators on the Hilbert space $\mathcal{H}$, and $H \in \mathcal{O}$ is the Hamiltonian. Then, any Galilei transformation $T_\alpha \in \mathcal{G}$ should apply to $\mathcal{S}$ as $\mathcal{S}$: $(\mathcal{O}\ H) \rightarrow \tilde{\mathcal{S}}$: $(\tilde{\mathcal{O}}, \tilde{H})$. However, as explained in Section 7.1 of the present chapter, any transformation $T_\alpha \in G$ is an automorphism $T_\alpha : \mathcal{X} \rightarrow \mathcal{X}$. As a consequence, the Galilei group $\mathcal{G}$ applies to the observables of the system in such a way that

$$\forall T_\alpha \in \mathcal{G}, \quad \text{if } O \in \mathcal{O} \text{ and } O \rightarrow \tilde{O}, \text{ then } \tilde{O} \in \mathcal{O}. \tag{7-30}$$

In other words, the space of observables of a quantum system is closed under the transformations of the Galilei group,

$$\forall\, T_\alpha \in \mathcal{G}, \quad \mathcal{O} \rightarrow \tilde{\mathcal{O}}. \tag{7-31}$$

This feature is physically reasonable, since one does not expect that the mere application of a Galilei transformation on the system $\mathcal{S}$ modifies its identity by modifying its space of observables $\mathcal{O}$. Therefore, the result of the application of the Galilei transformations to a quantum system only depends on the way in which the Hamiltonian is transformed:

$$\forall T_\alpha \in \mathcal{G}, \quad \mathcal{S}: (\mathcal{O}, H) \to \tilde{\mathcal{S}}: (\mathcal{O}, \tilde{H}), \tag{7-32}$$

where H, as discussed in the previous section, transforms *unitarily* as $\tilde{H} = U_\alpha H U_\alpha^{-1}$, with $U_\alpha = e^{iK_\alpha s_\alpha}$ and being K_α the generator of the transformation T_α.

As shown above, in the presence of external fields, the Schrödinger equation is not invariant under the Galilei group. This physically means that the system changes its behavior when merely displaced in space or time, rotated in space, or uniformly moved. Nevertheless, even in the presence of external fields, the Schrödinger equation preserves its covariance as the result of the transformation of the objects involved in it, in particular, of the Hamiltonian. According to the representative postulate RP1, strictly speaking, with external fields the original and the transformed systems are not the same system because the Hamiltonian has changed. This is completely consistent with the interpretative framework of the MHI. In fact, a system that preserves its identity when Galilei-transformed in the presence of fields must be a system endowed with a principle of identity that reidentifies it under the change of its properties. Such a principle should be given by a sort of substance, conceived as a characterless substratum to which properties are "stuck": the permanence of that substratum is what endows the system with its identity under property change. But, according to the MHI, a quantum system is not a substance acting as a bearer of properties, but a bundle of properties represented by the observables of the space $\mathcal{O}$ (these ontological matters will be discussed in depth in Chapters 9 and 10). To the extent that there is no identity-preserving substratum, the identity of the system changes as properties change.

With no external fields, on the contrary, the issue of the Galilei transformation of a system requires to be considered with care. In this case, the Schrödinger equation is invariant under the Galilei group, and this means that the application of a Galilei transformation does not modify the physical situation, but only expresses a change in the perspective from which the system is described. As a consequence, it is expected that, in this case, the system does not change its identity as a result of being Galilei-transformed: the system should be a Galilei-invariant object. The invariance of the system under time-displacements, space-displacements, and space-rotations follows directly from the invariance of the Hamiltonian under those transformations:

$$\mathcal{S}: (\mathcal{O}, H) \quad \to \quad \tilde{\mathcal{S}}: (\mathcal{O}, \tilde{H}) = \mathcal{S}: (\mathcal{O}, H). \tag{7-33}$$

But, again, the situation is completely different for boost-transformations: although the Schrödinger equation is invariant under boosts, the Hamiltonian is not. For this reason, this case needs to be analyzed in detail.

Let us consider a quantum system not affected by external fields, represented by $\mathcal{S}$: $(\mathcal{O}, H)$. In a generic reference frame F, the Hamiltonian reads $H = P^2/2m + W = K + W$, where the kinetic energy $K = P^2/2m$ only depends on the total momentum relative to F, and the internal energy W does not depend on the position and the momentum relative to F, but only depends on differences of positions and, eventually, on their derivatives. Therefore, it can be guaranteed that $[K,W] = 0$ and, as a consequence, H can be expressed as

$$H = K + W = H^K \otimes I^W + I^K \otimes H^W, \tag{7-34}$$

where H^K is the kinetic Hamiltonian acting on the Hilbert space $\mathcal{H}^K$, H^W is the internal energy Hamiltonian acting on the Hilbert space $\mathcal{H}^W$, and I^K and I^W are the identity operators of the respective tensor-product spaces. According to the System Decomposition postulate IP3 (see Section 4.1 of Chapter 4), eq. (7-34) implies that the system $\mathcal{S}$ is a composite system $\mathcal{S} = \mathcal{S}^W \cup \mathcal{S}^K$, whose elemental subsystems are:

- A system represented by $\mathcal{S}^W$: $(\mathcal{O}^W, H^W)$, where $\mathcal{O}^W$ is the space of the observables acting on $\mathcal{H}^W$ and $H^W \in \mathcal{O}^W$ represents the internal energy.
- A system represented by $\mathcal{S}^K$: $(\mathcal{O}^K, H^K)$, where $\mathcal{O}^K$ is the space of the observables acting on $\mathcal{H}^K$, and $H^K \in \mathcal{O}^K$ represents the kinetic energy.

If now a boost-transformation of velocity u_x is applied to the system $\mathcal{S} = \mathcal{S}^W \cup \mathcal{S}^K$, the unitarily transformed Hamiltonian is $\tilde{H} = H + T_B$ (see eq. (7-16)) and, then, it can be expressed as

$$\tilde{H} = H + T_B = \frac{P^2}{2m} + W + T_B = \tilde{K} + W, \tag{7-35}$$

where $\tilde{K}$ is the transformed kinetic energy (see eq. (7-18)):

$$K = \frac{P^2}{2m} \longrightarrow \tilde{K} = K + T_B = \frac{P^2}{2m} + T_B = \frac{(P + P_B)^2}{2m}. \tag{7-36}$$

For the same reasons as before, $[\tilde{K}, W] = 0$ and, as a consequence, $\tilde{H}$ can be written as

$$\tilde{H} = \tilde{K} + W = \tilde{H}^K \otimes I^W + I^K \otimes H^W, \tag{7-37}$$

where $\tilde{H}^K = H^K + T^B$ is the transformed kinetic Hamiltonian acting on $\mathcal{H}^K$. Therefore, the boost-transformed system is again a composite system $\tilde{\mathcal{S}} = \mathcal{S}^W \cup \tilde{\mathcal{S}}^K$, whose elemental subsystems are the original $\mathcal{S}^W$ and the system $\tilde{\mathcal{S}}^K$: $(\mathcal{O}, \tilde{H}^K)$ now defined by a kinetic energy $\tilde{H}^K$ that adds the kinetic energy T^B of the boost to the original kinetic energy H^K.

This argument shows that, with no external fields, a boost-transformation acts on a system represented by $\mathcal{S} = \mathcal{S}^W \cup \mathcal{S}^K$ as

$$\mathcal{S} = \mathcal{S}^W \cup \mathcal{S}^K \longrightarrow \tilde{\mathcal{S}} = \mathcal{S}^W \cup \tilde{\mathcal{S}}^K. \tag{7-38}$$

When, in particular, $\mathcal{S}$ is described in the reference frame at rest with respect to its center of mass, $P = 0$; then, $\mathcal{S}$ is an elemental system with Hamiltonian $H = W$, on which a boost acts as

$$\mathcal{S} = \mathcal{S}^W \longrightarrow \tilde{\mathcal{S}} = \mathcal{S}^W \cup \tilde{\mathcal{S}}^K, \tag{7-39}$$

where the subsystem $\tilde{\mathcal{S}}^K$ is now defined only by the kinetic energy of the boost. Therefore, the subsystem $\mathcal{S}^W$, carrying the internal energy of the system, is boost-invariant, in agreement with the fact that the internal energy W is a Casimir operator of the Galilei group. The application of a boost-transformation only affects the subsystem $\mathcal{S}^K$ by adding the kinetic energy of the boost to its Hamiltonian:

$$\mathcal{S}^W \longrightarrow \tilde{\mathcal{S}}^W = \mathcal{S}^W \qquad \tilde{H}^W = H^W, \tag{7-40}$$

$$\mathcal{S}^K \longrightarrow \tilde{\mathcal{S}}^K \qquad \tilde{H}^K = H^K + T^B. \tag{7-41}$$

This result leads to the question about the ontological status of both subsystems.

On the one hand, with no external fields, the action of a boost-transformation introduces a well-defined change in the energy of the composite system $\mathcal{S} = \mathcal{S}^W \cup \mathcal{S}^K$. However, it is also well known that energy is defined up to a constant value: the relevant information about the energy spectrum of a system is contained in its internal energy, and the kinetic energy only introduces a shift of that spectrum. Therefore, the boost-invariant subsystem $\mathcal{S}^W$ carries the physically meaningful structure of the energy spectrum, whereas $\mathcal{S}^K$ represents an energy shift that is not physically relevant because it is merely relative to the reference frame used for the description. On the other hand, even the composite or elemental character of the system $\mathcal{S}$ depends on the particular reference frame selected. In fact, in the reference frame F at rest with respect to the center of mass, $\mathcal{S} = \mathcal{S}^W$ is an elemental system; when, in turn, the system is described in a reference frame $\tilde{F}$ in uniform motion with respect to F, the system turns out to be composite, $\mathcal{S} = \mathcal{S}^W \cup \mathcal{S}^K$.

Both considerations point to the same direction: the *objective* content of the description is given by the internal energy. In other words, the objective description of the system is $\mathcal{S}^W$, that is, the description in the reference frame at rest with respect to the center of mass, where $H = W$: in this reference frame, $\mathcal{S}^W$ is completely invariant under the Galilei group. On the contrary, $\mathcal{S}^K$, which carries the kinetic energy, is a sort of "pseudo-system," whose identity is modified by a mere change in the descriptive perspective, and which may even "appear" or "disappear" as a consequence of such a change. In order to express this idea with precision, I will use the same terminology introduced in Chapter 4 to distinguish between the physical language and its ontological reference in the case of type-properties and case-properties: the symbol '[xx]' denotes the ontological item referred to by the word 'xx' of the physical language. On this basis, it can be said that $\mathcal{S} = \mathcal{S}^W \cup \mathcal{S}^K$ and $\mathcal{S}^W$ refer to the same *ontological system*:

$$[\mathcal{S}] = [\mathcal{S}^W \cup \mathcal{S}^K] = [\mathcal{S}^W], \tag{7-42}$$

where the symbol "=" denotes *logical identity* (that is, if $a = b$, then a and b are two names for the same item). Therefore, when *ontological* systems are free from external fields, they are invariant under all the transformations of the Galilei group, also under boosts:

$$[\mathcal{S}] = [\mathcal{S}^W \cup \mathcal{S}^K] = [\mathcal{S}^W] \longrightarrow [\tilde{\mathcal{S}}] = [\mathcal{S}^W \cup \tilde{\mathcal{S}}^K] = [\mathcal{S}^W] = [\mathcal{S}]. \tag{7-43}$$

As highlighted when discussing the relevance of symmetry to interpretation, the idea of a strong link between invariance and objectivity is very natural and appears in many works devoted to studying the meaning of symmetries in physics. The MHI conclusion about the objective description of a quantum system is in complete agreement with the general idea underlying those works: when the Galilei group leaves the Schrödinger equation invariant, the objective description of the system is also invariant and, consequently, the ontological system is not affected by the Galilei transformations.

7.8 Galilei-invariant versions of the Modal-Hamiltonian actualization rule

The modal-Hamiltonian actualization rule AR is not an object but a postulate, a sort of "interpretative law" that adds content to the theory. Then, the question about its covariance and its invariance makes perfect sense. Let us call $DVO\,(\mathcal{S})$ the set of the definite-valued observables of the elemental quantum system represented by $\mathcal{S}: (\mathcal{O}, H)$. According to AR, $DVO\,(\mathcal{S}) = \{H, A_i\}$, where the A_i are the observables commuting with H and having, at least, the same symmetries as H. Although AR is not a differential equation, it is easy to apply the concepts of covariance and invariance to it:

- AR is *Galilei-covariant* if, $\forall T_\alpha \in \mathcal{G}$, the set $DVO\,(\mathcal{S})$ transforms as

$$DVO(\tilde{\mathcal{S}}) = \{H, A_i\} \rightarrow DVO(\tilde{\mathcal{S}}) = \{\tilde{H}, \tilde{A}_i\}. \tag{7-44}$$

- AR is *Galilei-invariant* if, $\forall T_\alpha \in \mathcal{G}$, the set $DVO\,(\mathcal{S})$ transforms as

$$DVO\,(\mathcal{S}) \rightarrow DVO\,(\tilde{\mathcal{S}}) = DVO\,(\mathcal{S}). \tag{7-45}$$

On the one hand, the covariance of AR follows directly from the *unitarity* of the Galilei transformations. In fact, each Galilei transformation can be viewed as a definite rotation of the eigenvectors of all the observables $O \in \mathcal{O}$ in the Hilbert space $\mathcal{H}$. This rotation preserves all the commutation relations under the transformations; so, if $[H, A_i] = 0$, then $[\tilde{H}, \tilde{A}_i] = 0$ (see eq. (7-28)). But, as a mere rotation, such a transformation also preserves all the remaining relations between observables. As a consequence, if A_i has, at least, the same symmetries as H, then $\tilde{A}_i$ has, at least, the same symmetries as $\tilde{H}$. Therefore, if AR selects H and the A_i as the definite-valued observables of $\mathcal{S}$, these observables are transformed as $\tilde{H}$ and $\tilde{A}_i$, which are precisely the observables selected as definite-valued by AR applied to $\tilde{\mathcal{S}}$.

On the other hand, it is easy to see that, when the system is free from external fields, AR is invariant under time-displacements, space-displacements, and space-rotations, since the system is invariant under those transformations (see eq. (7-33)):

$$\mathcal{S} = \tilde{\mathcal{S}} \quad \Rightarrow \quad DVO\,(\mathcal{S}) = DVO\,(\tilde{\mathcal{S}}). \tag{7-46}$$

The difficulty comes, again, from the boost-transformation, which does not leave the Hamiltonian invariant. Since the set $DVO\,(\mathcal{S})$ depends on H, it changes with the change of H. At first glance, this appears to be a serious difficulty for the MHI: as a realist interpretation, the set DVO selected by its rule of definite-value assignment should not be modified by a mere change of reference frame, like that resulting from a boost on a system free from external fields. Nevertheless, the difficulty vanishes when the fact that AR applies to *elemental* quantum systems is recalled.

As seen in the previous section, when there are no external fields, the system represented by $\mathcal{S}$: $(\mathcal{O}, H)$ is, in general, a composite system $\mathcal{S} = \mathcal{S}^W \cup \mathcal{S}^K$, where the subsystem represented by $\mathcal{S}^W$: $(\mathcal{O}^W, H^W)$ carries the internal energy and the subsystem represented by $\mathcal{S}^K$: $(\mathcal{O}^K, H^K)$ carries the kinetic energy. Therefore, AR applies to the elemental subsystems independently:

- Since $\mathcal{S}^W$ is boost-invariant, $DVO(\mathcal{S}^W) = DVO(\tilde{\mathcal{S}}^W)$: the set of the definite-valued observables of $\mathcal{S}^W$ remains unaltered by the transformation.
- Under a boost, $\mathcal{S}^K$: $(\mathcal{O}^K, H^K)$ transforms as $\tilde{\mathcal{S}}^K$: $(\mathcal{O}^K, \tilde{H}^K)$, where $\tilde{H}^K = H^K + T^B$ (see eq. (7-41)); then $DVO(\mathcal{S}^K) \neq DVO(\tilde{\mathcal{S}}^K)$.

However, as argued in the previous section, the system $\mathcal{S}^K$ has no ontological reference by itself to the extent that it is a mere artifact of the descriptive perspective. In ontological terms, $[\tilde{\mathcal{S}}] = [\mathcal{S}^W] = [\mathcal{S}]$ (see eq. (7-43)). As a consequence, when applied to the objective description of the system, given by $\mathcal{S}^W$, AR turns out to be also boost-invariant.

Let us recall that, in the case that there are no external fields acting on the system, the Schrödinger equation is invariant under the Galilei group and, consequently, it is expected that AR can be reformulated in an explicitly Galilei-invariant form. By appealing to the ontological language introduced in the interpretative postulates of

Chapter 4, AR can be reformulated as follows (Lombardi, Castagnino, and Ardenghi 2010).

Actualization rule′ (AR′): Given an ontological system $[\mathcal{S}]$ free from external fields and represented by $\mathcal{S}$: $(\mathcal{O}, H) = \mathcal{S}^W \cup \mathcal{S}^K$, where the subsystems $[\mathcal{S}^W]$ and $[\mathcal{S}^K]$ are represented by $\mathcal{S}^W$: $(\mathcal{O}^W, H^W)$ and $\mathcal{S}^K$: $(\mathcal{O}^K, H^K)$, the set of the definite-valued type-properties DVO $([\mathcal{S}])$ of $[\mathcal{S}]$ is DVO $([\mathcal{S}]) = DVO$ $([\mathcal{S}^W]) = \{[W], [A_i]\}$, where $[W]$ is the type-property represented by $W = I^K \otimes H^W$, and the type-properties $[A_i]$ are represented by the observables A_i commuting with W and having, at least, the same symmetries as W.

Under this form, the MHI actualization rule is explicitly Galilei-invariant because, strictly speaking, $[\mathcal{S}]$ and $[\mathcal{S}^W]$ are only different names for the same ontological system, and $[\mathcal{S}^W]$ is Galilei-invariant:

$$[\mathcal{S}] = [\mathcal{S}^W] \quad \Rightarrow \quad DVO([\mathcal{S}]) = DVO\left([\mathcal{S}^W]\right) = DVO\left([\tilde{\mathcal{S}}^W]\right) = DVO\left([\tilde{\mathcal{S}}]\right). \quad (7\text{-}47)$$

This shows that, when going beyond the physical language and thinking in ontological terms, the seeming non-invariance of the actualization rule under boosts turns out to be a consequence of the descriptive physical language. Despite the change of the system's physical representation under boosts, with no external fields acting on the system, the set of its definite-valued type-properties is invariant under all the Galilei transformations, precisely due to the invariance of the ontological system. In other words, the identity and the behavior of the real quantum system behind the description are not modified by a mere change of the descriptive perspective, in agreement with the physical meaning of the Galilei group.

Although AR′ is conceptually meaningful from an ontological viewpoint, one can consider the possibility of reformulating it in such a way that its Galilei-invariance becomes explicit also in the physical language. The natural way to reach this goal is to appeal to the Casimir operators of the Galilei group: if the actualization rule has to select a Galilei-invariant set of definite-valued observables, such a set must depend on observables represented by those Casimir operators, which are invariant under all the transformations of the Galilei group. Precisely (Lombardi, Castagnino, and Ardenghi 2010),

Actualization rule″ (AR″): Given a quantum system free from external fields and represented by $\mathcal{S}$: $(\mathcal{O}, H)$, its definite-valued observables are the observables C_i represented by the Casimir operators of the Galilei group, and all the observables commuting with the C_i and having, at least, the same symmetries as the C_i.

Since the Casimir operators of the Galilei group are mass M, squared spin S^2, and internal energy W, this reformulation of the rule is in agreement with the original AR when applied to a system free from external fields:

- The definite-valuedness of M and S^2 postulated by **AR″** follows from AR: these observables commute with H and do not break its symmetries because, in non-relativistic quantum mechanics, both are multiples of the identity: $M = mI$, where m is the—scalar—value of mass, and $S^2 = s(s+1)I$, where s is the—scalar—value of spin. The fact that M and S^2 always acquire definite values is completely natural from a physical viewpoint, since mass and spin are properties that any quantum system is supposed to always possess and that are measurable in any physical situation.
- The definite-valuedness of W might seem to be in conflict with AR because W in general is not a multiple of the identity and it is not the Hamiltonian: whereas W is Galilei-invariant, H changes under the action of a boost. However, as seen above, this is not a real obstacle when the elemental subsystems to which AR applies are considered from an ontological viewpoint.

In addition to providing an explicitly invariant version of the MHI rule of definite-value assignment in the physical language, **AR″** leads us to a final thought. As already emphasized, the identity and the behavior of any ontological quantum system free from external fields are invariant under the Galilei group. In turn, from a realist stance, the fact that certain observables acquire definite values is an objective state of affairs that results from the very nature of the system and does not depend on the descriptive perspective. Therefore, the set of definite-valued observables selected by any realist interpretation must be Galilei-invariant. But the Galilei-invariant observables are always represented by the Casimir operators of the Galilei group. Consequently, it is natural to conclude that any realist interpretation that aims to preserve the objectivity of the set of the definite-valued observables and, then, of actualization, cannot stand very far from the MHI.

7.9 Galilei-invariant models of the hydrogen atom

As seen in Section 5.4 of Chapter 5, the physical model of a free hydrogen atom is a two-particle system consisting of an electron and a proton interacting through a Coulombian interaction. In this case, the Hamiltonian reads

$$H = \frac{P_e^2}{2m_e} + \frac{P_p^2}{2m_p} + \frac{e^2}{|Q_e - Q_p|}, \tag{7-48}$$

where the subindexes e and p refer to the electron and the proton, respectively, and e is the electric charge of the electron. Now, the coordinates of the center of mass and the relative coordinates can be defined as

$$Q_C = \frac{m_e Q_e + m_p Q_p}{m_e + m_p}, \qquad Q_R = Q_e - Q_p, \tag{7-49}$$

$$P_C = M\dot{Q}_C = P_e + P_p, \qquad P_R = \mu \dot{Q}_R = \frac{m_p P_e - m_e P_p}{m_e + m_p}, \tag{7-50}$$

where $m = m_e + m_p$ is the total mass, and $\mu = m_e m_p/(m_e + m_p)$ is the reduced mass. In this new coordinate system, the Hamiltonian can be written as

$$H = \frac{P_C^2}{2m} + \frac{P_R^2}{2\mu} - \frac{e^2}{|Q_R|}, \tag{7-51}$$

where the kinetic energy K and the internal energy W can be easily identified:

$$K = \frac{P_C^2}{2M}, \qquad W = \frac{P_R^2}{2\mu} - \frac{e^2}{|Q_R|}. \tag{7-52}$$

Since $[K, W] = 0$, H can be expressed as

$$H = K + W = H^K \otimes I^W + I^K \otimes H^W, \tag{7-50}$$

where H^K is the kinetic Hamiltonian acting on the Hilbert space $\mathcal{H}^K$, H^W is the internal energy Hamiltonian acting on the Hilbert space $\mathcal{H}^W$, and I^K and I^W are the identity operators of the respective tensor-product spaces. Therefore, the hydrogen atom is a composite system $\mathcal{S}^W \cup \mathcal{S}^K$ and the actualization rule AR must be applied to the elemental subsystems independently. However, as explained in Section 7.7 of the present chapter, the objective content of the description is given by the internal energy, that is, by $\mathcal{S}^W$, whereas $\mathcal{S}^K$ can be made to disappear if the reference frame is anchored to the center of mass of the system. Therefore, the preferred context of the hydrogen atom is defined by the application of AR to $\mathcal{S}^W$, with Hamiltonian H^W. The above procedure can easily be generalized to an n-body system whose total Hamiltonian can be expressed as the sum of two non-commuting terms: the internal energy term, W, which only depends on the relative coordinates, and the kinetic term, $K = P_C^2/2m$, which only depends on the momentum of the center of mass.

Quantum systems affected by external fields can also be studied: in the general case, the system cannot be decomposed into non-interacting quantum subsystems as in the previous case. However, this is possible in certain cases. Let us consider a system of two particles interacting through an internal potential and each one affected by an external potential depending on its position, $V_1(Q_1)$ and $V_2(Q_2)$. If the same change of coordinates is introduced as in the previous case, the total Hamiltonian is the sum of the kinetic term $K = P_C^2/2m$ and a Hamiltonian H_0 that, in general, is not the internal energy W and, therefore, $[K, H_0] \neq 0$. Nevertheless, if both $V_1(Q_1)$ and $V_2(Q_2)$ are linear in the positions—for example, $V_i(Q_i) = \alpha Q_i + \beta$—, the position Q_C of the center of mass and the relative position Q_{12} turn out to be uncoupled, and the system can be decomposed into two subsystems $\mathcal{S}^W$ and $\mathcal{S}^K$, as in the case of the free two-body system. Therefore, in this particular situation, the preferred context of the whole system is defined by the application of AR to $\mathcal{S}^W$ (see details in Ardenghi, Castagnino, and Lombardi 2009). It is interesting to note that, although this is a particular case, it correctly models the Zeeman effect, which arises in the hydrogen atom affected by a uniform magnetic field (as seen in Section 5.5 of Chapter 5).

7.10 A Casimir framework for quantum physics

The arguments developed in the previous sections show that the modal-Hamiltonian actualization rule mirrors the Galilei-covariance/invariance of the Schrödinger equation:

- In general, the rule is covariant, just like the Schrödinger equation.
- When the Schrödinger equation is invariant—no external fields acting on the system—, the rule is also invariant when expressed in ontological terms.

In turn, in the Galilei-invariant situation:

- The ontological system identified by the MHI is Galilei-invariant, in agreement with the idea that the Galilei transformations express a mere change of the perspective selected to describe the system.
- The modal-Hamiltonian actualization rule can be reformulated under an explicitly Galilei-invariant form in terms of the Casimir operators of the Galilei group, leading to results that agree with usual assumptions in the practice of physics.

The latter conclusion opens up a new interpretative avenue. In non-relativistic SQM, the external fields acting on a system are not quantized, and this fact is what breaks down the harmony of the free case: the Schrödinger equation loses its Galilei-invariance, and the Hamiltonian is no longer the generator of time-displacements in the Galilei group. In quantum field theory, on the contrary, fields are quantum items and not "external" fields affecting the behavior of the quantum system. As a consequence, the generators of the Poincaré group do not need to be reinterpreted in the presence of "external" factors, and the dynamical laws are always Poincaré-invariant. On this basis, it is not difficult to extrapolate the modal-Hamiltonian actualization rule—expressed in terms of the Casimir operators of the Galilei group in non-relativistic quantum mechanics—to quantum field theory by changing accordingly the relevant symmetry group: now the definite-valued observables of a system are those represented by the Casimir operators of the Poincaré group and the observables commuting with them and having, at least, the same symmetries. Since M and S^2 are the only Casimir operators of the Poincaré group, they are always definite-valued observables. This conclusion is in agreement with a usual physical assumption in quantum field theory: elementary "particles" always have definite values of mass and spin, and those values are precisely what define the different kinds of elementary particles in the theory (see Ardenghi, Castagnino, and Lombardi 2011).

The above remark makes it possible to go a step further in the extrapolation of the MHI to areas for which it was not originally devised. In fact, the central role played by the Casimir operators in the identification of the definite-valued observables leads to a general framework for realist interpretations of quantum theories, expressed by the following idea: *the definite-valued observables of a quantum system are defined by the Casimir operators of the symmetry group of the corresponding theory*. In this general framework, not only space-time symmetries are relevant: internal gauge symmetries can also play the same interpretative role. For example, in quantum electrodynamics, charge also turns out to be a definite-valued observable, which is a physically reasonable conclusion.

The importance of symmetries and invariances in physics has become an increasingly debated topic in the philosophy of physics in recent times (for an updated review, see Brading, Castellani, and Teh 2023). In this field, many different issues have been discussed, for example, the connection between symmetries, as studied in physics, and mathematical group techniques, the relation between symmetries and physical laws, the descriptive, explanatory, and normative roles of symmetries, among many others. In line with this general trend, this chapter has shown the relevance of the symmetry of a quantum theory for its interpretation, a question that has generally not taken a central place in the extensive debates on the interpretative challenges of quantum mechanics.

8

A Top-Down View of Quantum Mechanics

Since Galileo, physics has traditionally adopted a closed-system view, framed in the more generic strategy of describing highly idealized and generally non-existent situations. However, in the field of quantum mechanics—and also in other fields of physics—an open-system view has been gaining strength in recent decades, maybe under the influence of the practical success of the EID developments. This trend is at odds with the perspective of the MHI, which openly adopts a closed-system view.

The aim of this chapter is to discuss several aspects of quantum mechanics that reveal the difficulties faced by an open-systems approach. On this basis, a case will be made for a closed-system approach that leads to a top-down view of quantum mechanics. This top-down view is in agreement with the ontology of properties introduced in Chapter 4, which will be developed in detail in the following Chapters 9, 10, and 11.

8.1 Closed systems and their parts

As just mentioned, physics has traditionally adopted a closed-system view, and quantum mechanics is not an exception in this regard. However, quantum mechanics is very peculiar. Using Einstein's words (1948, translated by Howard 1985, 187), "the mutually independent existence (the 'being thus') of spatially distant things" is in general assumed in non-quantum physics; so, "things" can be individualized at least by their space-time trajectories. In contrast, precise trajectories cannot be assigned to quantum systems, and quantum correlations challenge the independent existence of the components of a quantum system. Therefore, in the quantum framework, the identification of the components of a closed system is by no means a trivial matter.

The so-called *quantum structural studies* deal with the different ways in which a quantum system can be decomposed into its parts (tensor product structures) (see, e.g., Dugić and Jeknić-Dugić 2008, Harshman and Ranade 2011, Kastner, Jeknić-Dugić, and Jaroszkiewicz 2016). The typical case is that of the hydrogen atom, which can be decomposed into the proton and the electron, but also into the center-of-mass system and the relative system, as shown in Section 7.9 of Chapter 7 (see Ardenghi, Castagnino, and Lombardi 2009, Harshman 2012). In the classical domain, there may be good reasons to privilege the partition of a classical system into individual particles, for example, the decomposition of the hydrogen atom into an electron and a proton. On the contrary, in the quantum realm, where the concept of individual goes into crisis, the theory itself does not provide a criterion for selecting a privileged partition:

The Modal-Hamiltonian Interpretation of Quantum Mechanics. Olimpia Lombardi, Oxford University Press.
© Olimpia Lombardi (2026). DOI: 10.1093/9780198951407.003.0008

"without further physical assumption, no partition has an ontologically superior status with respect to any other" (Zanardi 2001, 4). On this basis, different authors have formulated the most dissimilar criteria to select the privileged partition of a closed quantum system, which leads to identifying the open "subsystems" that compose it.

In the context of some Modal Interpretations, whose rule of definite-value ascription requires the factorization of the total Hilbert space of the closed system, the possibility of many different factorizations leads to a serious difficulty when attempting to assign definite values in a consistent way (see Bacciagaluppi 1995). To deal with this difficulty, Dennis Dieks (1998) proposes to select a privileged factorization that depends on symmetry considerations: under the assumption that there exist "atomic" systems corresponding to irreducible representations of the Galilei group, he suggests "the following necessary condition for factorizations to correspond to physically real systems. The factor Hilbert spaces should carry a representation of the space-time group (the Galilei group in our example of nonrelativistic quantum mechanics) in the same way as the factor spaces of the original factorization [in atomic systems], with the usual identification of generators of the space-time group and dynamical variables." (Dieks 1998, 148). Although with different motivations, Nathan Harshman and Sujeewa Wickramasekara (2007a) take a similar perspective by distinguishing between tensor product structures that are symmetry invariant and, therefore, also dynamically invariant, and those that are not: the authors select the invariant tensor product structures because they allow for measures of entanglement that do not depend on the frame of reference.

Other authors, by contrast, rely on pragmatic motivations for choosing a particular decomposition of the closed system. For example, Harshman (2016) is interested in tensor product structures that may be especially useful for storing and processing quantum information. In turn, Paolo Zanardi, Daniel Lidar, and Seth Lloyd (2004, 1) argue that "the partition of a quantum system into subsystems is dictated by the set of operationally accessible interactions and measurements."

From a completely different perspective and under the label "quantum mereology," Sean Carroll and Ashmeet Singh (2021) study the question of how to decompose a Hilbert space into a preferred tensor-product factorization without any pre-existing structure other than the Hamiltonian operator, in the particular case of a bipartite decomposition into "system" and "environment." The authors propose a quasi-classical criterion, according to which the preferred decomposition is defined by components that exhibit quasi-classical behavior: "The correct decomposition is one in which pointer states of the system are relatively robust against environmental monitoring (their entanglement with the environment does not continually and dramatically increase) and remain localized around approximately-classical trajectories." (Carroll and Singh 2021, 1).

In general, the problem of the identification of the components of a closed system is formulated in terms of tensor product structures, that is, the different ways in which a Hilbert space can be decomposed into a tensor product of Hilbert spaces. However, several authors have turned their attention from Hilbert spaces to algebras of observables. An example is John Earman (2015), who considers that subsystems of a composite system must enjoy some minimal degree of independence, which includes the mutual commutativity of the subsystems' algebras, so that the subsystems' observables can be simultaneously measured. However, this does not solve the issue of univocally selecting a privileged partition that identifies the "real" subsystems. Also from an algebraic perspective, Zanardi and collaborators take an algebra $\mathcal{M}$ of observables on a finite Hilbert space $\mathcal{H}$ as the starting point and prove that, given two independent subalgebras $\mathcal{A}$ and $\mathcal{B}$ of $\mathcal{M}$ that satisfy (i) independence ($[\mathcal{A}, \mathcal{B}] = 0$, that is, $[a, b] = 0$ for all $a \in \mathcal{A}$ and $b \in \mathcal{B}$) and (ii) completeness ($\mathcal{A} \otimes \mathcal{B} \cong \mathcal{A} \vee \mathcal{B} = \mathcal{M}$), then $\mathcal{A}$ and $\mathcal{B}$ induce a tensor product $\mathcal{H}_{\mathcal{A}} \otimes \mathcal{H}_{\mathcal{B}}$ (Zanardi 2001, Zanardi, Lidar, and Lloyd 2004). From their pragmatic perspective, the authors stress that, in this way, the partition of the algebra of observables can be made to depend on a particular set of experimental capabilities, that is, on the accessible observables in each situation. In turn, Harshman and Kedar Ranade (2011) provide an explicit constructive method to generate those subalgebras from a finite set of operators, which, although may look arbitrary from the viewpoint of the unstructured Hilbert space, have the correct properties to rigorously define locality, separability, and entanglement.

The above comments are only a sample of the different criteria used to identify the components of a closed system. This multiplicity of criteria shows that this is a subtle issue whose answer cannot be taken for granted when adopting an open-system view. The closed-system view of the MHI, by contrast, offers a straightforward solution to the problem by considering that only closed systems are legitimate quantum systems: they are *composite* if they can be decomposed into non-interacting—closed—quantum systems, and they are *elemental* otherwise (see Section 4.1 of Chapter 4). In the case of an elemental quantum system, all possible partitions are equally legitimate: however, these partitions do not lead to subsystems but rather to *parts* of the elemental whole. As a consequence, the quantum phenomena that rely on the partition of a closed system into parts turn out to be relative to that partition. This view is in agreement with the relativity of central quantum concepts, which will be discussed in the following sections.

8.2 The relativity of entanglement

As is well known, the phenomenon of entanglement is at the very heart of quantum mechanics. Moreover, at present it is the essential ingredient of the rapidly developing discipline of quantum information science, both in the case of quantum communication and quantum computation. However, the precise meaning of the concept of entanglement still deserves a deep inspection.

To begin, it is necessary to face the question about what the term "entanglement" is applied to. Does it refer to a property or to a relation? When it is said "x is entangled" or "x is entangled with y," what are x and y? Usually, the concept of entanglement is applied to particles, as if being entangled were a relation that links quantum objects to each other. Sometimes, such a relation is conceived as the consequence of an interaction between quantum systems, which persists even when the interaction has ceased. These common assumptions need to be carefully examined.

First, interaction is not a requirement for entanglement. A well-known case is that of the so-called *entanglement swapping* (first introduced in Żukowski et al. 1993), a phenomenon by means of which entanglement can be generated between systems that have never physically interacted with each other. In fact, the first experimental realization of entanglement swapping is described as "entangling photons that never interacted" (Pan et al. 1998, 3891). In other words, "one must avoid thinking that direct interaction is necessary for the appearance of entanglement." (Terra Cunha, Dunningham, and Vedral 2007, 2278).

Second, the idea of entanglement as a relation between particles must be set aside because there may be entanglement in a single photon or in a single atom. For example, a kind of "intraparticle entanglement" (Harshman and Wickramasekara 2007, 2) obtains between the momentum degrees of freedom and the spin degrees of freedom of a single atom (see also Harshman 2012) or between two spatial modes of a single photon passing through a beam splitter (Terra Cunha, Dunningham, and Vedral 2007). Furthermore, one-particle entanglement is as good as two-particle entanglement with respect to applications: a one-photon or one-atom state can be used to violate Bell's inequalities and to teleport a qubit (Lee and Kim 2000). On this basis, it has been claimed that a "common fallacy is to say that 'entanglement is a property of many particle systems.' It is, however, correct to say that 'entanglement is a property of composite systems,' i.e., systems that have more than one subsystem." (Terra Cunha, Dunningham, and Vedral 2007, 2280). This means that it is not necessary to talk about "particles" when entanglement is considered.

How is entanglement defined? As stressed by Alisa Bokulich and Gregg Jaeger (2010, xiv), "[t]he search for a fully general definition and measure of entanglement remains an active area of research." In fact, the issue of defining entanglement for mixed states is still an open question (see Earman 2015). Nevertheless, there is in general consensus regarding entanglement in the case of pure states: a state $|\psi_{12}\rangle$ of a quantum system is *entangled* if it is non-factorizable in terms of the tensor product of the states $|\psi_1\rangle$ and $|\psi_2\rangle$ of the component systems (see, e.g., Myrvold 2022a)

$$|\psi_{12}\rangle \neq |\psi_1\rangle \otimes |\psi_2\rangle. \tag{8-1}$$

This means that entanglement must be conceived not as a relation between particles, but as a property of systems, more precisely, of their states. Nevertheless, this does not mean that it is an absolute property of a state. As Earman (2015) clearly stresses, a given state is entangled or not only *in relation to the decomposition of the whole system into its components*. In fact, a given state may be entangled with respect to a certain decomposition and non-entangled with respect to another one. The typical example is the already mentioned case of the hydrogen atom, which can be decomposed at least in two ways: the entanglement of the atom's state is relative to the chosen decomposition. At present, there is an extensive agreement that *entanglement is a relative notion*. As Marcelo Terra Cunha, Jacob Dunningham, and Vlatko Vedral (2007, 2278) claim, "entanglement is a property of a quantum state relative to a given set of subsystems [...] a question like 'is there entanglement in such a state?' is meaningless in isolation." The authors even prove a theorem according to which, given a state vector $|\Psi\rangle$ in a finite-dimensional state space $\mathcal{H}$ with non-prime dimension $d = m\,n$, there always exists a tensor product structure $\mathcal{H} = \mathcal{H}^m \otimes \mathcal{H}^n$ with respect to which $|\Psi\rangle$ is factorizable. Since the theory itself does not provide a criterion for selecting a privileged partition, there are no grounds to privilege a statement about the entanglement of a quantum state over others.

In general, the relativity of entanglement is expressed in terms of different tensor product structures. However, in the algebraic formal framework, the concept of entanglement can be generalized to the case in which the decomposition of the closed system cannot be given in terms of a tensor product structure, as is general in the bosonic and fermionic cases. In this framework, Aiyalam Balachandran and his collaborators (2013a, 2013b) offer a completely algebraic approach to entanglement, according to which it can be said that a pure state is entangled with respect to a partition of the algebra of observables $\mathcal{A}$ representing the closed system into the subalgebras corresponding to $\mathcal{A}_1 \vee \mathcal{A}_2$ (I will come back to this point in Section 11.7 of Chapter 11, where the entanglement of indistinguishable particles will be discussed). Furthermore, Howard Barnum and collaborators (Barnum et al. 2003, 2004, Viola and Barnum 2010) take the observable viewpoint a step further: they introduce a subsequent generalization based on the idea that entanglement is relative to a specified subspace of observables rather than to a specified decomposition of the entire closed system.

From the closed-system perspective of the MHI, the relativity of entanglement turns out to be natural. If the quantum system, closed by definition, is composite, then the entanglement of its state is relative to its decomposition into subsystems; if it is elemental, the entanglement of its state is relative to its partition into parts. In both cases, the split of the closed system into smaller "pieces" is in general not unique, so entanglement is an essentially relative phenomenon. This relativity of entanglement has relevant consequences in the case of the so-called entanglement of indistinguishable particles: this issue will be discussed in detail in Section 11.7 of Chapter 11.

8.3 Quantum states versus reduced states

In physics, it is common to talk about composite quantum systems and their subsystems. As in the case of the discussion about decoherence in Section 5.7 of Chapter 5, in order to conform to the literature, in this section, I will use the common terminology that applies the term "system" also to the open case.

Commonly, given two systems $\mathcal{S}^1$ and $\mathcal{S}^2$, represented in the Hilbert spaces $\mathcal{H}^1$ and $\mathcal{H}^2$ respectively, the system $\mathcal{S} = \mathcal{S}^1 \cup \mathcal{S}^2$, represented in $\mathcal{H} = \mathcal{H}^1 \otimes \mathcal{H}^2$, is said to be composite: $\mathcal{S}^1$ and $\mathcal{S}^2$ are considered subsystems of $\mathcal{S}$. The relation between the composite system and its subsystems is usually defined independently of whether $\mathcal{S}^1$ and $\mathcal{S}^2$ interact with each other or not, and of whether the state of $\mathcal{S}$ is entangled or not relative to the corresponding decomposition: in any case, the common practice is to conceive the quantum states of the subsystems $\mathcal{S}^1$ and $\mathcal{S}^2$ as their reduced states, computed on the state of the whole system $\mathcal{S}$ by tracing over the degrees of freedom of $\mathcal{S}^2$ and $\mathcal{S}^1$ respectively. This is a basic assumption in the theory of environment-induced decoherence (Zurek 1981, 1982, 1991, 1993, 1994; for a review, see Zurek 2003), whose aim is to account for the emergence of classicality from the quantum realm in terms of the fast diagonalization of the reduced state of a system in interaction with its environment. As explained in Chapter 3, reduced states also play a central role in certain Modal Interpretations of quantum mechanics, such as the BDMI and the SDMI, in which it is assumed that operators resulting from partial traces represent the quantum states of the measured system and of the measuring apparatus after the measurement interaction.

The appeal to reduced states is certainly pervasive in practice, and physicists perfectly know how to use them. Nevertheless, the interpretation of the concept of reduced state is a subtle issue that has relevant consequences when the interpretation of quantum mechanics is the issue at stake. In general, a reduced state is simply conceived as the quantum state of an open system: "The projection operator $|\psi\rangle\langle\psi|$ is an observable of the total system [. . .]. But we are really interested in the individual properties of device and object taken by themselves. Therefore, we need substitutes for $|\psi\rangle\langle\psi|$ that represent the states of these individual systems. In the context of standard quantum mechanics such operators are readily available, namely the density operators for the partial systems" (Dieks 2007, 298). The justification for such an assumption is based on the fact that the reduced state allows to calculate the expectation values of all the observables of the open system, exactly like the state of the whole system allows to calculate the expectation values of all its observables: "we took [the reduced operator] W_1 for the state of system 1. This is standard practice in quantum mechanics; however, the usual justification relies on the probabilistic interpretation of the theory and the Born rule" (Dieks 2007, 299). These quotes suggest that the reduced state W_1 is conceived as the state of the open system in the same sense as $|\psi\rangle$ is the state of the total closed system. This is precisely the idea to be subjected to close scrutiny (see Fortin and Lombardi 2014).

Mathematically, the reduced state ρ_r^1 of a system $\mathcal{S}^1$, which is a component of a larger system $\mathcal{S} = \mathcal{S}^1 \cup \mathcal{S}^2$ represented in the Hilbert space $\mathcal{H} = \mathcal{H}^1 \otimes \mathcal{H}^2$, is a density operator since it satisfies the normalization condition ($Tr\,\rho_r^1 = 1$), the self-adjoint condition ($\rho_r^{1\dagger} = \rho_r^1$), and the non-negativeness condition ($\langle\varphi|\,\rho_r^1\,|\varphi\rangle \geq 0$ for all $|\varphi\rangle \in \mathcal{H}^1$). Nevertheless, it denotes an *improper* mixture (d'Espagnat 1976): although proper and improper mixtures are represented by the same kind of mathematical object—a density operator—, they refer to *physically different concepts*. In fact, if $\{|a_m\rangle\}$ is a basis of

$\mathcal{H}^1$ and $\{|b_n\rangle\}$ is a basis of $\mathcal{H}^2$, and ρ is the state of $\mathcal{S}$, the reduced states of $\mathcal{S}^1$ and $\mathcal{S}^2$ can be obtained as

$$\rho_r^1 = Tr_{(2)}\rho = \sum_n \langle b_n| \rho |b_n\rangle, \tag{8-2}$$

$$\rho_r^2 = Tr_{(1)}\rho = \sum_n \langle a_n| \rho |a_n\rangle, \tag{8-3}$$

and they are sufficient to compute the expectation value of any observable that belongs exclusively to $\mathcal{S}^1$ or $\mathcal{S}^2$, respectively. However, these two density operators are not sufficient, in general, to determine the state ρ of the composite system $\mathcal{S}$, since they provide no information about the correlations between $\mathcal{S}^1$ and $\mathcal{S}^2$. Thus, if one could only make measurements on $\mathcal{S}^1$ and could not make any on $\mathcal{S}^2$, then one would not be able to differentiate the improper mixture denoted by the reduced density operator ρ_r^1 from the analog proper mixture denoted by the density operator ρ^1. But, as Bernard d'Espagnat (1995) stresses, there is no theoretical reason that prevents us from having access to, at least, some of the traced over degrees of freedom, and this access would permit us to show that the proper mixture and the improper mixture are, in principle, testably different. In the context of the decoherence theory, the difference is also pointed out by Heinz-Dieter Zeh, one of the founding fathers of the approach: "The conceptually important difference between true and apparent ensembles was clearly pointed out by Bernard d'Espagnat (1976) when he distinguished between proper and improper mixtures. In the case of virtual (reversible) decoherence, this difference can even be observed as recoherence (a relocalization of the superposition, that would be impossible for a proper mixture)" (Zeh 2005, 2).

However, this difference between proper and improper mixtures has been considered an insufficient reason for depriving ρ_r^1 of the status of being the quantum state of the open system $\mathcal{S}^1$. In fact, some authors insist that improper mixtures are the quantum states of the components of a composite system, even though the quantum state of the composite system is not uniquely defined by the quantum states of its components: "[W]hile the states of subsystems of a composite system are uniquely determined by the state of the composite system via the reduced density operator, the state of a composite system is not, in general, uniquely determined by the states of its subsystems." (Bub 1997, 265; see also Hughes 1989). But this position does not consider the difference regarding time-evolutions: whereas quantum states evolve according to the Schrödinger (or the von Neumann) equation and, therefore, follow unitary evolutions, improper mixtures in general evolve non-unitarily. Therefore, the usual view about reduced states amounts to admit that, given the basic postulates of quantum mechanics, there are certain states that are not ruled by the dynamical postulate of the theory.

In the framework of this insistence on blurring the difference between proper and improper mixtures, it could still be argued that the non-unitary evolution of the reduced states does not imply the violation of one of the principles of the theory, to the extent that the dynamical postulate of quantum mechanics only applies to *closed* systems, but says nothing about the time-evolution of open systems. According to this position, reduced operators actually represent quantum states of *open* quantum systems, like those involved in measurement, whose time-dependence is ruled by non-unitary evolution laws. This amounts to claim that quantum mechanics is a mechanical theory with two different kinds of states, each one with its own evolution law. This would certainly be a novelty in Modern physics. To find such a case it is necessary to go back to Ancient times, when Aristotle's theory lacked a unified description of physical reality due to the difference between the sublunar and the supralunar world, each with its own dynamical law: the "law" of the natural places on Earth and the "law" of the uniform circular motion in Heaven.

The insistence on considering the reduced state (improper mixture) as the mixed quantum state of the open system seems to overlook the very definition of the concept of reduced state. The reduced state ρ_r^1 of $\mathcal{S}^1$ is defined as the density operator by means of which the expectation values of all the observables belonging to $\mathcal{S}^1$ can be computed. Precisely, if $\mathcal{H}^1$ and $\mathcal{H}^2$ are the Hilbert spaces of $\mathcal{S}^1$ and $\mathcal{S}^2$ respectively, $\mathcal{H} = \mathcal{H}^1 \otimes \mathcal{H}^2$ is the Hilbert space of $\mathcal{S} = \mathcal{S}^1 \cup \mathcal{S}^2$, $O^1 \in \mathcal{O}^1 \subset \mathcal{B}(\mathcal{H}^1)$ is an observable of $\mathcal{S}^1$, I^2 is the identity in $\mathcal{O}^2 \subset \mathcal{B}(\mathcal{H}^2)$, and ρ is the state of $\mathcal{S}$, then the reduced state of $\mathcal{S}^1$ is *defined* as the density operator ρ_r^1 such that

$$\forall O = O^1 \otimes I^2 \in \mathcal{O} \subset \mathcal{B}(\mathcal{H}), \quad \langle O \rangle_\rho = \langle O^1 \rangle_{\rho_r^1}. \tag{8-4}$$

On the basis of this definition, the reduced state can be *computed* by tracing over the degrees of freedom of $\mathcal{S}^2$, as $\rho_r^1 = Tr_{(2)}\, \rho$. As Maximilian Schlosshauer claims in his well-known book about decoherence, this means that, strictly speaking, a reduced density operator is only "a calculational tool" for computing expectation values at each time (Schlosshauer 2007, 48). For this reason, the author warns us "against a misinterpretation of reduced density matrices as describing a proper mixture of states" (2007, 69).

Perhaps the fact that proper and improper mixtures cannot be distinguished from a mathematical viewpoint is what leads many authors to suppose that the reduced state ρ_r^1 is the quantum state of the open system $\mathcal{S}^1$ in the same sense as ρ is the quantum state of the closed quantum system $\mathcal{S}$. However, that mathematical feature essentially depends on the formalism used to express the theory, in particular, the Hilbert space formalism. In fact, the situation is different in other cases. For instance, it has been proved (Masillo, Scolarici, and Sozzo 2009) that proper and improper mixtures are represented by different kinds of density operators in the so-called *quaternionic formulation of quantum mechanics* (see Adler 1995); hence they can be distinguished also from a mathematical viewpoint. This differentiated mathematical representation is in line with their different time-evolutions.

A different approach is to claim that the true states of quantum mechanics are the reduced states, and the unitarily evolving states are a limiting case of those generic states. This is the strategy adopted by Nancy Cartwright (1983), who proposed to replace the Schrödinger (or the von Neumann) equation by the generalized master equation governing the time-evolution of the reduced state as the dynamical postulate of quantum mechanics. However, this suggestion must face the difficulty that the generalized master equation has no general form, and must be derived case by case, depending on the specific features of each particular situation and always on the basis of assuming the unitary time-evolution of the closed system. Furthermore, the generalized master equation can be viewed as a convenient tool without independent theoretical meaning. In fact, as Juan Pablo Paz and Wojciech Zurek (2002, 97) admit, "[i]n principle, the evolution equation for [the reduced state] ρ could be obtained by solving Schrödinger (or von Neumann) equation for the total density matrix and then taking the trace. However, this task can be analytically completed in very few cases, and the study of the evolution of the reduced density matrix should be done by using some approximations." In other words, in principle the generalized master equation could be dispensed with by relying on the unitary evolution of quantum state of the closed system and taking the partial trace at each particular time to obtain the reduced state at that time.

Much more recently, Michael Cuffaro and Stephan Hartmann (2024) follow in Cartwright's footsteps in considering the unitarily evolving states as a limiting case of the generally non-unitarily evolving reduced states: the authors advocate for the ontic, epistemic, and explanatory fundamentality of open systems. They propose replacing the Schrödinger (or the von Neumann) equation with the Lindblad equation, a particular type of master equation that can be written in a simplified form as

$$\dot{\rho}_r = (\mathcal{L}_u + \mathcal{L}_{n-u})\,\rho_r, \tag{8-5}$$

where ρ_r is the reduced state of the open system, $\dot{\rho}_r$ is its time derivative, $\mathcal{L}_u = -i\,[H^S, \rho_r]$ is the unitary superoperator (an operator that acts on other operators) defined in terms of the Hamiltonian H^S of the open system $\mathcal{S}$, and $\mathcal{L}_{n-u}$ is the non-unitary superoperator. This open-system view, although initially compelling, must be analyzed carefully. The first point to take into account is that, like any master equation, the Lindblad equation does not have a general form: the non-unitary part must be calculated case by case depending on the particular system under consideration. A second aspect to notice is that, in the physical literature, the equation is always presented together with its derivation from the Schrödinger or the von Neumann equation, which sounds rather curious in the case of an equation that is proposed as a dynamical *postulate* of a theory. In a relatively recent paper, Daniel Manzano (2020) introduces two detailed derivations of the Lindblad equation: a physical and a formal derivation.

- The physical derivation starts from the von Neumann equation, which cancels the assumed fundamental character of the Lindblad equation. In addition, the derivation introduces several physical assumptions: (i) that the system interacts weakly with the environment, (ii) that the initial state of the complete system is separable, (iii) that the state of the environment is "thermal," and (iv) that the system and the environment are non-correlated during all the time evolution. These assumptions, indispensable for the derivation, show that the equation is not completely general as required of a basic theoretical postulate.

- The formal derivation comes from the question: What is the most general Markovian way of mapping density matrices onto density matrices? In this case, the map is required to be completely positive and trace-preserving. This derivation does not rely on the von Neumann equation. But precisely due to its formal character, its conclusion is a purely mathematical expression, without other physical reference than assuming that the Markovian map expresses a time evolution. In this case, the symbol ρ_r stands for a density matrix and the symbol H^S stands for a self-adjoint operator. In order to state that ρ_r refers to the reduced state of the open quantum system $\mathcal{S}$ and the symbol H^S refers to the Hamiltonian of $\mathcal{S}$ it is necessary to compare the formal result with the result obtained by starting from the von Neumann equation. And this leads us again to the non-fundamentality and the non-generality of the physical derivation.

These brief remarks show that, although the Lindblad equation is very useful in practice for the description of the time evolution of certain open "systems," like any other master equation it expresses a result that lacks the generality required by the dynamical postulate of a physical theory (see Lombardi 2025 for criticisms). Nevertheless, my rejection of the strong fundamentality of the open-system view does not imply that the closed-system view is endowed with such fundamentality. As I made clear at the outset, my purpose is to make sense of the quantum world and not to postulate a metaphysically privileged ontology that describes the structure of every domain of reality. From my perspective, the closed-system view and the open-system view can be seen as complementary approaches that need not be conceived as rivals.

8.4 Reduced states as coarse-grained states

Even if it is admitted that, strictly speaking, the reduced state ρ_r^1 cannot be said to be the *quantum* state of an open system $\mathcal{S}^1$, it is difficult to deny that it supplies a certain quantum description. In this section, it will be shown that reduced states are coarse-grained states. Although this claim has appeared in the literature on decoherence (see, e.g., Gell-Mann and Hartle 1993, Omnès 1994, Anastopoulos 2002), in general it does not go beyond pointing out the operation of tracing over the degrees of freedom of the environment. In this section, I will explain the precise sense in which reduced states can be understood by means of a generalized conception of coarse-graining.

Let us begin with a classical analogy, which will allow us to approach the issue from a more general perspective. In Gibbsian statistical mechanics, a statistical-mechanical state is represented by a density function $\rho(x, t)$, defined on a phase space Γ and evolving according to the Liouville equation. During the last decades, many discussions focused on the use of ensembles in the definition of the density function, the connection of the density function with the different interpretations of probability, etc. (see, e.g., Frigg 2007, Uffink 2007). One of the central problems in this theoretical context is how to explain the approach to equilibrium by means of the time evolution of the density function. In fact, as is well known, the classical evolution of $\rho(x, t)$ has no limit for $t \to \infty$: the volume of the support of $\rho(x, t)$ in Γ is time-invariant as a consequence of the Liouville theorem. When classical statistical mechanics is formulated in the Koopman formalism (Koopman 1931)—its Hilbert space formulation—, that result can be expressed by saying that the vector state ρ evolves unitarily according to $\rho(t) = U_t \, \rho_0 = e^{-iGt} \rho_0$, where ρ_0 is the initial state and G is the generator of the evolution, which represents the energy of the system: since U_t is a unitary operator, it does not change the angle of separation (the inner product) or the distance (the square modulus of the difference) between vectors representing two different states. In other words, due to its unitary evolution, $\rho(t)$ has no limit for $t \to \infty$. Therefore, the problem of irreversibility in classical statistical mechanics turns out to be how to account for a non-unitary irreversible approach to equilibrium in systems microscopically ruled by a unitary evolution law (see Frigg 2007).

The standard answer in the context of the Gibbsian approach consists in relying on coarse-graining: the phase space Γ is partitioned into cells C_i of the same volume c_i in Γ, and a coarse-grained distribution ρ_{cg} is defined such that, for each C_i,

$$\begin{cases} \text{if } \forall x \in C_i, \; \rho(x) = 0 \to \rho_{cg} = 0, \\ \text{if } \exists x \in C_i, \; \rho(x) \neq 0 \to \rho_{cg} = 1/c_i \displaystyle\int_{C_i} \rho(x)\, dx \neq 0. \end{cases} \tag{8-6}$$

It is quite clear that, since the coarse-grained state ρ_{cg} is not the original statistical-mechanical state, it is not governed by the dynamical law of classical statistical mechanics and, as a consequence, its evolution is not constrained by the Liouville theorem: $\rho_{cg}(t)$ may approach a definite limit for $t \to \infty$. This is what effectively happens when the system has a sufficiently high degree of instability: if the system is mixing, it can be proved that the coarse-grained state approaches an equilibrium state, $\rho_{cg}(t) \to \rho_{cg(eq)}$ (see, e.g., Berkovitz, Frigg, and Kronz 2006). Of course, discussions also focus on the meaning of the coarse-grained state: Does it supply the objective description of an emergent process? Or does it only describe our ignorance about the real underlying evolution? These questions express the different interpretations of irreversibility, in the context of a debate that is still as lively as it was at the birth of statistical mechanics (see Frigg and Werndl 2024 and references therein). Nevertheless, despite the heated discussions and the deep disagreements about interpretation, nobody ignores the difference between the statistical-mechanical state, which evolves unitarily according to the dynamical postulate of the theory, and the coarse-grained state, which may tend to a final stable state: no one intends to endow them with the same interpretation.

What happens in quantum mechanics? One of the hottest conceptual discussions about the theory is that referred to the interpretation of the quantum state. Does it describe a physical field on physical but empirically inaccessible degrees of freedom? Or does it represent our degree of knowledge? Or does it embody certain objective propensities to actualization? This kind of questions have led to the wide spectrum of interpretations of the theory and have attracted the interest of many researchers up to the present (see, e.g., Ney and Albert 2013, in particular, Maudlin's article "The nature of the quantum state" in that volume). Surprisingly, the interpretation of the reduced state is in general not discussed, under the implicit assumption that it is also a quantum state, but now of an open system. In other words, the unitary evolving quantum state and the reduced state are endowed with the same interpretation, despite their many differences such as those discussed in the previous section. In what follows it will be shown that, far from being the quantum state of an open system, as commonly conceived, the reduced state turns out to be *the coarse-grained state of a closed system* (see Fortin and Lombardi 2014).

In its traditional classical form, a coarse-grained description is based on a partition of a phase space into discrete and disjoint cells: this mathematical procedure defines a projector Π (see Mackey 1989). In other words, traditional coarse-graining amounts to a projection whose action is to cancel *some information about the exact values of all the dynamical variables* defining the fine-grained state. If this idea is generalized, coarse-graining can be understood as any operation that cancels a part of the information carried by a vector representing a state. The idea of expressing this generalized concept of coarse-graining in terms of projection was elegantly formulated by Robert Zwanzig (1961). Zwanzig projectors are idempotent operators that act either on classical statistical distributions or on quantum density operators: they are designed to formalize the procedure of neglecting certain "irrelevant" information from a complete state. From this generalized perspective, a partial trace is a particular case of generalized coarse-graining, since it involves the loss of *all the information about the values of some dynamical variables* defining the fine-grained state (Mackey 1989; see also Lombardi and Matta 2022).

Let us recall the definition of reduced operator given by eq. (8-4). Although for dimensional reasons the reduced state ρ_r^1 of $\mathcal{S}^1$ cannot be expressed as a projection $\Pi\,\rho$ acting directly on the quantum state ρ of the closed composite system $\mathcal{S} = \mathcal{S}^1 \cup \mathcal{S}^2$, the expectation value $\langle O_1 \rangle_{\rho_r^1}$ can also be expressed as the expectation value of the observable $O = O_1 \otimes I_2$ in a coarse-grained state ρ_{cg} of $\mathcal{S}$:

$$\langle O_1 \rangle_{\rho_r^1} = \langle O \rangle_{\rho_{cg}}. \tag{8-7}$$

The density operator ρ_{cg} represents a coarse-grained state because it can be obtained as $\rho_{cg} = \Pi\,\rho$, when the projector Π executes the following operation:

$$\Pi\,\rho = \left(Tr_{(2)}\,\rho\right) \otimes \widetilde{\delta}_2 = \rho_r^1 \otimes \widetilde{\delta}_2, \tag{8-8}$$

where $\widetilde{\delta}_2 \in \mathcal{O}^2$ is a normalized identity operator with coefficients $\widetilde{\delta}_{2\alpha\beta} = \delta_{\alpha\beta}/\sum_\gamma \delta_{\gamma\gamma}$. The operator Π is a projector since

$$\Pi\,\Pi\,\rho = \left(Tr_{(2)}\,\Pi\,\rho\right) \otimes \widetilde{\delta}_2 = \left(Tr_{(2)}\left(\rho_r^1 \otimes \widetilde{\delta}_2\right)\right) \otimes \widetilde{\delta}_2 = \rho_r^1 \otimes \widetilde{\delta}_2 = \Pi\,\rho. \tag{8-9}$$

It is quite clear that $\rho_{cg} = \Pi\,\rho$, although acting on $\mathcal{O} \subset \mathcal{B}(\mathcal{H})$, is not the quantum state of the system $\mathcal{S} = \mathcal{S}^1 \cup \mathcal{S}^2$ represented by $\mathcal{H} = \mathcal{H}_1 \otimes \mathcal{H}_2$: it is a coarse-grained state that disregards certain information of the quantum state of $\mathcal{S}$. It is interesting to notice that ρ_{cg} is a coarse-grained state of the whole *closed system* $\mathcal{S}$ that may evolve *non-unitarily*. Therefore, it is not the case that non-unitary evolutions are a specific feature of open quantum systems.

In turn, if the degrees of freedom of $\mathcal{S}^2$ are traced off from ρ_{cg}, the reduced state of $\mathcal{S}^1$ is recovered:

$$Tr_{(2)}\,\rho_{cg} = Tr_{(2)}\left(\rho_r^1 \otimes \widetilde{\delta}_2\right) = \rho_r^1. \tag{8-10}$$

This means that the coarse-grained state ρ_{cg}, which "erases" the components corresponding to $\mathcal{S}^2$, provides the same information about the open "system" $\mathcal{S}^1$ as the reduced state ρ_r^1, but now from the viewpoint of the composite system $\mathcal{S}$ (see Fortin and Lombardi 2014). Therefore, the reduced density operator ρ_r^1 can also be conceived as a kind of coarse-grained state of $\mathcal{S}$ that disregards certain degrees of freedom considered irrelevant in the particular situation under consideration.

It is worth emphasizing that the arguments presented here, pointing toward the careful distinction between proper quantum states and reduced states, do not amount to being committed to "The Church of the Large Hilbert Space." This is the name supposedly coined by John Smolin to describe the habit of always conceiving a mixed state as the result of a partial trace over a larger system, and any non-unitary evolution as being embedded in some unitary evolution of a larger system (see discussion in Brassard and Raymond-Robichaud 2013). Certainly, the present argumentation points to the fact that, as Paz and Zurek themselves admit, the fundamental dynamical law is the dynamical postulate of the theory that describes a unitary time-evolution, and any other evolution must be defined in terms of that one. Nevertheless, this does not amount to conceive pure states as the only proper quantum states of the theory. On the contrary, I consider, as other authors do (see, e.g., Ballentine 1998), that the quantum state of a closed system is represented by an operator that can denote pure or mixed states. In other words, not all mixed states need to be thought of as the result of a partial trace: there are mixed states that are legitimate quantum states not obtained by partial trace, which follow unitary evolutions and embody quantum correlations. The subtle point is to keep clear the difference between those quantum states and the reduced states, especially in interpretative contexts.

Once the reduced state is viewed as a coarse-grained state, its non-unitary evolution neither restricts the application of the dynamical postulate to the quantum states of closed systems nor does it require a new dynamical postulate to be described. The non-unitary evolution of reduced states turns out to be a situation analogous to the familiar case of classical statistical mechanics, in which obtaining a non-unitary coarse-grained evolution from the underlying unitary dynamics in the particular case of high instability is completely natural, with no need of restrictions or reformulations of the classical dynamical laws. An author who has emphasized the analogy between the classical statistical case and the quantum case is Roland Omnès (2001, 2002), by repeatedly claiming that decoherence is a particular case of the phenomenon of irreversibility (for a detailed comparison between the classical and the quantum case, see Fortin and Lombardi 2018, Lombardi and Matta 2022).

When the idea of the reduced state as the quantum state of an open system is set aside, the relativity with respect to the partition of the closed system reappears in this case as well. From the closed-system perspective, since there are many ways in which a closed system can be partitioned, many different coarse-grainings can be applied on it, and the reduced states of the components are always relative to the particular partition selected in each case. This has a relevant consequence for the understanding of the EID, in the context of which the reduced state of the decohering system plays a central role. Thus, the relativity of the reduced state is transferred to decoherence, which turns out to be a relative phenomenon (for this claim, see Castagnino, Laura, and Lombardi 2007). In fact, the many different coarse-grainings that can be applied on the same closed system represent the many ways in which it can be partitioned into a system of interest and its environment, and the occurrence of decoherence or not must be studied for each particular partition. Although already manifested in the environment-induced approach, the relativity of decoherence becomes clearer from a closed-system perspective, as will be explained below, in Section 8.7 of this chapter.

8.5 A closed-system approach to decoherence

If the description given by the reduced state can be recovered from the perspective of the closed system, it is not surprising that decoherence can be accounted for from a closed-system viewpoint. In this section, the closed-system approach to decoherence will be introduced on the basis of two assumptions (Castagnino, Laura, and Lombardi 2007, Fortin and Lombardi 2016):

- The object of study is always the *closed system*, which is considered from the viewpoint of some relevant observables. Therefore, the state used to describe decoherence is not the reduced state of an open system but a coarse-grained state of the closed system.
- The relevant evolution for decoherence is the evolution of the *expectation values* of the observables of the closed system.

The first assumption relies on the arguments developed in the previous section. Here, the second assumption will be explained.

The usual presentations of quantum mechanics place the concept of state at the center of the stage: the description of a system is given by the system's state and its time evolution. However, states in quantum mechanics do not provide the definite values of the observables of the system, as in the classical case. By contrast, they are the theoretical means for computing the expectation values of all the observables of the quantum system. Such expectations values carry the information that can be predicted by the theory and empirically tested through measurement. In turn, given the expectation values of all the observables of a system, it is theoretically possible to compute the system's state. Therefore, those expectation values can provide the complete theoretical description of the system and its time evolution, without resorting to the state.

This perspective centered on the observables is in resonance with the algebraic formalism of quantum mechanics: let us recall that, according to this formalism, a quantum system is mathematically characterized by the space $\mathcal{O}$ of observables, represented by the self-adjoint elements of an algebra of operators, and states are represented by functionals on $\mathcal{O}$. In this theoretical framework, the observables are the basic elements of the theory, and states are secondary elements, defined in terms of the basic ones. Therefore, in order to study the components of a closed system, it is necessary to consider the spaces of observables corresponding to those components.

Let us consider a closed system $\mathcal{S}$ partitioned as $\mathcal{S} = \mathcal{S}^1 \cup \mathcal{S}^2$. If $\mathcal{O}^S$ is the space of observables of $\mathcal{S}$, and $\mathcal{O}^1$ and $\mathcal{O}^2$ are the spaces of observables of $\mathcal{S}^1$ and $\mathcal{S}^2$ respectively, then $\mathcal{O}^S = \mathcal{O}^1 \otimes \mathcal{O}^2$. If ρ^S is the quantum state of $\mathcal{S}$, the reduced states of $\mathcal{S}^1$ and $\mathcal{S}^2$ can be computed by means of partial traces as $\rho_r^1 = Tr_{(2)}\rho^S$ and $\rho_r^2 = Tr_{(1)}\rho^S$, respectively. With these three states, the expectation values of all the observables of the closed system and of its components can be computed as

$$\langle O^S \rangle_{\rho^S} = Tr\left(\rho^S O^S\right) \quad \langle O^1 \rangle_{\rho_r^1} = Tr\left(\rho_r^1 O^1\right) \quad \langle O^2 \rangle_{\rho_r^2} = Tr\left(\rho_r^2 O^2\right), \qquad \text{(8-11)}$$

where $O^S \in \mathcal{O}^S$, $O^1 \in \mathcal{O}^1$, and $O^2 \in \mathcal{O}^2$. But, as is well known, there are always observables of the closed system $\mathcal{S}$ that are not observables of the components. For example, the expectation values of the observables with a generic form $O^1 \otimes O^2 \in \mathcal{O}^S$, where O^1 and O^1 are not identity operators, cannot be computed in terms of the components $\mathcal{S}^1$ and $\mathcal{S}^2$. For this reason, from the perspective centered on the observables, the viewpoint given by the closed system has conceptual priority: any partition into components leads to a description that is unavoidably partial, to the extent that it does not capture the information of all the observables of the closed system.

In the context of the closed-system approach, by contrast, the expectation values of the observables of the component systems can be computed in terms of the state and the observables of the closed system. In the case of the observables O^{S1} and O^{S2} of $\mathcal{S}$ that have the form $O^{S1} = O^1 \otimes I^2 \in \mathcal{O}^S$ and $O^{S2} = O^2 \otimes I^1 \in \mathcal{O}^S$, the expectation values of the observables of the subsystems can be computed as

$$\langle O^{S1}\rangle_{\rho^S} = Tr\left(\rho^S O^{S1}\right) = Tr\left(\rho^S\left(O^1 \otimes I^2\right)\right) = Tr\left(\rho_r^1 O^1\right) = \langle O^1\rangle_{\rho_r^1}, \tag{8-12}$$

$$\langle O^{S2}\rangle_{\rho^S} = Tr\left(\rho^S O^{S2}\right) = Tr\left(\rho^S\left(O^2 \otimes I^1\right)\right) = Tr\left(\rho_r^2 O^2\right) = \langle O^2\rangle_{\rho_r^2}. \tag{8-13}$$

These equations say the same as the second and the third eqs. (8-11). However, the closed-system approach introduces a relevant difference: it is completely general, since not restricted to the study of partitions of the closed system $\mathcal{S}$ in terms of tensor product structures. This approach permits any subset $\mathcal{O}^R \subset \mathcal{O}^S$ to be selected, where the $O^R \in \mathcal{O}^R$ do not need to be of the form $O^1 \otimes I^2$: the O^R are the observables considered relevant in the particular situation of interest. On this basis, here the closed system $\mathcal{S}$ is not split into a system of interest and its environment, but is partitioned into relevant and irrelevant observables: $\mathcal{O}^S = \mathcal{O}^R \cup \mathcal{O}^I$.

Let us consider the expectation value of a generic relevant observable $O^R \in \mathcal{O}^R$ when the system $\mathcal{S}$ is in the state ρ^S:

$$\langle O^R\rangle_{\rho^S} = Tr\left(\rho^S O^R\right) = \sum_i o_{ii}^R \rho_{ii}^S + \sum_{i\neq j} o_{ij}^R \rho_{ij}^S, \tag{8-14}$$

where the ρ_{ii}^S and the o_{ii}^R are the diagonal components, and the ρ_{ij}^S and the o_{ij}^R are the non-diagonal components of ρ^S and O^R, respectively, in the decoherence basis. The second sum of eq. (8-14) represents the specifically quantum interference terms of the expectation value. If those terms vanish, the expectation value adopts the structure of a classical expectation value,

$$\sum_i o_{ii}^R \rho_{ii}^S, \tag{8-15}$$

where, from a coarse-grained perspective, the o_{ii}^R might be interpreted as possible values and the ρ_{ii}^S might play the role of probabilities since they are positive numbers that are equal to or less than one and sum to one.

In the light of this formulation, from the closed-system approach based on expectation values (Fortin, Lombardi, and Castagnino 2014), there is decoherence for the relevant observables when the expectation values $\langle O^R\rangle_{\rho^S}$ tend, in an extremely short decoherence time t_D, to $\sum_i o_{ii}^R \rho_{ii}^{cgd}$, where $0 \leq \rho_{ii}^{cgd} \leq 1$ and $\sum_i \rho_{ii}^{cgd} = 1$. The ρ_{ii}^{cgd} can be conceived as the components of a kind of coarse-grained state ρ^{cgd}, diagonal in the decoherence basis. In other words, $\langle O^R\rangle_{\rho^S(t)}$ converges, after t_D, to a value that can be computed *as if* the quantum state of the system were represented by a diagonal density operator $\rho^{cgd}(t)$:

$$\langle O^R\rangle_{\rho^S(t)} \longrightarrow \sum_i o_{ii}^R \rho_{ii}^{cgd} = \langle O^R\rangle_{\rho^{cgd}(t)}, \tag{8-16}$$

in such a way that

$$\forall t \gg t_D,\ \langle O^R\rangle_{\rho^S(t)} = \langle O^R\rangle_{\rho^{cgd}(t)} \text{ and } \rho^{cgd}(t) \text{ is diagonal in the decoherence basis.} \tag{8-17}$$

In summary, the decoherence of certain relevant observables of the whole closed system depends on the behavior of the expectation values of those observables. Therefore, the phenomenon of decoherence can be explained in three general steps (see Fortin, Lombardi, and Castagnino 2014):

- *Step 1*: The space $\mathcal{O}^R \subset \mathcal{O}^S$ of relevant observables is defined. Regarding this step, the present proposal agrees with all the different approaches to decoherence, which always select a set of relevant observables in terms of which the time behavior of the system is described: gross observables (van Kampen 1954), macroscopic observables of the apparatus (Daneri, Loinger, and Prosperi 1962), relevant observables (Omnès, 2001, 2002), observables of the open system (environment-induced decoherence: Zurek, 1981, 1982), van Hove observables (self-induced decoherence: Castagnino and Lombardi 2004, 2005a, 2005b), etc.
- *Step 2*: The expectation value $\langle O^R \rangle_{\rho^S(t)}$, for any $O^R \in \mathcal{O}^R$, is obtained. This step can be performed in two different but equivalent ways:
  - $\langle O^R \rangle_{\rho^S(t)}$ is directly computed as the expectation value of O^R in the unitarily evolving state $\rho^S(t)$.
  - A coarse-grained state $\rho^{cg}(t)$, such that $\langle O^R \rangle_{\rho^S(t)} = \langle O^R \rangle_{\rho^{cg}(t)}$ for any $O^R \in \mathcal{O}^R$, is defined, and its non-unitary evolution (governed by a master equation) is computed.
- *Step 3*: It is proved that $\langle O^R \rangle_{\rho^S(t)} = \langle O^R \rangle_{\rho^{cg}(t)}$ reaches a value $\langle O^R \rangle_{\rho^{cgd}(t)}$:

$$\forall t \gg t_D, \quad \langle O^R \rangle_{\rho^S(t)} = \langle O^R \rangle_{\rho^{cgd}(t)}, \tag{8-18}$$

  where $\rho^{cgd}(t)$ remains diagonal in the decoherence basis for all times $t \gg t_D$.

This means that, although the off-diagonal terms of $\rho^S(t)$ never vanish through the unitary evolution, it might be said that the system decoheres *from the observational point of view* given by any relevant observable $O_R \in \mathcal{O}_R$.

8.6 Environment-induced decoherence from a closed-system approach

Since the 1980s, the environment-induced decoherence (EID) approach became the "new orthodoxy" in the community of quantum physicists (Leggett 1987, Bub 1997). In fact, decoherence is studied in many areas of physics, and has gained great relevance in quantum computation, where the aim is to take advantage of superpositions and, therefore, to avoid classicality. Despite the great success of the EID approach, from a conceptual viewpoint there are some challenges that must be faced if one wants to offer a self-consistent account of the phenomenon of decoherence. Here, I will focus on two of them: the "closed-system problem" and the "defining-system problem" (Castagnino, Laura, and Lombardi 2007, Fortin, Lombardi, and Castagnino 2014).

The *closed-system problem* consists in the fact that the EID approach cannot be applied to closed systems, in particular to the universe as a whole. According to its authors, the aim of the program "is to describe the consequences of the 'openness' of quantum systems to their environments and to study the emergence of the effective classicality of some of the quantum states and of the associated observables" (Zurek 1998, 1793). Therefore, the split of the universe into the degrees of freedom that are of direct interest to the observer—the system—and the remaining degrees of freedom—the environment—is absolutely essential to understand "the quantum origin of the classical world" (Paz and Zurek 2002, 77). Zurek has even claimed that the prejudice that seriously delayed the solution of the problem of the emergence of classicality relies on the fact that the role of the "openness" of quantum systems was ignored for a very long time (Paz and Zurek 2002, Zurek 2003). However, if decoherence explains the emergence of classicality, but only open systems can decohere, what about closed systems, in particular, the universe as a whole? (Pessoa Jr. 1998). In the literature, several models can be found that describe decoherence in systems without an environment understood in the traditional way. For example, in the Casati-Prosen model (Casati and Prosen 2005), decoherence is manifested by the vanishing of the interference pattern on a screen located in a closed cavity; independently of the details of the model, this is a case in which it is not possible to consider that the phenomenon is due to the interaction with an external environment. In the case of self-induced decoherence (Castagnino and Lombardi 2004, 2005a, 2005b), the loss of coherence is attributed to a very generic coarse-graining, and not to the intervention of an environment external to the system. Rodolfo Gambini and collaborators (Gambini, Porto, and Pullin 2007, Gambini and Pullin 2007, 2010) analyze the influence of an extra term in the evolution equation, which comes from quantum gravity considerations: such a term, responsible for decoherence, does not result from the interaction with an environment, but expresses a coarse-graining due to time-uncertainty. Some authors describe decoherence in the Heisenberg representation (Polarski and Starobinsky 1996, Kiefer and Polarski 2009), in such a way that the loss of coherence, treated by means of the Bogoliubov transformation, is due to the dynamics of the system itself. Given the peculiar features of the EID approach, these cases fall outside its scope of application. In fact, according to Zurek (1994, 181), since a closed system evolves deterministically forever, the issue of its classicality cannot even be posed.

The *defining-system problem* consists of the fact that the EID approach does not provide a criterion to decide where to place the "cut" between the system of interest and the environment. Sometimes "internal environments" are defined: the closed system is partitioned into some degrees of freedom that represent the system and the remaining degrees of freedom that play the role of the environment. For example, in the cosmological context, long wavelength modes are usually treated as the system of interest and short wavelength modes play the role of the environment (see, e.g., Lombardo and Mazzitelli 1996). However, this is not the only way to split the closed system. In the study of the fluctuations generated during the inflationary period of the cosmic evolution, it can be supposed that the tensor and the scalar fluctuations interact with each other, and the tensor fluctuations act as an environment that causes the loss of coherence of the scalar fluctuations, whose classicality is so justified (Franco and Calzetta 2011). These cosmological cases are only an example of the fact that, although the EID approach studies the correlations between system and environment and also between different subsystems (Paz and Roncaglia 2009), there is a conceptual difficulty regarding the definition of the systems involved in the phenomenon of decoherence: the approach does not furnish a general criterion to discriminate between them. In general, the classically behaving degrees of freedom are assumed in advance: the application of the EID formalism does not predict which observables will manifest a classical behavior but only confirms a previous assumption: "Decoherence cannot explain a quantum-classical divide without pain of some circularity, as any claim that a system has decohered is based on imposing a classical divide where it does not strictly apply, from the complete quantum perspective." (Tanona 2013, 3646). This problem is acknowledged by Zurek himself: "one issue which has been often taken for granted is looming big, as a foundation of the whole decoherence program. It is the question of what are the «systems» which play such a crucial role in all the discussions of the emergent classicality. This issue was raised earlier, but the progress to date has been slow at best" (Zurek 2000, 338; see also Zurek 1998).

As explained in Section 5.7 of Chapter 5, the EID approach proves that, in many physical models of systems in interaction with their environments, the non-diagonal terms of the reduced state $\rho_r^S(t)$ rapidly tend to vanish after an extremely short decoherence time t_D: the reduced state approaches a state $\rho_r^{Sd}(t)$ that is diagonal in the preferred basis. However, EID can be rephrased in such a way that it can be described from a closed-system perspective (see Castagnino, Laura, and Lombardi 2007, Fortin, Lombardi, and Castagnino 2014).

- *Step 1*: In this case, the whole composite system $\mathcal{S}$, whose space of observables is $\mathcal{O}^S$, is partitioned into an open component $\mathcal{Q}$, represented by the space of observables $\mathcal{O}^Q$, and an environment $\mathcal{E}$, represented by the space of observables $\mathcal{O}^E$. Therefore, the members of the space $\mathcal{O}^R \subset \mathcal{O}^S$ of relevant observables have the form

$$O^{SQ} = O^Q \otimes I^E \in \mathcal{O}^R, \tag{8-19}$$

  where $O^Q \in \mathcal{O}^Q$ is a generic observable of the system $\mathcal{Q}$ and $I^E \in \mathcal{O}^E$ is the identity operator corresponding to the environment $\mathcal{E}$.
- *Step 2*: The expectation value of any $O^{SQ} \in \mathcal{O}^R$ can be computed as

$$\langle O^{SQ} \rangle_{\rho^S} = Tr\left(\rho^S O^{SQ}\right) = Tr\left(\rho^S \left(O^Q \otimes I^E\right)\right) = Tr\left(\rho_r^Q O^Q\right) = \langle O^Q \rangle_{\rho_r^Q}. \tag{8-20}$$

This means that $\langle O^{SQ} \rangle_{\rho^S}$ can also be computed as the expectation value of any observable $O^Q \in \mathcal{O}^Q$ in the reduced state ρ_r^Q of $\mathcal{Q}$.

- *Step 3*: The evolution of the reduced state $\rho_r^Q(t)$ has its counterpart in the evolution of the expectation values. Therefore,

$$\langle O^{SQ} \rangle_{\rho^S(t)} = \langle O^Q \rangle_{\rho_r^Q(t)} \xrightarrow{t \gg t_D} \langle O^Q \rangle_{\rho_r^{Qd}(t)} = \langle O^{SQ} \rangle_{\rho^{cgd}(t)}, \tag{8-21}$$

where $\rho_r^{Qd}(t)$ and $\rho^{cgd}(t)$ remain diagonal in the decoherence basis for all times $t \gg t_D$. In other words, this step corresponds to the diagonalization of the reduced state.

This means that the results obtained by the EID approach can also be obtained in the framework of the closed-system approach. However, the latter is more general since it makes it possible to define decoherence for completely generic sets of observables, even sets of observables considered relevant but that do not define tensor product structures. In particular, it can be used to describe the phenomenon of decoherence in cases in which the EID approach cannot be applied.

On this basis, the challenges to the EID approach can be reassessed, since these difficulties derive precisely from what was considered the main advantage of the approach: its open-system perspective. For this reason, it is not surprising that they are resolved, or dissolved, from a closed-system view.

Regarding the closed-system problem, saying that closed quantum systems cannot decohere makes no sense: a closed quantum system may decohere from certain observational perspectives, that is, from the viewpoint of certain spaces of observables. Furthermore, the generality of the closed-system approach makes it particularly appropriate for cases that are beyond the scope of the EID approach. It is also interesting to recall that certain presentations of the EID approach suggest a close relation between decoherence and dissipation: since decoherence is a consequence of the interaction between an open system and its environment, it should be accompanied by other manifestations of openness, such as the dissipation of energy into the environment. Precisely for this reason, the "non-dissipative approaches" to decoherence were proposed as an alternative or rival to the orthodox EID approach (Polarski and Starobinsky 1996, Bonifacio *et al.* 2000, Ford and O'Connell 2001, Frasca 2003, Kiefer and Polarski 2009). Schlosshauer (2007) clearly stresses that energy dissipation is not a condition for decoherence: the loss of energy from the system is a classical effect, leading to thermal equilibrium in the relaxation time, whereas decoherence is a purely quantum effect that takes place in the decoherence time, many orders of magnitude shorter than the relaxation time. From the closed-system approach, to the extent that decoherence is not due to the interaction of a system with its environment, the possibility of confusing decoherence and dissipation disappears (Castagnino, Fortin, and Lombardi 2010b).

The closed-system approach implies, by its own nature, the dissolution of the "looming big" defining-system problem. In fact, the split of the closed system into an open subsystem and an environment is just a way to select the relevant observables of the closed system. As there are many different sets of relevant observables depending on the observational viewpoint adopted, the same closed system can be decomposed in many different ways: each decomposition represents a decision as to which degrees of freedom are relevant and which can be disregarded in the particular case under consideration. But since there is no privileged or "essential" decomposition, there is no need for a univocal criterion to decide where to place the cut between "the" system and "the" environment. Given that all the ways of selecting the relevant observables of the closed system are equally legitimate from a theoretical viewpoint, decoherence is a phenomenon *relative* to which observables of the whole closed system are considered relevant and which are disregarded in each case (see Lychkovskiy 2013). The relativity of decoherence will be discussed in detail in the next section.

These considerations cast new light on the relation between decoherence and energy dissipation. To the extent that decoherence is a relative phenomenon, no flow of a non-relative quantity from the open system to the environment should be associated to it. Although, as Schlosshauer points out, energy dissipation and decoherence can in general be easily distinguished because of their different timescales, the very reason for their difference is that energy dissipation is not a relative phenomenon but results from the effective dynamics of a physical magnitude, whereas decoherence is relative to the selected observational partition of the whole closed system.

It is worth insisting on the difference between the open-system and the closed-system approaches to decoherence by emphasizing the distinction between the concepts of subjective and relative. The open-system approach relies on the cut between system and environment. However, since there is no univocal criterion to decide where to place the cut, the decision about the "right" partition seems to rest on the observer, that is, it seems to be a *subjective* matter. From the closed-system perspective, by contrast, the discrimination between relevant and irrelevant observables is not unique from the outset. The only system univocally defined is the closed system. The choice of a set of relevant observables is a selection of a kind of observational reference frame *relative* to which decoherence is evaluated: the interference terms of the expectation values vanish for certain sets of relevant observables and not for others. This flexibility is what endows the closed-system approach with the capability of studying the behavior of any set of observables, without the need to conceive one or some of them as the privileged ones.

8.7 The relativity of decoherence

As just argued, the relativity of the phenomenon of decoherence is clearly manifested when different sets of relevant observables of the same closed quantum system are selected. In this section, I will show how such relativity appears in a particular model, the spin-bath model in its traditional form and in a generalized version, and how some seemingly paradoxical consequences can be avoided from a closed-system approach.

The traditional spin-bath model is a very simple model that has been exactly solved in previous papers (Zurek 1982). Here I will present it from the closed-system perspective discussed in the previous sections (see Castagnino, Fortin, and Lombardi 2010a). Let us consider a closed system $\mathcal{U} = \mathcal{P} \cup \mathcal{P}^1 \cup \mathcal{P}^2 \cup \cdots \cup \mathcal{P}^N = \mathcal{P} \cup (\cup_{i=1}^N \mathcal{P}^i)$, where $\mathcal{P}$ is a spin-1/2 "particle" represented by the Hilbert space $\mathcal{H}^P$, and each $\mathcal{P}^i$ is a spin-1/2 "particle" represented by its Hilbert space $\mathcal{H}^i$. The Hilbert space of the closed system $\mathcal{U}$ is, then,

$$\mathcal{H} = (\mathcal{H}^P) \otimes \left(\bigotimes_{i=1}^{N} \mathcal{H}^i \right). \tag{8-22}$$

In the particle $\mathcal{P}$, the two eigenstates of the spin operator $S^P_{\vec{v}}$ in direction $\vec{v}$ are $|\Uparrow\rangle$ and $|\Downarrow\rangle$. In each particle $\mathcal{P}^i$, the two eigenstates of the spin operator $S^i_{\vec{v}}$ in direction $\vec{v}$ are $|\uparrow_i\rangle$ and $|\downarrow_i\rangle$. Therefore, a pure initial state of $\mathcal{U}$ before interactions reads

$$|\psi_0\rangle = (a\,|\Uparrow\rangle + b\,|\Downarrow\rangle) \otimes \left(\bigotimes_{i=1}^{N} (\alpha_i\,|\uparrow_i\rangle + \beta_i\,|\downarrow_i\rangle) \right), \tag{8-23}$$

where $|a|^2 + |b|^2 = |\alpha_i|^2 + |\beta_i|^2 = 1$. If the self-Hamiltonians H^P of $\mathcal{P}$ and H^i of the $\mathcal{P}^i$ are taken to be zero, and there is no interaction among the $\mathcal{P}^i$, then the total Hamiltonian H of the system $\mathcal{U}$ is given by the interaction between the particle $\mathcal{P}$ and each particle $\mathcal{P}^i$. For example (see Zurek 1982),

$$H = \frac{1}{2}(|\Uparrow\rangle\langle\Uparrow| - |\Downarrow\rangle\langle\Downarrow|) \otimes \sum_{i=1}^{N} \left(g_i\,(|\uparrow_i\rangle\langle\uparrow_i| - |\downarrow_i\rangle\langle\downarrow_i|) \otimes \left(\bigotimes_{\substack{j=1 \\ j\neq i}}^{N} I^j \right) \right), \tag{8-24}$$

where $I^j = |\uparrow_j\rangle\langle\uparrow_j| + |\downarrow_j\rangle\langle\downarrow_j|$ is the identity operator on the subspace $\mathcal{H}^j$ and the g_i are the coupling constants.

(a.1) Decomposition 1. In the typical situation studied by the EID approach, the open system $\mathcal{S}$ is the particle $\mathcal{P}$ and the remaining particles $\mathcal{P}^i$ play the role of the environment $\mathcal{E}$: $\mathcal{S} = \mathcal{P}$ and $\mathcal{E} = \cup_{i=1}^N \mathcal{P}^i$. Then, the TPS (tensor product structure) for this case is

$$\mathcal{H} = \mathcal{H}^S \otimes \mathcal{H}^E = (\mathcal{H}^P) \otimes \left(\bigotimes_{i=1}^{N} \mathcal{H}^i \right), \tag{8-25}$$

and the relevant observables O^R of $\mathcal{U}$ are those corresponding to the particle $\mathcal{P}$:

$$O^R = O^S \otimes I^E = (s_{\Uparrow\Uparrow}\,|\Uparrow\rangle\langle\Uparrow| + s_{\Uparrow\Downarrow}\,|\Uparrow\rangle\langle\Downarrow| + s_{\Downarrow\Uparrow}\,|\Downarrow\rangle\langle\Uparrow| + s_{\Downarrow\Downarrow}\,|\Downarrow\rangle\langle\Downarrow|) \otimes \left(\bigotimes_{i=1}^{N} I^i \right). \tag{8-26}$$

The expectation value of these observables in the state $|\psi(t)\rangle = |\psi_0\rangle\, e^{-iHt}$ is given by

$$\langle O^R \rangle_{\psi(t)} = |a|^2 s_{\Uparrow\Uparrow} + |b|^2 s_{\Downarrow\Downarrow} + 2\,\mathrm{Re}\left(ab^* s_{\Downarrow\Uparrow}\, r(t)\right) = \Sigma^d + \Sigma^{nd}(t), \tag{8-27}$$

where

$$r(t) = \langle \varepsilon_{\Downarrow}(t) | \varepsilon_{\Uparrow}(t) \rangle = \prod_{i=1}^{N} \left(|\alpha_i|^2 e^{-ig_i t} + |\beta_i|^2 e^{ig_i t} \right). \tag{8-28}$$

By means of numerical simulations it can be shown that, for $N \gg 1$, in general $|r(t)|^2 \to 0$ and, therefore, the non-diagonal part of the expectation values also tend to vanish: $\Sigma^{nd}(t) \to 0$. This means that the particle $\mathcal{P}$ decoheres in interaction with a large environment $\mathcal{E}$ composed by N particles $\mathcal{P}^i$.

(a.2) Decomposition 2. Although in the usual presentations of the model the system of interest is the particle $\mathcal{P}$, there are different ways of splitting the whole closed system $\mathcal{U}$. For instance, one can decide to observe a particular particle $\mathcal{P}^k$ of what was previously considered the environment, and to consider the remaining particles as the new environment: $\mathcal{S} = \mathcal{P}^k$ and $\mathcal{E} = \mathcal{P} \cup \left(\cup_{i=1, i\neq k}^{N} \mathcal{P}^i\right)$. The total Hilbert space of the closed system $\mathcal{U}$ is still given by eq. (8-22), but now the TPS is

$$\mathcal{H} = \mathcal{H}^S \otimes \mathcal{H}^E = \left(\mathcal{H}^k\right) \otimes \left(\mathcal{H}^P \otimes \left(\bigotimes_{i=1, i\neq k}^{N} \mathcal{H}^i\right)\right), \tag{8-29}$$

and the relevant observables O^R are those corresponding to the particle $\mathcal{P}^k$:

$$O^R = O^S \otimes I^E = \left(\zeta_{k\uparrow\uparrow} |\uparrow_k\rangle\langle\uparrow_k| + \zeta_{k\uparrow\downarrow} |\uparrow_k\rangle\langle\downarrow_k| + \zeta_{k\downarrow\uparrow} |\downarrow_k\rangle\langle\uparrow_k| + \zeta_{k\downarrow\downarrow} |\downarrow_k\rangle\langle\downarrow_k|\right)$$
$$\otimes \left(I^P \otimes \left(\bigotimes_{i=1, i\neq k}^{N} I^i\right)\right). \tag{8-30}$$

The expectation value of these observables in the state $|\psi(t)\rangle$ is given by

$$\langle O^R \rangle_{\psi(t)} = \left(|\alpha_k|^2 \zeta_{k\uparrow\uparrow} + |\beta_k|^2 \zeta_{k\downarrow\downarrow}\right) + 2\,\mathrm{Re}\left(\alpha_k \beta_k^* \zeta_{k\downarrow\uparrow} e^{ig_k t}\right) = \Sigma^d + \Sigma^{nd}(t). \tag{8-31}$$

In this case, numerical simulations are not necessary to see that the time-dependent term of eq. (8-31) is an oscillating function which, therefore, has no limit for $t \to \infty$. This result is not surprising, but rather completely reasonable from a physical point of view. In fact, with the exception of the particle $\mathcal{P}$, the remaining particles of the environment $\mathcal{E}$ are uncoupled from each other: each $\mathcal{P}^i$ evolves as a free system and, consequently, cannot reach a final stable reduced state.

The traditional spin-bath model can be generalized in the following sense (see Castagnino, Fortin, and Lombardi 2010a). Let us consider a closed system $\mathcal{U} = \mathcal{A} \cup \mathcal{B}$ where:

(i) The subsystem $\mathcal{A}$ is composed of M spin-1/2 particles $\mathcal{A}^i$, with $i = 1, 2, \cdots, M$, each one represented in its Hilbert space $\mathcal{H}^{A^i}$: in each $\mathcal{A}^i$, the two eigenstates of the spin operator $S_{\vec{v}}^{A^i}$ in direction $\vec{v}$ are $|\Uparrow_i\rangle$ and $|\Downarrow_i\rangle$.

(ii) The subsystem $\mathcal{B}$ is composed of N spin-1/2 particles $\mathcal{B}^k$, with $k = 1, 2, \cdots, N$, each one represented in its Hilbert space $\mathcal{H}^{B^k}$: in each $\mathcal{B}^k$, the two eigenstates of the spin operator $S_{\vec{v}}^{B^k}$ in direction $\vec{v}$ are $|\uparrow_k\rangle$ and $|\downarrow_k\rangle$.

The Hilbert space of the composite system $\mathcal{U} = \mathcal{A} \cup \mathcal{B}$ is, then,

$$\mathcal{H} = \mathcal{H}^A \otimes \mathcal{H}^B = \left(\bigotimes_{i=1}^{M} \mathcal{H}^{A^i} \right) \otimes \left(\bigotimes_{k=1}^{N} \mathcal{H}^{B^k} \right), \tag{8-32}$$

and a pure initial state of $\mathcal{U}$ reads

$$|\psi_0\rangle = |\psi_A\rangle \otimes |\psi_B\rangle = \left(\bigotimes_{i=1}^{M} \left(a_i |\Uparrow_i\rangle + b_i |\Downarrow_i\rangle\right) \right) \otimes \left(\bigotimes_{k=1}^{N} \left(\alpha_k |\uparrow_k\rangle + \beta_k |\downarrow_k\rangle\right) \right), \tag{8-33}$$

with $|a_i|^2 + |b_i|^2 = |\alpha_k|^2 + |\beta_k|^2 = 1$. As in the original spin-bath model, the self-Hamiltonians H^{A^i} and H^{B^k} are taken to be zero, and there is no interaction between the particles $\mathcal{A}^i$ nor between the particles $\mathcal{B}^k$. As a consequence, the total Hamiltonian $H = H^A \otimes H^B$ of the composite system $\mathcal{U}$ is given by

$$\begin{aligned} H = &\left(\sum_{i=1}^{M} \left(\frac{1}{2} \left(|\Uparrow_i\rangle\langle\Uparrow_i| - |\Downarrow_i\rangle\langle\Downarrow_i| \right) \otimes \left(\bigotimes_{j=1,\, j\neq i}^{M} I^{A^j} \right) \right) \right) \\ &\otimes \left(\sum_{k=1}^{N} \left(g_k \left(|\uparrow_k\rangle\langle\uparrow_k| - |\uparrow_k\rangle\langle\downarrow_k| \right) \otimes \left(\bigotimes_{l=1,\, l\neq k}^{N} I^{B^l} \right) \right) \right), \end{aligned} \tag{8-34}$$

where $I^{A^j} = |\Uparrow_j\rangle\langle\Uparrow_j| + |\Downarrow_j\rangle\langle\Downarrow_j|$ and $I^{B^l} = |\uparrow_l\rangle\langle\uparrow_l| + |\downarrow_l\rangle\langle\downarrow_l|$ are the identity operators on the subspaces $\mathcal{H}^{A^j}$ and $\mathcal{H}^{B^l}$ respectively. Let us notice that the eq. (8-24) of the original model is the particular case of eq. (8-34) for $M = 1$.

(b.1) Decomposition 1. Let us consider that $\mathcal{A}$ is the open system $\mathcal{S}$ and $\mathcal{B}$ is the environment $\mathcal{E}$. This is a generalization of Decomposition 1 in the traditional spin-bath model: the only difference is that here $\mathcal{S}$ is composed of $M \geq 1$ particles instead of only one. Then, the TPS for this case is

$$\mathcal{H} = \mathcal{H}^S \otimes \mathcal{H}^E = \left(\bigotimes_{i=1}^{M} \mathcal{H}^{A^i} \right) \otimes \left(\bigotimes_{k=1}^{N} \mathcal{H}^{B^k} \right), \tag{8-35}$$

and the relevant observables O^R are those corresponding to $\mathcal{A}$:

$$O^R = O^S \otimes I^E = O^A \otimes \left(\bigotimes_{i=1}^{N} I^i \right). \tag{8-36}$$

When the expectation values $\langle O^R \rangle_{\psi(t)} = \Sigma^d + \Sigma^{nd}(t)$ of the observables O^R in the state $|\psi(t)\rangle$ are computed, two cases can be distinguished:

- *Case (α)*: $M \ll N$. Numerical simulations show that $\Sigma^{nd}(t) \to 0$ very fast for increasing time (see Figure 2 of Castagnino, Fortin, and Lombardi 2010a). This means that, as expected, a small open system $\mathcal{S} = \mathcal{A}$ of M particles decoheres in interaction with a large environment $\mathcal{E} = \mathcal{B}$ of $N \gg M$ particles.
- *Case (β)*: $M \gg N$ or $M \simeq N$. Numerical simulations show that $\Sigma^{nd}(t)$ exhibits an oscillating behavior and, then, it does not approach zero for increasing time (see Figures 3 and 4 of Castagnino, Fortin, and Lombardi 2010a). This means that, when the environment $\mathcal{E} = \mathcal{B}$ of N particles is not large enough when compared with the open system $\mathcal{S} = \mathcal{A}$ of M particles, $\mathcal{S}$ does not decohere.

(b.2) Decomposition 2. In this case one decides to observe only one particle of $\mathcal{A}$. This amounts to splitting the closed system $\mathcal{U}$ into two new components: the open system $\mathcal{S}$ is, say, the particle $\mathcal{A}^M$ and the environment is $\mathcal{E} = \left(\cup_{i=1}^{M-1} \mathcal{A}^i\right) \cup \mathcal{B} = \left(\cup_{i=1}^{M-1} \mathcal{A}^i\right) \cup \left(\cup_{k=1}^{N} \mathcal{B}^k\right)$. Let us notice that the Decomposition 2 of the traditional spin-bath model is a particular case of this one, for $N = 1$ (where N plays the role of the M of this case). Here, the TPS is

$$\mathcal{H} = \mathcal{H}^S \otimes \mathcal{H}^E = \left(\mathcal{H}^{\mathcal{A}^M}\right) \otimes \left(\left(\bigotimes_{i=1}^{M-1} \mathcal{H}^{\mathcal{A}^i}\right) \otimes \left(\bigotimes_{k=1}^{N} \mathcal{H}^{\mathcal{B}^k}\right)\right), \tag{8-37}$$

and the relevant observables O^R are those corresponding to $\mathcal{A}^M$:

$$O^R = O^S \otimes I^E = O^{\mathcal{A}^M} \otimes \left(\left(\bigotimes_{i=1}^{M-1} I^i\right) \otimes \left(\bigotimes_{k=1}^{N} I^k\right)\right). \tag{8-38}$$

When the expectation value $\langle O^R \rangle_{\psi(t)} = \Sigma^d + \Sigma^{nd}(t)$ is computed, numerical simulations show that, if $N \gg 1$, $\Sigma^{nd}(t) \to 0$ very fast for increasing time (see Figures 5, 6, and 7 of Castagnino, Fortin, and Lombardi 2010a). This means that the particle $\mathcal{A}^M$ decoheres when $N \gg 1$, independently of the value of M. But since the particle $\mathcal{A}^M$ was arbitrarily selected, the same argument holds for any particle $\mathcal{A}^i$ of $\mathcal{A}$. Then, when $N \gg 1$ and independently of the value of M, any particle $\mathcal{A}^i$ decoheres in interaction with its environment $\mathcal{E}$ of $N + M - 1$ particles. In turn, on the basis of the symmetry of the whole system $\mathcal{U}$, analogous consequences obtain when the system $\mathcal{S}$ is one of the particles of $\mathcal{B}$: when $M \gg 1$ and independently of the value of N, any particle $\mathcal{B}^k$ decoheres in interaction with its environment $\mathcal{E}$ of $N + M - 1$ particles.

When the two possible partitions of the whole closed system are taken into account, interesting conclusions can be drawn (see Castagnino, Fortin, and Lombardi 2010b). Let us consider the generalized spin-bath model with $M \simeq N \gg 1$. In this case, the subsystem $\mathcal{A} = \cup_{i=1}^{M} \mathcal{A}^i$ does not decohere (Decomposition b.1), but the particles $\mathcal{A}^i$, considered independently, do decohere (Decomposition b.2). In other words, in spite of the fact that certain particles decohere and may behave classically, the subsystem composed by all of them retains its quantum nature. In turn, since $M \gg 1$, all the particles $\mathcal{B}^k$, considered independently, decohere. Then, in this case not only all the

$\mathcal{A}^i$, but also all the $\mathcal{B}^k$ decohere. This means that all the particles of the closed system $\mathcal{U} = \left(\cup_{i=1}^{M} \mathcal{A}^i\right) \cup \left(\cup_{k=1}^{N} \mathcal{B}^k\right)$ may behave classically when considered independently, although the whole closed system $\mathcal{U}$ certainly does not decohere and, therefore, retains its quantum character.

The fact that certain particles may decohere or not and, as a consequence, may behave as classical or quantum particles depending on how they are considered might sound paradoxical; however, this is an unavoidable result in the context of an approach as EID, which explains decoherence as the effect of an interaction between open systems (see discussion in Lombardi, Fortin, and Castagnino 2012). This difficulty can also be seen as a manifestation of the problem of defining the open systems involved in decoherence. The irony of this story is that such a problem is the consequence of what has been considered to be the main advantage of the decoherence program, its open-system perspective, according to which particles interacting with other particles are well-defined open systems, and the collections of those particles are also open systems. So, the "looming big" problem is to decide which one of all these open systems is the system $\mathcal{S}$ that decoheres or, in other words, where to place the cut between the system $\mathcal{S}$ and its environment $\mathcal{E}$.

These paradoxical consequences can be avoided from a closed-system approach, according to which decoherence is a relative phenomenon. Since there are many ways in which a closed quantum system can be decomposed, and there is no privileged decomposition, it is not necessary to single out a criterion to place the split between "the" decohering system and "its" environment.

8.8 The observable viewpoint

Although admitting that the relativity of decoherence dissolves the defining-system problem, this does not yet answer the issue of the emergence of classicality. In fact, if classicality (or quasi-classicality) is conceived as an objective property of systems, the fact that a system behaves classically or not cannot depend on the way in which the original closed system is decomposed into a system of interest and its environment. The conceptual difficulty is even more serious due to the already pointed out result: in certain situations the fact that classicality emerges or not in a component system seems to depend on the larger system in which that component is embedded. More precisely, given two partitions of a closed system $\mathcal{U}$, $\mathcal{U} = \mathcal{S}^1 \cup \mathcal{E}^1$ and $\mathcal{U} = \mathcal{S}^2 \cup \mathcal{E}^2$, it seems possible that $\mathcal{S}^1$ and $\mathcal{S}^2$ decohere and behave classically when considered independently, but $\mathcal{S}^1 \cup \mathcal{S}^2$ does not decohere and, so, classicality does not emerge in it. When expressed in this way, this conclusion is difficult to be accepted: even if emergent, classicality should not be a relative property. In other words, it is expected that a given "open system" behaves as classical or not, and the theory should account for this behavior.

The closed-system approach, while accepting that the phenomenon of decoherence is essentially relative to the decomposition of the closed system, solves the problem of the emergence of classicality because of its focus on observables. Given the closed system $\mathcal{U}$, saying that it decoheres from the perspective of the relevant observables $O^R \in \mathcal{O}^R$ amounts to saying that the interference terms of the expectation values of all the observables belonging to $\mathcal{O}^R$ vanish with the (unitary) time-evolution of $\mathcal{U}$, and this is not a relative fact. The same physical fact can be expressed by saying that all the relevant observables $O^R \in \mathcal{O}^R$ decohere and, as a consequence, they behave classically. Furthermore, the observables of $\mathcal{U}$ can be studied *one by one* to analyze whether they decohere or not, that is, whether the interference terms of the their expectation values tend to vanish with time: in this way, the space of observables of $\mathcal{U}$ can be partitioned into the set of observables that do not decohere and those that decohere. The latter set $\mathcal{O}^{CL}$, which includes all the classically behaving observables of $\mathcal{U}$, can be defined with no ambiguity or relativity.

The closed-system approach perfectly agrees with the closed-system viewpoint of the MHI. Moreover, the perspective based on observables rather than on open systems fits perfectly with the algebraic formalism adopted by the MHI proposal, and finds its ontological counterpart in the ontology of properties to be developed in the following chapters. But for now it can be concluded that, according to the explanation of the emergence of the classical world given by the closed-system approach, strictly speaking *classicality is not a property of systems*: thinking in systems that become classical leads to the already mentioned confusions and perplexities. The difficulties can be overcome once it is recognized that *classicality is a property of observables.* The emergent classical world is, then, the world described by the observables that behave classically with respect to their expectation values.

8.9 Quantum mechanics from a closed-system perspective

The previous sections of this chapter have provided a number of arguments in favor of a closed-system perspective, which lead to a "top-down" approach to quantum mechanics, according to which the theory applies primarily to closed systems and only derivatively to their constituents. Furthermore, strictly speaking only closed systems are quantum systems, because their behavior is described by the unitary dynamical postulate of the theory. Here, I summarize the most relevant aspects of this approach:

- There are many different ways in which a quantum system can be decomposed into parts, and SQM itself does not provide a criterion for selecting a privileged decomposition. This multiplicity of criteria used to identify the components of a closed system shows that such identification is a subtle matter and cannot be taken for granted when an open-system view is adopted. The closed-system view of quantum mechanics, by contrast, offers a straightforward solution to the problem by considering that only closed systems are legitimate quantum systems: they are composite if they can be decomposed into non-interacting—closed—quantum *subsystems* or elemental otherwise. All the possible partitions of an elemental quantum system are equally legitimate: they do not lead to subsystems but rather to *parts* of the whole system. As a consequence, the quantum phenomena that rely on the partition of a closed system into parts turn out to be relative to that partition.

- Entanglement is a property of the quantum state of a closed quantum system. From the closed-system view, if the system is composite, then the entanglement of its state is relative to its decomposition into subsystems; if it is elemental, the entanglement of its state is relative to its partition into parts. In both cases, the split of the closed system into smaller "pieces" is in general not unique, so *entanglement is an essentially relative phenomenon.*
- From the closed-system perspective, reduced states are not quantum states of open "subsystems" in the same sense as quantum states are states of the whole closed system. By relying on a generalized concept of coarse-graining, it can be proved that *reduced states are coarse-grained states of closed systems*, which disregard certain degrees of freedom considered as irrelevant. Therefore, their non-unitary evolutions do not need the reformulation of the dynamical postulate of SQM nor do they require a new dynamical postulate in order to be described. The non-unitary evolution of a quantum reduced state turns out to be a situation analogous to the familiar case of classical statistical mechanics, where it is natural to obtain a non-unitary coarse-grained evolution from the underlying unitary dynamics of the unstable closed system, with no need for reformulations of the classical dynamical laws. And, as in the classical case, there are many different coarse-grained states that can be defined on the same closed system, depending on which degrees of freedom are retained and which are disregarded.
- If the description given by the reduced state can be recovered from the perspective of the closed system as a coarse-grained description, it is not surprising that *decoherence can be accounted for from a closed-system view.* In fact, it is possible to study the loss of coherence in a closed system by considering the internal structure of the expectation values of its observables: decoherence is conceived in terms of the vanishing of the interference terms of the expectation values of certain observables of the closed system. This closed-system approach to decoherence can account for the results of the environment-induced approach, but also dissolves the problem of identifying the system that decoheres: since the split of the closed system into an open "subsystem" and an environment is just a way of selecting the relevant observables among many possible decompositions, and since there is no privileged or "essential" decomposition, then there is no need for an univocal criterion to decide where to place the cut between "the" system and "the" environment.
- The observables of a closed system can be studied one by one to analyze whether they decohere or not. Therefore, the set of all the classically behaving observables of the closed system can be defined with no ambiguity or relativity. This means that *classicality is not a property of systems but a property of observables.*

- From a closed-system perspective, the problem of non-locality does no longer consist in accounting for correlations between the properties of non-interacting and possibly distant objects. Since the closed system is a single whole, *EPR-correlations are correlations between properties of a single system.* In turn, the problem of indistinguishability does no longer consist in explaining why a permutation of elementary particles of the same kind does not lead to a different complexion in quantum statistics. From this perspective, *indistinguishability* is not a relation between particles but turns out to be the result of *an internal symmetry of the closed system* conceived as an indivisible whole.
- The open-system approach not only leads to the defining-system problem and to ignoring the relativity of some central quantum phenomena, but also disregards the well-known holism of quantum mechanics, understanding *holism* as the feature of some *physical systems that are not composed of physical components but are indivisible wholes.* In order to retain its holistic nature, a quantum system must be considered as a whole: the open "subsystems" are only partial descriptions of the whole closed system.

Whereas the first five points have been covered extensively in the present chapter, the two last points are related to the ontological challenges of quantum mechanics, and will be analyzed in depth in the following three chapters, in particular, in Chapter 11.

9
Metaphysical preliminaries

The issue about the nature of quantum systems usually hinges on the category of individual: the question is whether this category can be applied to understand the peculiarities of quantum behavior. However, some metaphysical subtleties are commonly overlooked in this discussion. What is the difference between the notions of object and of individual? What metaphysical principles are quantum systems supposed to fulfill? What ontological kinds of properties are involved in the quantum domain? These questions should be taken into account when the issue at stake is the ontological nature of quantum systems and, more generally, the structure of the quantum realm.

In order to address these and other metaphysical questions, different ontological distinctions that are scarcely taken into account in the debate will be brought to the fore in this chapter. In the light of these distinctions, the ontological challenges that quantum mechanics poses to the ontological picture presupposed by traditional metaphysics will be discussed in the following chapter.

9.1 Ontological categories

In the discussions about the interpretation of quantum mechanics, it is commonly acknowledged that the theory introduces a deep break with respect to the classical view of reality. In order to understand the extent of such a break, the first step is to recall what an ontological category is.

A category is not a class defined by a concept, like "yellow" or "round," which gathers certain objects together because they possess a certain property—or a cluster of properties—in common. A category is not a *taxon*, like "vegetal" or "mammal," which classifies objects into well-defined kinds. Categories are prior to any classification, since they are what endow reality with a certain structure. For this reason, they are conditions of possibility for any classification (Lewowicz 2005).

In his *Tractatus Logico-Philosophicus*, Ludwig Wittgenstein (1921) introduces the distinction between *saying* and *showing*. A state of affairs external to language itself is "*sayable*" and is depicted by a proposition. A proposition has a content that is fully intelligible to a person who is fluent in the language, even if they do not know whether it is true or false. However, although propositions can depict the whole of reality, they cannot depict their own logical form, since this would require adopting a perspective outside of language itself. Thus, a proposition does not represent its logical form but *shows* it in its own structure. In other words, the logical structure

The Modal-Hamiltonian Interpretation of Quantum Mechanics. Olimpia Lombardi, Oxford University Press.
© Olimpia Lombardi (2026). DOI: 10.1093/9780198951407.003.0009

of language cannot be *said*; it can only be *shown*. As Wittgenstein states: "What can be shown, cannot be said" (1921, Proposition 4.1212). These "unsayable things" are shown by the *form* of the propositions: they are there, in language, even though they cannot be said. In turn, the structure of language, shown by language itself, is also the structure of reality: "Propositions show the logical form of reality. They display it" (Wittgenstein 1921, Proposition 4.121). Therefore, the analysis of the logical structure of language allows us to understand the ontological structure of reality. A proposition such as 'the flower is pink' says something—namely, it describes the fact that a certain flower is pink—but it cannot *say* that 'flower' is a noun representing an object and 'pink' is an adjective representing a property. the proposition can only *show*, through its structure, that it speaks of a reality populated by objects and properties.

Once this Wittgensteinian distinction is taken into account, it becomes clear that categories are *said* neither with nouns nor with predicates nor with any other kind of word. *Categories are shown by language*: each language manifests, in its own structure, the categories that shape and organize the ontology to which it refers. For example, the structure of language will tell us whether the ontology is inhabited by objects, properties, and relations, whether ontological items can be categorized as one versus multiple, whether events are temporally ordered, and whether there are causal links between them. In turn, one task of metaphysics is to organize reality on the basis of categories. From Aristotle to Kant, many authors have proposed categories to structure reality. Here, only those categories that play a relevant role in the ontological interpretation of quantum mechanics will be considered, in particular, the categories of *object*, *property*, and *event*. I will use the term 'item' to refer to anything that exists, regardless of the category to which it belongs: objects, properties, and events are ontological items.

The ontological category of object is mirrored in language by the linguistic category of subject. As Ernst Tugendhat clearly explains: "There is a class of linguistic expressions which are used to stand for an object; and here we can only say: to stand for something. These are the expressions which can function as the sentence-subject in so-called singular predicative statements and which in logic have also been called singular terms" (Tugendhat 1982, 23). This means that the category of object has its linguistic correlate in the so-called *singular terms*, which play the role of logical subjects of propositions and have singular references.

Since the category of object is closely related to the idea of subject of predication, it is always complemented by the ontological category of property: properties are attributed to objects and are represented by linguistic predicates. In other words, if an object corresponds to a subject of predication at the linguistic level, at the ontological level, it is the bearer of the properties attributed to it (see Rettler and Bailey 2022). It is relevant to emphasize that monadic properties, usually called "properties," and *n*-adic properties, usually called "relations," both belong to the ontological category of "property" and are distinguished by their arity. Or, in the other conceptual direction, a monadic property can be conceived as a particular case of a relation, corresponding to $n = 1$.

On the basis of these distinctions, the most traditional metaphysical picture is that of an ontology of objects and properties. This picture is present, under different forms, in the Aristotelian dichotomy between primary substance and attributes and in Locke's doctrine of the substance in general as the underlying substratum in which properties inhere, among many other cases in the history of philosophy. It is this kind of ontology that underlies Western ordinary languages and most systems of logic. In fact, in the propositions 'Socrates is mortal' and 'water is liquid,' both of the form 'S is P,' the linguistic distinction between the subject S and the predicate P expresses the ontological distinction between object and property. The same can be said in propositions with a non-copulative verb, such as the case 'The baby cries': here, too, there is a subject, 'the baby,' representing an object, and a predicate, 'to be crying,' representing a property. Moreover, the predicate does not need to be monadic; in a relational proposition such as 'Aristotle was the teacher of Alexander,' the names 'Aristotle' and 'Alexander' refer to objects, and the dyadic predicate 'to be the teacher of' denotes a dyadic property—that is, a relation.

The application of a monadic property to one object or of an *n*-adic property to *n* objects leads to a *fact*, which is logically expressed by a true proposition (Armstrong 1993). Another ontological category is that of *event*, which is endowed with a strong temporal connotation. Some philosophers consider that the link between events and facts is close enough to justify the assimilation of the two categories (Wilson 1974, Tegtmeier 2000). However, facts are commonly conceived as having a certain temporal stability: they last for a certain amount of time, in general, not too short. By contrast, events are thought of as occurrences that are instantaneous or last for a very short time. On the basis of this temporal connotation, if events are assumed as a primitive ontological category, then temporal instants or intervals can be obtained as derived items, for instance, as maximal sets of simultaneous or partially simultaneous events (Russell 1914, Whitehead 1929), or time itself can be constructed as a linear ordering of events, induced by the binary relation "*x wholly precedes y*" between them (Thomason 1989).

9.2 Objects: Individuals versus stuff

The traditional examples of objects are individual things, such as persons, tables, or flowers. However, if an object is understood strictly as a bearer of properties, the category of object is more general than that of individual: there are bearers of properties, such as water, soup, and jelly, which are not individuals but belong to a different subcategory. I will call it "*stuff*" in order to avoid the philosophical and physical connotations of terms like 'matter' and 'mass.'

An *individual* is an object that bears properties, but it needs additional features to be so. An individual is a *whole unity* in the sense that, as such an individual, it is indivisible. This means that either it cannot be divided (it is "atomic") or, if it can be divided, the results of the division are other individuals or parts of a different kind from the original one. In turn, an individual is subject to the Kantian category of *quantity* (unity-plurality): individuals are either one (each one of them) or many, that is, a plurality. In the plural case, individuals form aggregates, in which they can be *counted*. As Henry Laycock says, the key to the character of this general category "evidently rests in the notions of unity and singularity—and thereby perhaps, more generally, in the concepts of number and countability" (Laycock 2010, 8).

In the Western philosophical tradition, the properties of an individual can be either (i) accidental, which are those that can change over time because the individual may or may not possess them, or (ii) essential, that is, those that the individual necessarily possesses and that, in many cases, allow the individual to be reidentified over time. In this sense, space-time properties always play a central role, either as essential properties of the individual or as the properties that confer individuality to it, in both cases under the assumption of impenetrability, which guarantees that two individuals cannot occupy the same place at the same time.

Most systems of logic include individual constants and individual variables, which represent items belonging to the category of individual. For instance, in first-order logic, a proposition 'Pa' states that the property referred to by the predicate 'P' applies to the individual denoted by the individual constant 'a'; in turn, in the expressions '$\forall xPx$' and '$\exists xPx$,' the range of the individual variable x is a domain of individuals. The presence of individual constants and variables is not specific to traditional logic: the vast majority of systems of logic, even extensions of traditional logic and deviant logics (see Haack 1974, 1978), all include symbols to represent individuals, so that an ontology populated by individuals is presupposed. In turn, in traditional set theory, the elements of a set are also individuals: when we say '$a \in A$,' we mean that the individual denoted by 'a' belongs to the set represented by 'A,' and this holds even in the case where the element denoted by 'a' is itself a set, since in this case the set behaves as an individual.

Although it may not be easy to define what an individual is, it seems quite clear that the ontology we commonly talk about includes individuals, precisely because the symbols used to denote them are ubiquitous in our ordinary and formal languages. Perhaps for this reason, the assumption of an ontology of individuals and properties has been the dominant view in Western philosophical thought. It has also shaped physics since modern times, from the corpuscular philosophy of Galileo and Boyle up to present-day physics, with the standard model of fundamental particles. On the contrary, the answer to the question 'What is stuff?' is far from easy, as there is no strong tradition to help us. Laycock introduces the issue with an everyday example: "Removing a fly from a bowl of soup inevitably involves removing some soup as well; but it seems grammatically inappropriate to say that in such a case, there is another thing which is removed, alongside the fly. [...] the spatio-temporal isolation of any such soup will be arbitrary or adventitious [...] that, in essence, is why soup must be served in discrete bowls." (Laycock 2010, 15). The key difference between the two cases is that a fly is an individual, whereas soup is stuff.

As in the case of individuals, a *stuff* must have an identity principle—that is, a principle that distinguishes it from other stuffs or kinds of stuff. However, in this case, such a principle has nothing to do with space and time: what distinguishes water from alcohol has no relation to spatio-temporal properties. Despite this, *portions of stuff* do exist in space and time: a portion of water is spatio-temporally located, for example, now and here, in my glass. Unlike individuals, a portion of stuff can be further split into portions of the same stuff—that is, it *can be divided* without losing its identity: if a piece of chalk is broken into smaller pieces, the resulting parts are also pieces of chalk. However, a stuff is not each of its portions: the meaning of the term 'chalk' cannot be established by pointing to one piece of chalk. But neither is a stuff the mere aggregate of its portions: the reference of the word 'water' is something beyond all the portions of water existing in the universe, and such reference would not change if some or many portions of water disappeared. This means that a stuff embodies *unity and multiplicity* at the same time: it is one stuff, but it has multiple manifestations in its portions. Despite this, portions of stuff do not behave as individuals because, when put together, they *cannot be counted*: the joining of two portions of water is not "two waters" but "more water." Moreover, whereas individuals preserve their identity in the aggregate, portions of stuff cannot be reidentified once they are put together: it cannot be said that "this" is one and "that" is the other of the original portions of, for instance, water (see Lewowicz and Lombardi 2013).

Analytic philosophy has confronted the trouble of understanding the category of stuff from a linguistic perspective, through the discussion of the so-called *problem of mass terms*. In the words of Donald Davidson (1967), the problem is to understand the difference between countable nouns and mass nouns or non-countable nouns (see Pelletier 1979). In fact, a mass term refers to something that cannot be counted. It is in this sense that mass terms are said to have the semantic property of referring cumulatively: "any sum of parts which are water is water" (Quine 1960, 91). On the one hand, like plural nouns, mass terms are semantically *non-singular*, a fact reflected in their non-acceptance of singular determiners: one can speak of "all water," "some water," and "more water," but not of "a water," "each water," or "one water." This means that mass terms do not denote individual portions of stuff. On the other hand, they have in common with singular nouns the fact that they are semantically *non-plural*: whereas we can say "all oranges are sweet," we can only say "all water contains impurities." These features of mass terms show that their reference is neither singular nor plural, since they designate neither one nor many individual things: "we should not expect a successful reduction to singular reference and singular predication, something that the application of traditional first-order logic would require [...] when we say that water surrounds our island [...] our discourse is not singular discourse (about an individual) and is not plural discourse (about some individuals); we have no single individual or any identified individuals that we refer to when we use 'water'" (McKay 2008, 310–311).

Summing up, it is necessary to take into account that properties do not always apply to individuals, but they also apply to stuff: the category of object is more general than that of individual.

9.3 Objects: Metaphysical identity

The question of identity is a traditional topic of metaphysics. A distinction is commonly made between *qualitative identity* and *numerical identity* (Noonan and Curtis 2022). Qualitative identity is a relation between two objects that have certain properties in common and is a matter of degree. Numerical identity, by contrast, is a relation that an object has only with itself and involves absolute qualitative identity. In the present work, I am concerned exclusively with numerical identity, so the term 'identity' will always be used in this sense. It is in this context that different forms of identity are proposed. *Synchronic identity* is what makes it possible to identify an object by distinguishing it from all the others at a certain instant. *Diachronic identity* is what makes it possible to reidentify an object over time. In addition, what can be called *trans-setting identity* is what makes it possible to reidentify an object in different situations or circumstances. Commonly, identity criteria are discussed in the case of individuals; here, I will consider them in the more general case of objects.

When the criterion of synchronic identity depends on an ontological category that transcends the properties of the object, a *transcendental identity* (transcendental individuality in terms of Post 1963) is established. Duns Scotus's *haecceitas*, a notion still present in contemporary metaphysics, is an example of this approach (see Williams 2022); substance in Locke's sense is another typical example (see Kaplan 1975). By contrast, for the *bundle theory*, the criterion of synchronic identity depends exclusively on the properties of the object (Armstrong 1989), since an object is nothing but a bundle of properties. In this case, identity is an equivalence relation (reflexive, symmetric, and transitive) satisfying Leibniz's law, which is the result of the conjunction between the principle of identity of indiscernibles and the principle of indiscernibility of identicals.

The principle of indiscernibility of identicals states that if x and y are identical (they name the same object), then exactly the same properties apply to them (this is a case of absolute qualitative identity). Assuming that the identity relation satisfies this condition is not controversial. However, this principle alone cannot act as a definition or criterion of identity, since the identity relation appears in its antecedent. The converse of this principle, the *principle of identity of indiscernibles* (PII), states that if exactly the same properties apply to x and y, then they are identical. The PII could work as an identity criterion because the identity relation appears in its consequent. However, in the specification of the minimal subset of properties that suffices to obtain identity, a new problem arises. In fact, three versions of the PII can be distinguished on the basis of the subset of properties considered relevant (French and Krause 2006): (i) PII(1), if two objects have all their monadic and relational properties in common, then they are identical; (ii) PII(2), if two objects have all their monadic and relational properties in common, except space-time properties, then they are identical; and (iii) PII(3), if two objects have all their monadic properties in common, then they are identical. These three versions are ordered from lowest to highest logical strength. PII(2) and PII(3), being stronger, have counterexamples in the classical (non-quantum) domain. The traditional one is the case of Black's spheres (Black 1952): two exactly equal spheres in an empty Newtonian space share all their monadic and relational properties, except the space-time ones, and yet they are two objects. PII(1), being the weakest, is the only version that is a candidate for metaphysical necessity.

Diachronic identity, although not as intensely discussed as synchronic identity, is not less relevant. What is it that makes an object the same at different times? Advocates of transcendental identity may claim that substance or haecceitas is what persists over time and thus reidentifies the object despite the change in its properties. Supporters of the bundle theory are committed to selecting certain essential properties that do not change over time, whereas accidental properties may be replaced. In the case of the bundle conception of individuals, space-time properties are traditionally conceived as the essential properties that confer diachronic identity in terms of their continuity (see Gallois 2016).

9.4 Properties: Universals versus tropes

Properties are items that are present in almost any ontological picture, since they are necessary for classification: objects that are numerically different can, nevertheless, be similar in terms of a certain feature or characteristic and, as a consequence, can be grouped in the same class. However, despite their ubiquity, the metaphysical nature of properties has been the subject of controversy since the origin of philosophical thought.

Since Plato, who used the term 'εἶδος' to designate them, properties have traditionally been conceived as *universals*—that is, items that can be shared by different objects—in contrast to particulars, which have a single existence. As Duns Scotus, following Aristotle, asserted, the peculiarity of universals is that they are "one-in-many" (see, e.g., King 1992): a universal is one (e.g., redness), but it is instantiated in a multiplicity of cases (red in this case, red in that case). A universal is fully present in each of its instances, and the existence of a universal in one case is unrelated to its simultaneous existence in another case (see MacLeod and Rubenstein 2006). The so-called *problem of universals* has permeated the history of philosophy from its beginnings, with questions about whether and how universals exist. There are three long-standing answers to these questions: *realism*, *nominalism*, and *conceptualism*. Whereas realists include universals as elements of the ontology, conceptualists and nominalists refuse to accept their independent existence. Conceptualists explain the similarity between objects by appealing to general concepts that exist only in minds. Nominalists, in turn, leave the relation of qualitative similarity as a brute and primitive fact. According to realists, by contrast, universals exist as mind-independent items. For *transcendent* realism, such as that proposed by Plato, they exist even though they are not instantiated, and they are thus "transcendent" or "*ante res*" ("before the things"). For *immanent* realism, such as that defended by Aristotle, universals are "immanent" or "*in rebus*" ("in things"), since they exist only if they are instantiated by objects. Whatever the differences between the two forms of realism, in both cases the *instances of a universal property* are many but absolutely *indistinguishable*: they differ "*solo numero*," that is, they are *only numerically different*. For example, given two instances of sphericity, the mere idea of distinguishing between them as properties makes no sense.

Since a universal has multiple instances, it is necessary to elucidate the concept of *instantiation*. Instantiation is traditionally conceived as the relation that a universal property maintains with the substratum to which it is "attached" (see Orilia and Paolini Paoletti 2022). However, from a bundle-theory viewpoint, there are no substances, substrata, or individuals in which the properties inhere. Therefore, in this case, instantiation becomes a subtler notion. One possibility is to conceive instantiation as the relation that a property maintains with its multiple manifestations, called "instances." For example, the property "red" has many instances in different situations, that is, when it appears in different bundles. In the physical realm, for example, energy and momentum are properties that are instantiated in their multiple empirical manifestations. Another possibility is to consider that the notion of instantiation does not really refer to a relation, since the universal and its multiple instances are two indivisible sides of a single coin. Here I will not delve into these metaphysical matters; my proposal only requires accepting the natural idea that properties can be absolutely indistinguishable.

Another issue that I will not discuss here is the alternative between transcendent and immanent realism, since the metaphysical choice between them is irrelevant to the ontological proposal to be developed in the following chapters. Nevertheless, I consider immanent realism, according to which there are no non-instantiated universals, to be more reasonable when the task is to devise a physical ontology, since the very idea of universal properties without physical manifestations is meaningless from a scientific viewpoint.

Empiricist-minded philosophers have considered that the appeal to universals is unnecessary and even undesirable to account for properties. Of course, clever arguments have been proposed to support this position. However, resorting to universals provides the simplest explanation for the fact that the properties "shape" and "color" are different and can be applied to the same object or, in physics, that "energy" and "position" are different properties and can be applied to the same system. Although, in metaphysics, the criteria of identity for universals are a matter of discussion (see, e.g., Alvarado 2020), in the physical sciences, the different universal properties are distinguished in a straightforward way by assigning different units of measurement to them: nobody confuses energy and position because energy is measured in joules and position is measured in meters. On this basis, the empiricist philosopher could adopt the distinction between different physical quantities, such as energy and position, as a brute fact which, therefore, does not require the appeal to universals to be explained. I find this position philosophically uninteresting, but I leave the discussion of this issue to expert metaphysicians.

In twentieth-century metaphysics, a new approach to properties entered the scene: properties as *tropes*. This approach intends to occupy an intermediate position between realism and nominalism with respect to universals. Tropes are *particular* properties, such as the particular shape, weight, and texture of an object (see Maurin 2018). That two objects "share" a property (e.g., a particular shade of redness) means that they each exemplify a redness-trope, where the two redness-tropes, though numerically distinct, nevertheless resemble each other exactly. Because tropes are particulars, trope theorists face the problem of providing a principle of individuation for tropes (see Schaffer 2001). A natural answer is to individuate tropes by reference to the objects that instantiate them. However, this strategy does not work in the context of a bundle view: if objects are bundles of tropes, that principle of individuation would become circular. A different approach appeals to spatio-temporal individuation, according to which two tropes are different when they are located at different space-time positions: the redness-trope now and here is different from the redness-trope then and there (see, e.g., Campbell 1990). But this standpoint needs to exclude space-time position as a property in the same sense as the rest of the trope properties. For these and other reasons, many trope theorists have opted for a primitivist perspective, according to which the fact that two tropes are distinct is a primitive fact lacking any further metaphysical explanation (see, e.g., Keinänen and Hakkarinen 2014).

9.5 Properties: Determinables and determinates

Another distinction regarding properties is the traditional difference between determinables and determinates, that is, properties that stand in a distinctive specification relation; let us call it *determination* (see Wilson 2022). For example, color is a determinable, having red, blue, and other specific shades of color as determinates; shape is a determinable, having rectangular, oval, and other specific shapes (including many irregular ones) as determinates; mass is a determinable, having specific mass values as determinates.

The determination relation differs from other specification relations. Unlike the genus-species and conjunct-conjunction relations, in which the more specific property can be understood as a conjunction of the less specific property and some independent property or properties, a determinate cannot naturally be treated in conjunctive terms (whereas man can be conceived as the conjunction between animal and rational, red is not a conjunctive property of having color and some other property or properties as conjuncts). And, unlike the disjunct-disjunction relation, in which disjuncts may be dissimilar and compatible (as in the case of 'red or round'), the determinates of a determinable are both similar and incompatible (red and blue are similar in the sense that they are both colors, but nothing can be simultaneously and uniformly both red and blue).

It is extremely important to emphasize that the determinable-determinate relation should not be confused with the universal-instance relation. The latter is the relation between a universal property and its many instances: for example, the universal property "color" has countless instances of colored items. The first, by contrast, is the relation between a property and other more specific properties that are specifications of it: for example, the determinable "color" has red, green, yellow, and so on, as its determinates. The distinction between the two kinds of relations is clear in a classical ontology. In fact, the position Q and the momentum P of a particular classical object, say, this billiard ball, are instances of the universal properties "position" and "momentum," respectively. In turn, the position q_1 = "10 cm from the corner of the table" and the momentum p_1 = "20 g·cm/sec with respect to the table" (in both cases, the billiard table is taken as the reference frame) are determinates of the determinables Q = "position of the billiard ball" and P = "momentum of the billiard ball," respectively. Furthermore, those determinates define the classical state s_1 of the billiard ball, conceived as a classical system, at a given time $t_1 : s_1 = (q_1, p_1)$.

According to Jessica Wilson (2023), except for minimally specific determinables and maximally specific determinates, the characterization of a property as determinable or determinate is relative. In other words, determinables can be determined at different "levels of specificity" (see a similar idea under the name of "level of determination" in Calosi and Wilson 2019). For example, red is a determinate of color but a determinable of scarlet. Red, blue, and yellow are determinates of the determinable color at one level of specification; scarlet, navy, and emerald are determinates of the determinable color at another (more finely grained) level of specification.

The *principle of omnimode determination* is a principle that was generally accepted in modern philosophy. For example, it already appears in the works of Wolff: "*Apparet hinc, individuum esse ens omnimode determinatum*" ("Hence it appears that an individual is a completely determined being") (Wolff 1728, 152). It can also be found in Bernoulli's famous treatise on the calculus of probabilities: "*Sed quicquam in se et sua natura tale esse [viz. incertum et indeterminatum], non magis a nobis posse concipi, quam concipi potest, inde simul ab Auctore naturæ creatum esse et non creatum*" ("That anything is uncertain and indeterminate in itself and by its very nature is as inconceivable to us as it would be inconceivable for that thing both to have been created and not created by the Author of nature") (Bernoulli 1713, 227). It is also repeated several times by Kant in his lectures on metaphysics: "*Alles, was existiert, ist durchgängig determinirt*" ("Everything that exists is continuously determined") (1902, AA 18:332, 5710; AA 18:346, 5759; see also LM XXVIII 554) (I am grateful to Roberto Torretti for bringing these quotes to my attention). The central idea of the principle is that, in any object, all determinables are determinate: if the determinable "color" applies to an object, the object necessarily has some determinate color, say red, independently of its other determinable properties, and also independently of our knowledge about what that determinate color is. In other words, it is not possible for an object to have a determinable property that is not determinate: an object cannot be colored without being of some particular color, say, red, blue, white, and so on. Perhaps the intuitiveness of the principle of omnimode determination is the reason why it has not been as extensively discussed in the history of philosophy as other metaphysical principles, such as Leibniz's law.

9.6 Modality: Different types of possibility

Already in the Middle Ages, it was recognized that true propositions fall into two groups: "necessarily true" and "contingently true." In other words, there are two "modes" in which a proposition can be true (or false): the mode of necessity and the mode of contingency—hence the term '*modality*.' At present, the term is retained to mean "pertaining to possibility and related concepts" (see Kment 2021). In fact, there are other concepts besides "possible" involved in modal discourse: impossible, necessary, and contingent. However, all of them can be defined in terms of *possibility*: X is impossible ≡ not (X is possible); X is necessary ≡ not (non-X is possible); X is contingent ≡ not (X is necessary) ≡ (non-X is possible). Therefore, the attention here will be focused on the concept of possibility.

In addition to the above modal concepts, another notion is essential to understanding what the term 'possibility' means: *actuality*. In principle, something that is possible may be realized or not; it may actualize, that is, it may become actual, or not. In other words, the realm of possibility is partitioned into possible-and-actual and possible-and-non-actual. The relation between possible and actual is expressed by certain modal intuitions that any conception of possibility should incorporate: whatever is actual is possible, and whatever is necessary is actual.

The very first question is *what the predicate 'possible' applies to*. At the beginning of the twelfth century, Peter Abelard seems to be the first to realize, while studying syllogistics, that possibility may be said in two ways: (i) *de dicto* possibility when it applies to what is expressed by a non-modal sentence (e.g., possibly 'every A is B'), or (ii) *de re* possibility when it expresses the mode in which the predicate belongs to the subject and, therefore, is associated with the thing (e.g., 'every A is possibly B') (see Lagerlund 2021, Knuuttila 2021). In contemporary terms (see, e.g., Bunge 1977) it is said that *de dicto modality* applies to linguistic items, in particular, to propositions and, thus, it is studied by *logics*, whereas *de re modality* applies to ontic items (objects, properties, states of affairs), so it belongs to *ontology*.

Related to, but different from, the distinction *de dicto-de re*, another classification expresses the different forms of modality: semantic, subjective, and objective modality. *Semantic modality* is relative to a logical system S and is defined in terms of derivability in the context of S: a proposition p is semantically possible relative to a logical system S iff $\neg p$ is not derivable from S, and it is semantically necessary relative to S iff it is derivable from S (Bunge 1977). *Subjective or epistemic modality* is relative to an agent and the evidence he possesses, and is defined in terms of evidence and ideal reasoning: a proposition p is subjectively necessary for an agent A iff the evidence E_A that A possesses and ideal reasoning are sufficient to rule out $\neg p$ (Kment 2021). *Objective modality* is an ontological category, which refers to how reality may be and, then, is closely related to modality *de re*: modal concepts apply to ontic items (objects, properties, states of affairs) independently of knowledge and agents, because objective modality is a mode of being of reality in itself. In general, the *de dicto* use of modality is linked with semantic and subjective modality. Nevertheless, it may be the case that the *de dicto* use of modality is the linguistic expression of *de re* modality: this happens when a proposition p is *de dicto* possible because the fact semantically referred to by p is objective-*de re* possible.

Despite the short life of logical positivism, its great influence spread throughout the entire twentieth century. Its rejection of metaphysics in favor of logic led most philosophers of science belonging to the analytic tradition to avoid ontological discussions by translating them into linguistic and logical issues. It is not surprising, then, that this general position showed a profound skepticism toward any appeal to objective modality, manifested by the *de re* use of modal concepts. Perhaps the most conspicuous enemy of objective modality was Willard V. O. Quine (1960). According to him, if modality were to make any sense, it should be understood in terms of the concept of analyticity, so it should be identified with semantic modality; but since objective modality cannot be understood in terms of analyticity, it cannot be understood at all.

Many authors reject objective possibilities on the grounds that there is no non-trivial identity criterion for non-actual items; this is again the case of Quine (1969) with his slogan "no entity without identity." In order to meet this challenge, different strategies have been developed for the purpose of identifying possibilities: possible worlds, subsistence as different from existence, etc. These discussions consider examples such as the possibility of the existence of Vulcan as a planet closer to the Sun than Mercury, or the possibility that Julius Caesar had a sixth finger on his right hand. But these are not situations that are relevant to the treatment of the problem of possibility in physics: in this context, there is no need for an all-encompassing theory of possibility that can be applied to any kind of possible fact, even those outside the field of physics as those just mentioned—under the assumption that such an all-encompassing theory makes sense. The notion of objective possibility relevant here is one that refers to the facts described by physics: it is physical theories that define the space of objective possibility. In what follows, I will confine the discussion to this type of possibility, which can be called *physical possibility*: a *de re* objective possibility, embodied in the laws of physics.

An example of the rejection of metaphysics in favor of logic is the case of Richard Montague (1960). According to this author, p is physically necessary iff p is deducible from a certain class of physical laws L. This view, which attempts to elucidate the concept of physical necessity in logical terms, conflates necessity with lawfulness. However, regardless of how natural laws are philosophically conceived, this is not the way in which the relation between laws and modality is understood in the practice of physics: laws do not fix what necessarily happens or what actually happens, but only what may happen. What actually happens depends not exclusively on the laws but also on certain particular circumstances of the phenomenon in question that do not depend on the laws. The paradigmatic case is that of dynamical laws: they only define the evolutions that a system may describe; but which of all the possible evolutions will become actual depends on the initial conditions and the boundary conditions of the particular situation. In other words, a physical law distinguishes the realm of possibility from the realm of impossibility. For example, special relativity makes it impossible for a body to have a velocity greater than the velocity of light. Within the realm of possibility, physical laws fix some necessary facts, for example, those referring to certain elemental properties such as the electric charge of the electron or the mass of certain subatomic particles; but this does not imply that lawfulness must be identified with necessity.

It was only in the last decades of the twentieth century that the philosophy of science began to get free from the strong ties imposed by logical positivism and its descendants. In particular, philosophers of physics started to examine the problems related to the ontological reference of physical theories. When this task focused on modality, the need to recover an objective meaning of modality came to the fore.

9.7 Modality: Actualism versus possibilism

The problem of the nature of objective possibility is as old as philosophy itself. It already appears in Aristotle's discussion on future contingents. The question was how to assign a truth value today to statements about the future, such as 'There will be a sea-battle tomorrow' and 'There will not be a sea-battle tomorrow': according to Aristotle, whereas the past is fixed, the future is open to possible alternatives. In the generation after Aristotle, the Megarian logician Diodorus Cronus devised an argument to show that there can be no future contingency at all: the future is as fixed as the past, so there are no real possibilities (see Øhrstrøm and Hasle 2020). This ancient discussion brings to light two different conceptions of modality.

Actualism is the position according to which everything that exists, everything that in any sense can be said to be, is actual. In other words, actualism denies that there is any kind of being beyond actual existence: to be is to exist, and to exist is to be actual (see Menzel 2021). This was precisely the position of Diodorus Cronus; in Cicero's words, "Diodorus defines the possible as that which either is or will be" (quoted in Kneale and Kneale 1962, 117). According to Diodorus, what actually happens is all that *may* happen. Therefore, any claim about possibility must be reduced to a claim about actuality. For Arthur Prior, Diodorus "seems to have been an ancient Greek W. V. Quine, who regarded the Aristotelian logic of possibility and necessity with some skepticism, but offered nevertheless some «harmless» senses that might be attached to modal words" (Prior 1967, 16). This actualist view survived through the centuries until contemporary times; for instance, for Bertrand Russell, 'possible' means 'sometimes,' 'necessary' means 'always,' and 'impossible' means 'never' (Russell 1919, 165). In other words, for actualists, the categories of actuality and possibility are coextensive: possibilities that never happen do not exist; thus, possibility must be reduced to actuality (for a criticism of the reductive accounts of modality, see Bueno and Shalkowski 2015).

Possibilism is the denial of actualism. It is the thesis that there are ontic items (things, properties, events, facts) that are not actual. As a consequence, possibility is an irreducible ontological category. This conception is exemplified in Philo of Megara, for whom possibility is an intrinsic capability of propositions of being or not being true. In Boethius' words, for Philo, "possible is that which is capable of being true by the proposition's own nature" (Boethius 1880, 2, II 234). In his dispute with Diodorus, the Stoic philosopher Chrysippus also adopts a possibilist position when he maintains that "even though something is not true and may never be true, it may nevertheless be possible" (quoted in Gould 1970, 77–78); so, he defines possible as "that which is not prevented by anything from happening, even if it does not happen" (quoted in Bunge 1977, 172). These quotes show that, for possibilists, possibility is an irreducible ontological category: reality unfolds in two domains, the domain of possibility and the domain of actuality.

The debate actualism versus possibilism is still alive in present-day metaphysics, but in general, it is centered on the possible existence of objects (see Yagisawa 2023): according to actualists, all possible objects are actual, whereas possibilists admit the existence of *possibilia*, that is, non-actual possible objects, such as the planet Vulcan and Sherlock Holmes. However, when the interest is focused on modality in physics, particular objects play no role: physics does not deal with this or that, say, electron, but with electrons in general. In turn, in the contemporary debate, *possibilia* are conceived as things that are not actual but could have been actual: to be a *possibile* is to be unactualized (see Menzel 2024). Here I will not take this stance because, in physics, it is necessary to be able to describe possible states of affairs that become actual when certain conditions hold. Therefore, actuality will be understood as a subclass of possibility: for possibilism, as understood here, there are possibilities that may never become actual, but there are also possibilities that enter the realm of actuality while remaining possible. In this sense, the notion of possibility is closer to the Aristotelian idea of *potentia* than to the way possibility is treated in most contemporary metaphysical debates.

When the two conceptions of the relation between possibility and actuality are clearly distinguished, it is not difficult to perceive again the influence of logical positivism on twentieth-century philosophical thought. According to the verificationist theory of meaning, a sentence is meaningful only when its truth or falsity can, at least in principle, be empirically determined. But it is quite clear that if a modal sentence refers to a possible state of affairs, its truth value cannot be determined by empirical means. This explains the rejection of possibilist possibility in traditional philosophy of science, which was strongly biased by logical positivism.

This actualist conception of possibility can be easily found in the treatment of determinism during the twentieth century. The basic idea of determinism can be expressed in William James's words: "The future has no ambiguous possibilities hidden in his womb" (James 1897, 150). However, as soon as actualist philosophers attempt to elucidate the concept, they fall into positions that oscillate between emptiness and triviality. On the one hand, on the basis of the verificationist theory of meaning, the notion of determinism is meaningless if it is not defined in terms of an empirical criterion that unequivocally establishes whether a system is determinist or indeterminist. For this reason, the notion of determinism has usually been closely linked to that of predictability. For instance, Karl Popper proposes a "scientific" concept of determinism as "the doctrine that the state of any closed physical system at any given future instant of time can be predicted, even from within the system, with any specified degree of precision" (Popper 1982, 36). This claim, which conflates epistemology with ontology, renders the notion of determinism empty because the state of any physical system becomes virtually unpredictable after a sufficiently long time and under sufficiently unstable conditions. By contrast, in an attempt to purify the notion of determinism from epistemic components, Bertrand Russell states that "[a] system is said 'deterministic' when, given certain data $e_1, e_2, \ldots, e_n$ at times $t_1, t_2, \ldots, t_n$ respectively, concerning this system, if E_t is the state of the system at any time t, there is a functional relation of the form $E_t = f(e_1, t_1, e_2, t_2, \ldots, e_n, t_n)$" (Russell 1953, 398). However, as the author himself acknowledges, this definition trivializes the notion of determinism, since it is always possible to define a function that "passes" through a finite number of points. To avoid this conclusion, Russell needs to add further requirements, such as the simplicity of the function f or the condition that f does not explicitly depend on t, which require independent justifications.

Perhaps the most recent manifestation of actualism in the philosophy of science is Humeanism regarding laws of nature, according to which laws are just patterns in the patchwork of actual events. This approach is inspired by David Lewis's "Humean supervenience," that is, "the doctrine that all there is to the world is a vast mosaic of local matters of particular fact, just one little thing and then another" (Lewis 1986a, ix). According to Lewis (1994), among all the deductive systems that make true claims about the "*Humean mosaic*," the "*best system*" is the one that achieves the best balance between simplicity, strength, and fit. A natural law is a theorem in the best system (for criticisms, see, e.g., Maudlin 2007). It is quite clear that, in this picture, there is no room for non-actualized possibilities.

Despite the strong presence of actualism in the current philosophy of science (see, e.g., Hoefer 2019), the question is whether this doctrine agrees with the way in which practitioners of physics conceive physical laws. The answer is that a concept of objective possibility, irreducible to actuality, seems to be pervasively present in the discourse of physics. When physicists assert the impossibility of superluminary signals, they are not merely asserting their de facto non-existence: the physical idea is that it is impossible for such signals to exist in a reality described by relativistic laws. In turn, it cannot be guaranteed that everything that is physically possible will become actual at some point in time: for a possible event to actually occur, certain circumstances must be present, without which the event remains within the realm of possibility forever. For example, according to statistical mechanics, it is possible for all the molecules of a gas confined in a container to spontaneously agglomerate at one of its corners, although it may be the case that this never occurs in the entire history of the universe. The transition from possibility to actuality, from the perspective of physics, has much more in common with the Aristotelian categories of potency and act than it shares with most of the present-day philosophy of science and its actualist interpretation of possibility.

9.8 Possibility and probability

One of the peculiarities of quantum mechanics is that the quantum state does not specify definite values of the observables of a system, but rather assigns probabilities to those definite values. Perhaps for this reason, in the field of the interpretation of quantum mechanics, the interest has focused mainly on understanding the concept of probability, whereas the concept of possibility has been much less addressed. However, if probability is conceived, in some sense, as a measure of possibility, it is easy to see that there is a correspondence between the different interpretations of the concept of probability and those of the concept of possibility.

According to Ian Hacking's well-documented study on the emergence of probability (Hacking 1975), our modern conception of probability has had a dual nature since its birth in the mid-seventeenth century: in its *epistemic* sense, it measures the degree to which evidence supports a given hypothesis; in its *ontological* sense, it describes indeterministic regularities exhibited in nature. In the twentieth century, two widely recognized analyses of epistemic probability have been proposed (see Hájek 2023). According to the *logical* interpretation—the tradition of John Keynes (1921) and Rudolf Carnap (1950)—probabilities can be determined by the examination of the space of possibilities and express the degree of support that a piece of evidence confers to a given hypothesis. According to the *subjective* interpretation—represented by Frank Ramsey (1931) and Bruno de Finetti (1937)—probabilities measure degrees of confidence or belief of suitable agents. In turn, ontological probability appeared originally under two different forms. The *frequentist* interpretation—for example, Hans Reichenbach (1949) and Ludwig von Mises (1957)—identifies probabilities with relative frequencies in infinite sequences. In turn, since the 1950s, the idea of interpreting ontological probabilities as *propensities*—Karl Popper (1959)—began to be considered by some philosophers as an alternative to the frequentist interpretation. Recently, the neo-Humean so-called *best-system* interpretation of probability shows some similarities with the frequentist approach but tries to avoid its main difficulties: inspired by Lewis's conception of laws of nature (Lewis 1994), probabilities are defined by those probabilistic laws that turn out to be the best account of the Humean mosaic according to simplicity, strength, and fit.

When these interpretations of probability are compared with the different views on possibility as introduced in the previous subsections, it is not difficult to see a conceptual match. On the one hand, *epistemic* probability can be conceived as the counterpart of *de dicto* possibility. In this epistemic context, the *logical* interpretation corresponds to *semantic* possibility, whereas the *subjective* interpretation corresponds to *subjective* possibility. On the other hand, *ontological* probability is the counterpart of *de re* and *objective* possibility. In this context, the *frequentist* and the *best-system* interpretations implicitly presuppose an *actualist* view of possibility, whereas the *propensivist* interpretation relies on a *possibilist* conception of possibility.

The strong influence of the positivists' rejection of metaphysics, expressed by an excessively strict conception of meaning, led to a non-possibilist conception not only of possibility but also of probability. This position could be tenable to a certain extent if physical theories were still formulated within a deterministic framework with reducible probabilities. But this is not the case since the advent of quantum physics. In Section 10.8 of the next chapter, I will show that, despite the inescapable presence of probability in the field of quantum mechanics, many approaches to quantum mechanics are still based on that non-possibilist view.

9.9 Metaphysical indeterminacy

In Latin, the word '*determinatio*' means boundary, end, limit. This term enters philosophy as 'determinacy' to designate the feature of what is fully defined and, therefore, can be characterized in an unequivocal way; 'indeterminacy' refers to the absence of determinacy.

The influence of logical positivism is also noticeable in this regard. In general, the traditional philosophy of the twentieth century conceived of indeterminacy as a linguistic or epistemic concept: language is not precise enough to capture reality, or human knowledge is inherently limited. From this widespread perspective, the very idea of metaphysical indeterminacy, as indeterminacy in the world, is incoherent or unintelligible (see, e.g., Russell 1923, Dummett 1975, Evans 1978, Salmon 1981, Lewis 1986b). It was only in recent decades that the idea of *metaphysical indeterminacy* began to be rehabilitated, with several authors arguing for its philosophical relevance (see Torza 2023 for an overview).

The approach to metaphysical indeterminacy that first appeared in recent philosophical debates was supervaluationism, according to which metaphysical indeterminacy consists in a fundamental kind of unsettledness in the world: "When *p* is metaphysically indeterminate, there are two possible (exhaustive, exclusive) states of affairs—the state of affairs that *p* and the state of affairs that not-*p*—and it is simply unsettled which in fact obtains. No further explication is possible or needed" (Barnes and Williams 2011, 113–114; see also Akiba 2004, Williams 2008, Barnes 2010). More recently, other approaches to metaphysical indeterminacy have been proposed. Here I will focus on the determinable-based approach developed by Jessica Wilson

(2013, 2016). According to this perspective, metaphysical indeterminacy involves states of affairs in which an object has a determinable property, but *not a unique* determinate of that determinable. The difference between the supervaluationist and the determinable-based approaches to metaphysical indeterminacy is clearly explained by Claudio Calosi and Wilson (2019, 2601): "On a metaphysical supervaluationist approach, MI [metaphysical indeterminacy] involves its being indeterminate which determinate (precise) state of affairs obtains; on a determinable-based approach, MI involves its being determinate (or just plain true) that an indeterminate (imprecise) state of affairs obtains."

Different situations have been presented as cases of metaphysical indeterminacy—for example, objects with fuzzy boundaries like clouds, the openness of the future, the phenomenon of iridescence, a borderline-bald man, among others. Nevertheless, there is no consensus about what constitutes a legitimate case of metaphysical indeterminacy. In fact, it is always possible to argue that those cases can be interpreted as involving merely linguistic or epistemic vagueness. By contrast, quantum mechanics seems to provide a source of examples of genuine metaphysical indeterminacy. This idea has led some authors to speak of "quantum indeterminacy," which will be analyzed in Section 10.10 of the following chapter.

10
The Structure of the Modal-Hamiltonian Ontology

In Chapter 4, the MHI was introduced in terms of interpretative postulates that define the ontological reference of each physical term of quantum mechanics in its standard formulation. On the basis of that presentation, an ontology of properties, framed in the algebraic formalism of quantum mechanics, was proposed. The aim of this chapter is to set out in detail the structure of the ontology proposed by the MHI, now with the technical metaphysical resources developed in the previous chapter. In particular, the nature of the properties that constitute the fundamental ontological items will be discussed, and it will be argued that quantum systems are not objects at all, but mere collections of properties that do not retain their identity either synchronically or diachronically. More precisely, quantum systems turn out to be non-objectual bundles of properties.

10.1 Different proposals for the quantum ontology

At present, hardly anyone denies the underdetermination of metaphysics by physics: quantum mechanics is compatible with distinct metaphysical "packages." In other words, the quantum domain has been conceived as structured on the basis of very different fundamental ontological categories.

According to Bohmian Mechanics (see Dürr, Goldstein, and Zanghì 2013), the universe is a configuration of particles in precise positions relative to each other. Therefore, elemental quantum systems are *individual particles* with synchronic and diachronic individuality: they are ontologically—not epistemically—discernible by their relative positions in the configuration, and they can be ontologically—not epistemically—reidentified over time by the continuous trajectories traced by their motion. The difference with respect to the classical case is that the dynamics of quantum individuals is described by the law of motion known as "guiding equation," which makes the evolution of each particle to depend on the position of all the others through the wavefunction. Proponents of Bohmian Mechanics believe that their theory provides the smallest deviation from the ontology of classical mechanics that is necessary to accommodate quantum phenomena (see, e.g., Esfeld 2019).

The Ghirardi-Rimini-Weber (GRW) Spontaneous Collapse Theory (Ghirardi, Rimini, and Weber 1986) states that, in order to solve the quantum measurement problem, the dynamical postulate of SQM must be modified by adding stochastic and

The Modal-Hamiltonian Interpretation of Quantum Mechanics. Olimpia Lombardi, Oxford University Press.
© Olimpia Lombardi (2026). DOI: 10.1093/9780198951407.003.0010

nonlinear terms to the Schrödinger equation. The resulting new equation describes the spontaneous jumps undergone by the wavefunction in configuration space at random times. This theory has been ontologically interpreted in terms of two different pictures: *the matter-density ontology* (GRWm) and the *flash ontology* (GRWf). In GRWm, the wavefunction describes a continuous matter-density field (see Ghirardi, Grassi, and Benatti 1995), which varies at different points of the three-dimensional physical space and changes in time. This means that there is a single fundamental object in the universe, a *stuff* that "fills" the entire space. The definite outcomes of measurements are the result of the spontaneous contraction of the matter-density field at certain points or regions of space: the collapse of the wavefunction represents such a contraction. Discrete individual objects are mere appearances, which arise in certain regions of the physical space where the matter density is higher. According to GRWf, by contrast, each jump of the wavefunction in configuration space represents an *event* occurring at a point of physical space. These point-events are known as "flashes" (Tumulka 2006). The time evolution of the wavefunction in configuration space represents the probability of the occurrence of future flashes, given an initial distribution of flashes. Therefore, the physical space is not filled with objects, neither individuals nor stuff, but with a sparse distribution of discrete events. Objects are, from this viewpoint, nothing but clusters of a huge number of flashes.

These cases are examples that show how the literature focusing specifically on the ontology of quantum mechanics has appealed to different fundamental ontological categories in order to design the structure of the quantum domain: *individuals and properties*—strictly speaking, one property—in Bohmian Mechanics, *stuff* in the matter-density ontology, *events* in the flash ontology. However, the possibility of an ontology in which properties are the only fundamental items has scarcely been considered. As far as I know, the idea of bundles of properties has appeared only a few times in the literature on quantum physics. It has been proposed as the ontology for quantum field theory in its algebraic version by Meinard Kuhlmann (2010), barely suggested by Cord Friebe (2014) in his objections to GianCarlo Ghirardi's criterion for entanglement of indistinguishable particles, and more recently used by Andrea Oldofredi (2021) to provide a metaphysical interpretation of Relational Quantum Mechanics. In this scarcity lies the novelty of the idea of an ontology of properties as that proposed by the MHI.

10.2 Type-properties and case-properties

As pointed out in Section 1.1 of Chapter 1, although a formalism does not fix its interpretation, formalisms are not ontologically neutral: by appealing to some examples, I argued that different formal systems, even if mathematically equivalent, may suggest different ontologies. Another example, which is relevant to the present proposal, is the comparison between the Hilbert space formalism and the algebraic formalism, traditionally considered mathematically equivalent. The Hilbert space formalism suggests an ontology of objects and properties: objects labeled by their states represented by vectors of a Hilbert space—states that, in quantum mechanics, are not mere collections of the properties of the system as in classical mechanics—and properties represented by observables "applied" to states—represented by operators acting onto vectors. In the algebraic approach, by contrast, the fundamental elements of the theory are observables; states are logically posterior since they are represented by functionals over the space of observables.

Let us recall that the MHI adopts an algebraic approach as its formal starting point. Following the Wittgensteinian idea that the structure of language mirrors the structure of reality, the structure of the quantum reality can be designed by establishing the ontological counterpart of the algebraic formalism, that is, by providing an ontological interpretation for each physical/mathematical term of that formalism. Thus, if the logical priority of observables over states is transferred to the ontological domain, observables turn out to embody the representation of the fundamental items of the ontology, which belong to the category of property. In other words, whereas a realm of objects and properties seems to be the natural reference of the theory in the Hilbert space formalism, the algebraic approach favors the adoption of an ontology of properties, in which the ontological category of object is not fundamental. Now it is necessary to discriminate between different specifications of the category of property.

The term "*observable*" is used in SQM to refer to certain quantifiable magnitudes of physical relevance, which are mathematically represented by self-adjoint operators. Ontologically, they correspond to items belonging to the category of property. In turn, since in general observables have different values, their ontological counterparts are *determinable properties*, which here will be referred to as "*type-properties*." But it is also necessary to decide about the very nature of properties: here they are conceived as universals. Therefore, it is necessary to distinguish between *universal type-properties* (*U-type-properties*) and *instances of universal type-properties* (*I-type-properties*). The ontological counterparts of general physical magnitudes are *U-type-properties*, and of observables are *I-type-properties*. I will symbolize a U-type-property as $\langle A\rangle$, and its I-type-properties as $[A]$. An example of U-type-property is spin in the x-direction $\langle S_x\rangle$, which can be instantiated as the spin in the x-direction $[S_x]$ in system $\mathcal{S}$. Let us stress that, although this talk seems to refer to an ontology of objects, below I will define the concept of quantum system as a non-objectual ontological item.

On this basis, the first ontological postulate **OP1** can be introduced (da Costa, Lombardi, and Lastiri 2013; see also da Costa and Lombardi 2014):

> **OP1**: There exist *universal type-properties, U-type-properties*, symbolized as $\langle A\rangle$, $\langle B\rangle$, $\langle C\rangle$, etc., each one of which with countless instances. An *instance of a universal type-property* $\langle A\rangle$, *I-type-property*, is symbolized as $[A]$.

In physics, each I-type-property $[A]$ of the U-type-property $\langle A\rangle$ is represented by an observable A belonging to a certain space of observables $\mathcal{O}$ ($A \in \mathcal{O}$). Mathematically, each observable is represented by a self-adjoint operator A belonging to a space of self-adjoint operators $\mathcal{O}$ ($A \in \mathcal{O} = \mathcal{B}(\mathcal{H})$). In what follows, the difference between the physical and the mathematical meaning of the terms will be clear from the context.

Since a physical observable is a quantifiable magnitude, it has different *possible values*, which are mathematically represented by the eigenvalues of the corresponding self-adjoint operator. Their ontological counterparts are *determinate properties*, which here will be referred to as "*case-properties*" of the corresponding type-property. At this level, it is also necessary to distinguish between *universal case-properties* (*U-case-properties*) and *instances of universal case-properties* (*I-case-properties*): given an I-type-property $[A]$ of a U-type-property $\langle A \rangle$, its I-case-properties will be symbolized as $[a_i]$, each one of which is an instance of the respective U-case-property symbolized as $\langle a_i \rangle$. Following with the above example, we can talk of the I-case-properties $[\uparrow_x]$ and $[\downarrow_x]$ of the spin in the x-direction $[S_x]$ in system $\mathcal{S}$, where $[S_x]$ is an I-type-property of the U-type-property spin in the x-direction $\langle S_x \rangle$, and $[\uparrow_x]$ and $[\downarrow_x]$ are instances of the U-case-properties $\langle \uparrow_x \rangle$ and $\langle \downarrow_x \rangle$, respectively.

On this basis, the second ontological postulate **OP2** can be introduced (da Costa, Lombardi, and Lastiri 2013):

> **OP2**: Any I-type-property $[A]$ of a U-type-property $\langle A \rangle$ has *I-case-properties*, symbolized as $[a_i]$, each one of which is an instance of the respective *U-case-property* symbolized as $\langle a_i \rangle$.

In physics, the I-case-properties $[a_i]$ are represented by the possible values a_i of the observable A. Mathematically, the possible values a_i are represented by the eigenvalues of the self-adjoint operator $A = \sum_i a_i \Pi_i$, where Π_i is the eigenprojector corresponding to a_i, such that $\Pi_i = |a_i\rangle\langle a_i|$ in the non-degenerate case.

In this ontological proposal, type-properties and case-properties are conceived as universals with their corresponding instances. A legitimate question is why not to appeal to tropes. The answer is strongly linked to the problem of indistinguishability. As explained in Section 9.4 of the previous chapter, although tropes may be absolutely similar, they are neither absolutely indistinguishable nor only numerically different, precisely because they can be individuated and distinguished by their space-time position (redness here and now), by the object to which they apply (red of this individual balloon), or because the distinction between them is taken to be primitive. Since the elemental items of the quantum ontology should be adequate to provide the foundations of quantum indistinguishability, an ontology of tropes would face the same difficulties as an ontology of objects, since in both cases they are distinguishable items. By contrast, the instances of a universal are absolutely indistinguishable because they are manifestations of the same property: the roundness of a billiard ball and the roundness of a water drop are both instances of the universal roundness, and trying to distinguish them as different properties makes no sense. For this reason, an ontological approach based on universals and their instances paves the way toward an ontologically reasonable answer to the problem of indistinguishability, as will be discussed in Section 11.4 of the next chapter.

As remarked in Section 9.5 of the previous chapter, according to the orthodox view in metaphysics, a determinable property can have different levels of specification (Wilson 2022). However, this idea does not reflect how observables are conceived in physics, where it is commonly accepted that self-adjoint operators mathematically represent what in philosophy are identified as determinable properties. Thus, the mathematical fact that a self-adjoint operator is defined by its eigenvalues with their corresponding eigenprojectors formally expresses the fact that a determinable property is univocally defined by its determinate properties. This means that, in quantum mechanics, physically determinable properties do not have different levels of specification. Samuel Fletcher and David Taylor (2024) take a different position by assuming that a determinable property is represented, not by a self-adjoint operator, but by the domain of the operator; since different operators (with their different collections of eigenprojectors and eigenvalues) can have the same domain, they can represent the same determinable property. On this basis, the authors need to reformulate the notion of compatibility between properties and the well-known eigenstate-eigenvalue link by taking into account a level of specification. I consider that, by rejecting the usual link between property and operator, this position introduces unnecessary complications regarding the concept of compatibility and the eigenstate-eigenvalue link. Moreover, it is forced to deny that different properties/observables correspond to different experimental questions with their specific answers, an assumption that is very common in physics. In addition, it needs to accept that a single determinable property can have determinate properties with different physical units at different levels of specification. For example, the determinates of the determinable energy have energy units in one level of specification, defined by the Hamiltonian's spectrum, and do not have energy units in another level of specification, defined as "ground state" and "non-ground state." In short, Fletcher and Taylor's position, while agreeing with the orthodoxy in metaphysics, conflicts with many common assumptions in the physical practice of quantum mechanics. For this reason, in the proposed MHI metaphysical picture, I prefer to agree with those assumptions by accepting that, in the quantum ontology, determinable properties, represented by self-adjoint operators, do not have different levels of specification.

10.3 Possible properties and actual properties

When the postulates of the MHI were introduced in Chapter 4, "possibility" and "actuality" were predicated to facts: what is possible or actual is a fact. However, in an ontology of properties, facts are not conceived in terms of the assignment of a certain property to an object, but are characterized as the occurrence of a property, in particular, of a case-property of a certain type-property. In particular, a possible/actual fact is the possible/actual occurrence of the corresponding case-property. Therefore, in order to make the presentation more fluent, hereafter I will dispense with the terms "the occurrence of" and speak of possible or actual case-properties, with the caveat that this does not imply conflating the categories of fact and property.

In Section 9.5 of the previous chapter, the importance of not confusing between the universal-instance relation and the determinable-determinate relation was emphasized by means of an example coming from classical mechanics: the position and the momentum of a particular billiard ball as determinable instances of the universal properties "position" and "momentum," respectively, having determinate values with respect to a billiard table. The difference between the two relations, universal-instance and determinable-determinate, is completely analogous in the quantum framework. The divergence between the classical and the quantum cases resides not in those relations, but in the role played by possibility and actualization. In the classical case, at a given time, a single possible determinate corresponds to each determinable and, as a consequence, such a determinate becomes actual. For example, at time t_1, the determinable Q = "position of the billiard ball" has a possible determinate, say, q_1 = "10 cm from corner of the table," and this is the actual determinate position of the billiard ball at t_1. In the quantum case, by contrast, at a given time, determinables may have many different possible determinates, among which at most only one becomes actual. For example, at a given time, the determinable S_z = "spin in direction z in the quantum system $\mathcal{S}$" has two possible determinates, s_{z1} = "spin up in direction z" and s_{z2} = "spin down in direction z": it may be the case that one of them becomes actual; but, according to the Kochen-Specker theorem, it may also be the case that none of them becomes actual.

The terms 'type-property' and 'case-property' introduced in the previous section stand for determinable and determinate, respectively; they are used to emphasize that the corresponding items belong to the ontological category of property. However, there is another reason why the terminology "type-property" and "case-property" has been preferred over the traditional terminology "determinable" and "determinate." The latter, coined in the light of a classical ontology, suggests that possibility is confined to the side of the determina*ble* (possible to be determinate), whereas the determinate is linked to actuality: according to the principle of omnimode determination, in a particular object, a single and actual determinate corresponds to each determinable. In the quantum case, on the contrary, the determinate properties are also possible properties, which may be actualized or not. The terminology 'type-property'/'case-property' does not confine possibility to the side of the determinable, and is not anchored to the actual character of the determinate that is proper to the realm of traditional ontology.

In order to express the fact that the case-properties of an instance of a universal type-property belong to the realm of possibility, and that one of them may enter the world of actuality, the third ontological postulate **OP3** can be introduced (da Costa, Lombardi, and Lastiri 2013):

OP3: Given any I-type-property $[A]$ of a U-type-property $\langle A \rangle$, among all its I-case-properties $[a_i]$, *no more than one* of them becomes actual. When one of the $[a_i]$, say $[a_k]$, becomes actual, it will be said that the I-type-property $[A]$ *actualizes* and acquires the *actual I-case-property* (*AI-case-property*) $[\boldsymbol{a_k}]$.

Although in their everyday language physicists are used to speaking of the value that an observable "effectively" acquires—generally as a result of a measurement—, quantum mechanics does not account for actualization. As a consequence, neither in the physical language nor in the mathematical language is there a representation of the AI-case-property $[\boldsymbol{a}_k]$, belonging to the realm of actuality, as different from the I-case-property $[a_k]$, belonging to the realm of possibility.

Notice that the clause "no more than one" in **OP3**, which corresponds to "exactly one" in the classical case, can be "zero" in the quantum case. In ontological words, an I-type-property $[A]$ of the U-type-property $\langle A \rangle$ may or may not actualize. This is a consequence of the Kochen-Specker theorem, which establishes one of the central differences between the quantum world and the classical world: in the quantum case, omnimode determination does not hold.

10.4 Systems as bundles of properties

The problem of the nature of objects remains one of the main topics of controversy in contemporary metaphysics: is an object a substratum supporting properties or a mere "bundle" of properties? (see, e.g., Robinson 2021, Rettler and Bailey 2022). The conception of an object as a substratum acting as a carrier of properties has pervaded the history of philosophy: it is present under different forms, for example, in Aristotle's "primary substance" and in Locke's "substance in general." However, following Hume's rejection of the notion of substance, many philosophers belonging to the empiricist tradition, such as Bertrand Russell, Alfred Ayer, and Nelson Goodman, have regarded the postulation of a characterless substratum as a metaphysical abuse, and have adopted some version of the *bundle theory*. According to this view, an object is nothing but a bundle of properties: properties have metaphysical priority over objects and are, therefore, the fundamental items of the ontology.

In the metaphysical literature, it has been argued that the difference between the substratum theory and the bundle theory is only verbal: in the bundle theory, an object is the result of a "compresence relation" that serves the same purpose as the substance in the traditional substratum-plus-properties theory. Therefore, the decision as to whether an object is a substratum supporting properties or simply a bundle of properties remains a matter of metaphysical taste (see Benovsky 2008). Whereas this may be the case in the classical domain, quantum mechanics calls this conclusion into question. For this reason, although in the present proposal quantum systems are ontologically characterized as bundles of properties, it is important to emphasize the peculiarity of this perspective.

The first point to consider is related to the difference between determinables and determinates, which was scarcely taken into account in the traditional works about the interpretation of quantum mechanics, but which has begun to be discussed in the field of the metaphysics of physics focused on quantum mechanics (see, e.g., Calosi and Wilson 2019, Calosi and Mariani 2021, Fletcher and Taylor 2024). According to the traditional versions of the bundle theory, an object is the bundle of certain determinate properties, under the implicit assumption that the object's determinable properties are all determinate. For example, a billiard ball is the confluence of a definite value of position, say here, a definite shape, say round, a definite color, say white, etc. So, the problem is to decide whether this object is a substratum in which determinate position, roundness, and whiteness inhere, or is the mere bundle of those determinate properties. But in the two cases the properties that compose the bundle are *actual* determinate properties. In the quantum case, by contrast, not all the determinable properties of a system are determinate; as a consequence, the system cannot be identified with a bundle of determinate properties. For this reason, in the present proposal, a quantum system is conceived as a bundle of determinables, that is, a *bundle of I-type-properties*, each one of them with its possible determinates, that is, its *I-case-properties*. This is the first reason why this interpretation of the nature of quantum systems cannot be assimilated to the notion of object, not even according to the traditional bundle theory. Later, in Section 10.9 of this chapter, it will be pointed out why dispensing with the non-actualized possibility is not an adequate option.

Secondly, in its traditional versions, the bundle theory is a theory about particular objects, according to which objects are composed of items of a different ontological category (namely, properties). In other words, the bundle theory is designed to account for objects without appealing to a substratum on which properties inhere (see, e.g., O'Leary-Hawthorne 1995, French and Bigaj 2024). To this end, some properties must be selected to play the role of the principle that endows objects with synchronic and diachronic identity. The proposed quantum bundle view, by contrast, completely dispenses with the ontological category of object: bundles of properties do not behave as objects in any way. This means that, when two bundle-systems are combined, the system resulting from the combination is also a bundle. And since bundles are not objects, there is no principle that preserves their identity through the combination: in the resulting system, the identity of the original systems is not preserved, precisely because they are not objects at all. In this sense quantum systems are conceived as *non-objectual bundles* of properties.

On the basis of the above discussion, the fourth ontological postulate **OP4** can be introduced (da Costa, Lombardi, and Lastiri 2013):

> **OP4**: A *bundle* $\mathfrak{B} = \{[A], [B], [C], \ldots\}$ is a collection of I-type-properties $[A], [B], [C], \ldots$ corresponding to the U-type-properties $\langle A \rangle, \langle B \rangle, \langle C \rangle, \ldots$.

Since in the algebraic formalism the physical concept of quantum system $\mathcal{S}$ is mathematically represented by a space of observables, the concept of bundle is the ontological correlate of the physical concept of *system*. This means that the bundle $\mathfrak{B}$ is physically represented by a system $\mathcal{S}$, which is identified with a space of observables $\mathcal{O}$. In turn, mathematically, the physical system $\mathcal{S}$ is represented by the space of operators $\mathcal{O}$, or by the Hilbert space $\mathcal{H}$ if $\mathcal{O} = \mathcal{B}(\mathcal{H})$.

Precisely due to their non-objectual nature, bundle-systems require a different kind of logic. An ontological picture in which properties are the fundamental items and do not constitute objects is not adequately captured by any formal theory whose elemental symbols are individual variables referring to classical objects. But, as remarked in Section 9.2 of the previous chapter, most systems of logic are designed to treat individual objects. A way out of this problem would be to develop a "logic of predicates" in the spirit of the "calculus of relations" proposed by Alfred Tarski (1941), in which individual constants and variables are absent. A different strategy is to apply the so-called *quasi-set theory* (see, e.g., Krause 1992, da Costa and Krause 1999): although it was originally devised to provide a formalism for indiscernible quantum objects, due to its formal character it can be adapted to deal with aggregates of items that do not belong to the ontological category of object but to that of property, so that bundles turn out to be represented by quasi-sets of properties (for a detailed presentation, see Holik, Jorge, Krause, and Lombardi 2022). I will come back to this point in Section 11.6 of the next chapter.

The idea of a realm without objects has been proposed by other authors for different reasons. A very active program in present-day philosophical literature is structural realism, which appeared in its epistemological version in the context of the problem of theory change (Worrall 1989). This view turned into an ontic version in the hands of James Ladyman (1998), who was driven by the ontological problems derived from quantum mechanics to the adoption of an eliminativist stance about objects. Ontic structural realism proposes, then, a reconceptualization of the ontology at the most basic metaphysical level, which leads to a shift from objects to structures. This kind of realism follows Ernst Cassirer (1938) in the claim that elementary particles are not objects but "points of intersection" of certain relations: physical objects are "reduced to mere 'nodes' of the structure, or 'intersections' of the relevant relations" (French 2006, 173). Ontic structural realists also stress the formal limitations to express their position, by pointing out "the descriptive inadequacies of modern logic and set theory which retains the classical framework of individual objects represented by variables and which are the subject of predication or membership respectively" (French and Ladyman 2003, 41). As noticed by Steven French (2020), the view of non-objectual bundles proposed by the MHI finds a natural resonance with ontic structural realism. In fact, the ontology of properties as introduced here can be conceived as an attempt to complete the picture suggested by ontic structural realism by identifying with precision the items belonging to the category of property that constitute the structure of the quantum realm.

10.5 Physical properties and the structure of quantum bundles

Up to this point, the quantum ontology is described as a property-only realm, in which systems are non-objectual bundles of properties. However, this characterization remains formal, as it says nothing about which properties effectively populate the quantum domain. In fact, the properties involved in the quantum ontology are not generic properties, but *physical* properties.

As Leslie Ballentine (1998) points out, although the formal structure of quantum mechanics is a necessary basis for the formulation of the theory, it has by itself very little physical content. When concrete physical problems are to be faced, the relevant observables of the system, endowed with a clear physical meaning, must be identified. These observables are closely related to space-time symmetry transformations. In particular, the space-time symmetries represented by the Galilei group identify the fundamental physical magnitudes of the theory (see Section 2.3 of Chapter 2): the energy $H = \hbar K_{\tau}$, the three momentum components $P_i = \hbar K_{p_i}$, the three angular-momentum components $J_i = \hbar K_{\theta_i}$, and the three boost components $G_i = \hbar K_{u_i}$. The rest of the physical magnitudes can be defined in terms of these basic ones: for example, the three position components are $Q_i = G_i/m$, the three orbital angular momentum components are $L_i = \varepsilon_{ijk} Q_j P_k$, and the three spin components are $S_i = J_i - L_i$. In summary, space-time symmetries endow the formalism of SQM with the content that makes it a well-specified physical theory. From the ontological viewpoint, they play the role of identifying the fundamental physical properties of the quantum realm.

Another important question is what holds properties together in the bundle-system. According to the traditional bundle theory, objects are composed of items of a different category: they are bundles of properties. However, not just any collection of properties forms a bundle that is an object: following Russell (1940), properties must hold a relation that binds them together in order to constitute an object. Consequently, objects are either bundles of coinstantiated universals in the universal-view or bundles of compresent tropes in the trope-view. Both *coinstantiation* and *compresence*, which tie properties together, are commonly regarded as primitive relations, serving the same purposes as substance or bare particulars in the traditional object-and-properties ontological view.

In the quantum case, not every type-property of the bundle-system has an actual case-property. For this reason, in the MHI ontological proposal, a quantum system is not a bundle of AI-case-properties, but a bundle of I-type-properties, that is, instances of universal type-properties, and the bundle itself is formally represented by the space of self-adjoint elements of an algebra of operators. Bundling, in this case, does not require a coinstantiation or compresence relation that plays the role of the substance in equipping the bundle with a feature that distinguishes it from other bundles and reidentifies it over time. Such relations are not necessary precisely because bundles are not objects.

Does this mean that a quantum bundle is a mere collection or aggregate of properties, with nothing relating them to each other? The answer to this question is negative. Bundle-systems have a well-defined *internal structure*, given by the specific relations that link the I-type-properties of the bundle to each other. Those ontological relations are formally represented by the *commutation relations* of the form $[A, B] = C$ between the observables of the algebra. In other words, what "binds" the I-type-properties in a bundle is a physically meaningful *structure of relations*, which is accurately represented by the mathematical structure of the space of self-adjoint operators included in the corresponding algebra. That structure of relations is neither vaguely defined nor arbitrary, but follows from the symmetry group of quantum mechanics. As a Lie group, the Galilei group is defined by the commutation relations between its generators, which, as explained above, represent the basic physical observables of the theory (again, see Section 2.3 of Chapter 2).

In summary, the symmetry group of quantum mechanics endows the structure of the quantum ontology with a precise physical reference. In this sense, the Galilei group plays a dual role. On the one hand, it defines the physical content of the properties that make up the bundle-systems. On the other hand, it establishes the relations between the properties of the bundle; these relations are what give cohesion to the bundle without turning it into an object as conceived by traditional metaphysics. From a more general perspective, this subsection shows that physics and metaphysics are not at odds at all. On the contrary, when metaphysical notions are elucidated from the outset, they can acquire physical content in a natural way.

10.6 Indistinguishability in an ontology of properties

In the discussions about the ontological commitments of quantum mechanics, several authors have pointed out the serious challenge posed by the theory to the notion of individual. Already in the 1960s, Heinz Post (1963) argued that elementary particles cannot be conceived as individuals, but must be viewed as "non-individuals." Paul Teller (1998) addressed the problem in terms of "haecceity" or "thisness": an unanalyzable feature that makes an object to be different from all others in some way that transcends all properties. According to this author, quantum mechanics provides good reasons for rejecting any aspect of quantum entities that might be thought to do the work of haecceity: "I suggest that belief in haecceities, if only tacit and unacknowledged, plays a crucial role in the felt puzzles about quantum statistics" (Teller 1998, 122). These positions led to the so-called *Received View* about quantum indistinguishability, according to which elementary particles cannot be regarded as individual objects (see French and Bigaj 2024). However, in general the Received View gives no metaphysical characterization of those items beyond their non-individuality: they are only negatively characterized. An exception is the case of Teller (1998), who proposes an account of quantum indistinguishability in terms of stuff: if quantum objects are stuff, permutation invariance follows naturally, because the permutation of two portions of the same stuff over two states does not give rise to a different complexion. Nevertheless, this account does not solve the challenge posed by the Kochen-Specker theorem, which affects any object, not only individuals, but also stuff (this problem will be analyzed in detail in Section 11.2 of the next chapter).

The quantum feature that has given rise to a deep skepticism about the notion of individual is the indistinguishability of the so-called *identical particles*, that is, particles of the same kind, which have the same state-independent properties.

Indistinguishability is introduced in the formalism of SQM as a restriction on the set of states: non-symmetric states are rendered inaccessible. French (1998) considers that such a restriction is consistent with the ontological view of particles as individuals: quantum statistics is recovered by regarding those states as possible but never actually realized. However, the cutout of non-symmetric states has an unavoidable ad hoc flavor in the context of the theory. In this sense, Michael Redhead and Teller (1992) reject the talk of individuals by claiming that the positing of inaccessible non-symmetric states amounts to the introduction of a surplus structure in the formalism.

Despite the fact that physicists are used to speak of "indistinguishable particles," quantum mechanics is not endowed with a formal relation for indistinguishability: there is no mathematical representation for this relation. Some authors developed logical systems specifically designed to deal with indistinguishability in a formal context: the semi-extensional quasi-sets theory, introduced by Décio Krause (1992; see also da Costa and Krause 1994, 1997, 1999, da Costa, French, and Krause 1992), and the intensional quasets theory, developed by Maria Luisa dalla Chiara and Giuliano Toraldo di Francia (1993, 1995), both describe collections having cardinality but not ordinality, that is, collections of objects to which the concept of individual of classical logic does not apply.

These are just a few of the many authors who have addressed the problem of quantum indistinguishability (this problem will be discussed in Section 11.4 of the next chapter). However, their positions make manifest the assumption underlying the usual way of conceiving the problem: quantum particles are perhaps non-individuals, but they still belong to the ontological category of objects, and indistinguishability is a relation between objects. In the context of the MHI ontology of properties, this assumption must be revised from the very beginning.

According to the MHI, properties are the elemental items of the ontology. Therefore, *indistinguishability is primarily a relation between properties*, which, not being objects, are not subject to any restriction that prevents them from being only numerically different. In particular, if a case-property is a "one-in-many" universal, its instances differ "*solo numero*" and, consequently, they are absolutely indistinguishable. On this basis, the fifth ontological postulate **OP5** can be introduced:

OP5: Any two I-case-properties $[a_1]$ and $[a_2]$ are *indistinguishable* when they are I-case-properties of the same U-case-property $\langle a \rangle$. Indistinguishability will be symbolized with the symbol '$\triangleq$': $[a_1] \triangleq [a_2]$.

According to this postulate, indistinguishability is primarily a relation between I-case-properties. Nevertheless, the relation is immediately inherited by I-type-properties (da Costa, Lombardi, and Lastiri 2013):

OP6: Any two I-type-properties $[A^1]$ and $[A^2]$ are *indistinguishable*, $[A^1] \triangleq [A^2]$, when they are I-type-properties of the same U-type-property $\langle A \rangle$ and their respective I-case-properties are indistinguishable: $[a_i^1] \triangleq [a_i^2]$.

Concerning the issue of "indistinguishable particles," the other concept that needs to be revised in the framework of the MHI is that of particle, in particular, of *elementary particle*. Mathematically speaking, the Galilei group has irreducible representations, in which the Casimir operators are multiples of the identity: $M = mI$, $W = wI$, and $S^2 = s(s+1)I$; as a consequence, each irreducible representation is labeled by a triplet (m, w, s). In quantum physics, it is assumed that each irreducible representation of the symmetry group of a theory corresponds to a kind of elementary particle. In the case of SQM, each irreducible representation of the Galilei group represents a kind of elemental bundle, characterized by mass m, internal energy w, and spin s. On this basis, the concept of *atomic bundle* can be defined as the ontological correlate of the physical concept of elementary particle (da Costa, Lombardi, and Lastiri 2013):

> **OP7**: A bundle $\mathfrak{B}_{at}$ is *atomic* if (i) no more than one I-type-property of each U-type-property belongs to it, (ii) the I-type-properties $[M_{at}]$ of the U-type-property mass $\langle M\rangle$, $[W_{at}]$ of the U-type-property internal energy $\langle W\rangle$, and $[S^2_{at}]$ of the U-type-property squared spin $\langle S^2\rangle$ always belong to it, and (iii) these I-type-properties are represented by the observables M_{at}, W_{at}, and S^2_{at}, which, in turn, are mathematically represented by operators that are multiples of the identity.

On the basis of the ontological postulate **OP7**, an atomic bundle is mathematically represented by the Casimir operators M_{at}, W_{at}, and S^2_{at}, which are multiples of the identity and, as a consequence, have a single eigenvalue: m_{at}, w_{at}, and $s_{at}(s_{at}+1)$, respectively. Ontologically, this means that each one of the I-type-properties $[M_{at}]$, $[W_{at}]$, and $[S^2_{at}]$ of the U-type-properties $\langle M\rangle$, $\langle W\rangle$, and $\langle S^2\rangle$ has a single I-case-property: $[m_{at}]$, $[w_{at}]$, and $[s_{at}(s_{at}+1)]$, respectively. According to the MHI actualization rule **AR″**, formulated in terms of the Casimir operators of the Galilei group (see Section 7.8 of Chapter 7), in the atomic bundle $\mathfrak{B}_{at}$ the I-type-properties $[M_{at}]$, $[W_{at}]$, and $[S^2_{at}]$ actualize, that is, they acquire an AI-case-property. But since all of them have a single I-case-property, this is the I-case-property that becomes actual. In other words, the following corollary can be inferred (da Costa, Lombardi, and Lastiri 2013):

> **Cor**: In any atomic bundle $\mathfrak{B}_{at}$, the I-type-properties $[M_{at}]$, $[W_{at}]$, and $[S^2_{at}]$ of the U-type-properties $\langle M\rangle$, $\langle W\rangle$, and $\langle S^2\rangle$ actualize, and they acquire the AI-case-properties $[m_{at}]$, $[w_{at}]$, and $[s_{at}(s_{at}+1)]$, respectively.

Once indistinguishability is defined as a relation that links primarily I-case-properties to each other and derivatively I-type-properties to each other, it can be extrapolated to be applied to atomic bundles: two atomic bundles "inherit" the indistinguishability between their respective instances (da Costa, Lombardi, and Lastiri 2013):

> **OP8**: Two atomic bundles $\mathfrak{B}^1_{at}$ and $\mathfrak{B}^2_{at}$ are *indistinguishable*, symbolized as $\mathfrak{B}^1_{at} \triangleq \mathfrak{B}^2_{at}$, when their respective I-type-properties are indistinguishable: $(\forall [A^1] \in \mathfrak{B}^1_{at}, \forall [A^2] \in \mathfrak{B}^2_{at}, [A^1] \triangleq [A^2]) \Rightarrow \mathfrak{B}^1_{at} \triangleq \mathfrak{B}^2_{at}$.

From OP6 and OP8, it is easy to see that the corresponding AI-case-properties of indistinguishable atomic bundles are also indistinguishable: $\mathfrak{B}^1_{at} \triangleq \mathfrak{B}^2_{at} \Rightarrow [m^1_{at}] \stackrel{\wedge}{=} [m^2_{at}]$, $[w^1_{at}] \stackrel{\wedge}{=} [w^2_{at}]$, and $[s^1_{at}(s^1_{at}+1)] \stackrel{\wedge}{=} [s^2_{at}(s^2_{at}+1)]$. Consequently, indistinguishable atomic bundles are only *numerically different*. The ontological indistinguishability between two atomic bundles $\mathfrak{B}^1_{at}$ and $\mathfrak{B}^2_{at}$ expresses the physical indistinguishability between the corresponding "elementary particles," and agrees with the physical fact that all the elementary particles of the same kind—that is, with the same values of mass, internal energy, and spin—are indistinguishable.

Summing up, in the ontology of properties proposed by the MHI, indistinguishability is not a relation between particles, because there are no objects in the ontology. Primarily, indistinguishability is a relation between I-case-properties, which is inherited by I-type-properties and then transferred to relate atomic bundles. This view, although heterodox, provides a natural solution to the problem of indistinguishability, as will be explained in Section 11.4 of the next chapter.

10.7 Combination of bundles

Up to this point, I have considered bundle-systems independently of each other. But bundle systems also combine to constitute new bundles. Therefore, the combination of bundles must be characterized in ontological terms (da Costa, Lombardi, and Lastiri 2013):

OP9: Given two bundles $\mathfrak{B}^1$ and $\mathfrak{B}^2$, physically represented by the spaces of observables $\mathcal{O}^1$ and $\mathcal{O}^2$ respectively, and mathematically represented by the corresponding spaces $\mathcal{O}^1$ and $\mathcal{O}^2$ of self-adjoint operators, the *combination* between bundles is defined as an operation $\mathfrak{B}^1 \diamond \mathfrak{B}^2 = \mathfrak{B}^{comb}$, where $\mathfrak{B}^{comb}$ is a new bundle physically represented by the space of observables $\mathcal{O}^{comb}$ such that it is mathematically represented by the corresponding space of self-adjoint operators $\mathcal{O}^{comb} = \mathcal{O}^1 \otimes \mathcal{O}^2$.

It is important to recall again that, in the context of the MHI ontology of properties, bundles are not objects and, as a consequence, they do not have a principle that preserves their identity through change. As a consequence, any combination of bundles of properties is a new bundle of properties, in which the identity of the components is not retained, precisely because they are not objects.

The practice of quantum mechanics usually assumes certain relations between the observables of two systems and the observables of the system resulting from their combination. For example, the observable A^1 of a system $\mathcal{S}^1$ and the observable $A = A^1 \otimes I^2$ of the combined system $\mathcal{S} = \mathcal{S}^1 \cup \mathcal{S}^2$ represent the same property. This assumption can be expressed in the ontological language as follows (da Costa, Lombardi, and Lastiri 2013):

OP10: Given two bundles $\mathfrak{B}^1$ and $\mathfrak{B}^2$, which combine in the bundle $\mathfrak{B}^{comb} = \mathfrak{B}^1 \diamond \mathfrak{B}^2$, the I-type property $[A^1] \in \mathfrak{B}^1$ of the U-type-property $\langle A \rangle$, represented by the operator $A^1 \in \mathcal{O}^1$, is the same I-type property as $[A^1 \otimes I^2] \in \mathfrak{B}^{comb}$ of the same U-type-property $\langle A \rangle$, represented by the operator $A^1 \otimes I^2 \in \mathcal{O}^1 \otimes \mathcal{O}^2$. In short, $[A^1] = [A^1 \otimes I^2]$.

It is clear that, when it is said that $[A^1] = [A^1 \otimes I^2]$, the symbol "=" represents *logical identity*: "a=b" means that the terms "a" and "b" refer to the same item in the ontological realm.

Bundles can combine with each other in different ways, depending on whether they interact or not (da Costa, Lombardi, and Lastiri 2013):

OP11: Let us consider two bundles $\mathfrak{B}^1$ and $\mathfrak{B}^2$, which combine in the bundle $\mathfrak{B}^{comb} = \mathfrak{B}^1 \diamond \mathfrak{B}^2$, where the I-type-property $[H^{comb}]$ of the U-type-property energy $\langle H \rangle$ belongs to $\mathfrak{B}^{comb}$. If $[H^{comb}]$ is physically represented by an observable H^{comb}, which in turn is mathematically represented by an operator $H^{comb} = H^1 \otimes I^2 + I^1 \otimes H^2 + H_{\text{int}} \in \mathcal{O}^{comb} = \mathcal{O}^1 \otimes \mathcal{O}^2$, there are two types of combination between $\mathfrak{B}^1$ and $\mathfrak{B}^2$:

- If $H_{\text{int}} \neq 0$, $\mathfrak{B}^{comb}$ is the result of an *interaction*,
- If $H_{\text{int}} = 0$, $\mathfrak{B}^{comb}$ is an *aggregate*, symbolized as $\mathfrak{B}^{agg} = \mathfrak{B}^1 * \mathfrak{B}^2$.

The behavior of the aggregates of atomic bundles will be essential to face the problem of indistinguishability, as will be discussed in Sections 11.4 and 11.5 of the next chapter.

10.8 Possibilist possibility

As remarked in Section 9.8 of the previous chapter, during the twentieth century, the strong influence of the logical-positivist rejection of metaphysics, expressed by an excessively strict conception of meaning, led to the prevalence of a non-possibilist, actualist conception not only of possibility, but also of probability. Logical positivism has been losing strength for a long time now; nevertheless, many approaches to quantum mechanics are still based on that non-possibilist view.

Although the Copenhagen Interpretation is not a single, definite doctrine (for a critical historical overview, see Howard 2004), its different versions share certain points about how possibility and probability enter the scene. In particular, quantum mechanics is considered an indeterministic theory, in the context of which the wavefunction expresses a probability amplitude. However, probabilities still have a positivistic flavor. On the one hand, in general they are referred to measurements: probabilities are not simply manifested by the frequencies of the measurements' outcomes, but are defined in terms of those frequencies. On the other hand, as a result of an operationalist influence, all variables are considered ill-defined unless they refer to measurement outcomes; therefore, speaking of probabilities independently of the experimental setup is meaningless because it is devoid of operational meaning.

In contrast to the Copenhagen Interpretation, Bohmian Mechanics is a deterministic theory. In this framework, the complete description of a quantum system requires not only the wavefunction, which evolves according to the Schrödinger equation, but also the specification of the actual positions of the particles, which evolve according to a guiding equation that expresses the velocities of the particles in terms of the wavefunction (for updated presentations, see Goldstein 2021, Tumulka 2022). Therefore, the ultimate justification for probability distributions is given in terms of the statistical patterns exhibited by ensembles of the actual subsystems of the universe, which are probabilistically distributed according to the Born rule. In other words, Bohmian Mechanics endows probabilities with an epistemic, "ignorance" meaning, as in the case of the probabilities involved in classical statistical mechanics.

At present, there are many versions of the Many Worlds Interpretation; nevertheless, some features are common to all of them (see Wallace 2012 for an updated presentation). The state vector never collapses: with each measurement (or interaction), the entire world splits into many copies, each of which "contains" one of the components of the superposition, and from then on each copy follows its own evolution. Sometimes it is not sufficiently emphasized that, under this interpretation, quantum mechanics is a deterministic theory, according to which all possibilities are actual in the branching "multiverse" (see Vaidman 2021). Consequently, possibility and probability must be interpreted as epistemic, in terms of the ignorance of an observer confined to only one of the many worlds.

Quantum Bayesianism or QBism, currently led by Christopher Fuchs, originated as a way of conceiving probabilities in quantum mechanics (Caves, Fuchs, and Schack 2002a, 2002b; see Fuchs, Mermin, and Schack 2014 for an updated presentation). QBists claim that the quantum state does not refer to the situation of a physical system, but rather expresses the subjective state of belief of an agent regarding possible future experiences. Therefore, they explicitly adopt a subjectivist interpretation of probability inspired by the work of Bruno de Finetti (1937). This perspective makes the Born rule not a law of nature, but an empirically motivated rule that a rational agent must follow in order to win when betting on quantum systems.

According to Wavefunction Realism, the wavefunction is a concrete entity, a physical field on configuration space (Albert 1996; see also Ney 2013). However, in this case the configuration space is not an abstract space used to represent possible configurations of particles in the three-dimensional space, but rather it is the physical space of the universe. In this multidimensional space, material objects are classical configurations of particles that correspond to regions in which the amplitude of the wavefunction is high. This means that our impression of living in a three-dimensional space is somehow entirely illusory. Furthermore, since the wavefunction evolves deterministically in the fundamental multidimensional space, probabilities also turn out to be subjective "illusions," derived from the fact that we perceive only three spatial dimensions.

This very quick survey of some of the currently discussed interpretations of quantum mechanics shows how difficult it remains in this philosophical field to conceive of a possibilist possibility and a correlative propensity probability. Some of these interpretative approaches design different strategies to make probability a measure of the observer's ignorance about an underlying reality that is inherently deterministic or, in the case of QBism, a mere playing field for betting. Only the most traditional Copenhagen Interpretation endows probability with an objective content, but with an actualist-frequentist conception in terms of measurement outcomes. The challenge seems to be, then, to devise an interpretation of quantum mechanics based on a possibilist conception of possibility. Leaving aside Heisenberg's intuitions about "*potentia*," as far as I know, only three interpretations decidedly show that there is room for a possibilist view that endows modality with an ontological content.

In the Spontaneous Collapse Theory, the collapse of the wavefunction is interpreted as a spontaneous process occurring at random times. More precisely, each microscopic quantum system is subjected to random localization processes around certain positions at random times. In this theory, those spontaneous "hittings" occur with higher probability where, according to the Copenhagen Interpretation, there is a higher probability of finding the system if a measurement were performed on it. However, here there is no measurement involved: hittings are completely spontaneous processes (see Ghirardi and Bassi 2024). Therefore, in the Spontaneous Collapse Theory, probabilities are objective and not tied to operational manipulations.

According to the Transactional Interpretation of quantum mechanics, originally proposed by John Cramer (1986, 1988), quantum interactions are *transactions*. An "emitter" generates a retarded wavefunction forward in time (an offer wave OW, represented by the usual quantum state $|\Psi\rangle$), and also an advanced wavefunction backward in time. An "absorber" responds by generating a retarded wavefunction forward in time and an advanced wavefunction backward in time (a confirmation wave CW, represented by the dual state $\langle\Phi|$). A transaction event occurs when the confirmation wave CW from the absorber exactly reinforces the offer wave OW from the emitter, and the advanced field from the emitter and the retarded field from the absorber exactly cancel each other; thus, all that remains is a fully retarded wave carrying energy from the emitter to the absorber. This proposal was endowed with a clear possibilist content by Ruth Kastner: the Possibilist Transactional Interpretation (PTI) explicitly announces "the reality of possibility" from the very title of Kastner's 2013 book. According to this position, when an emitter generates an OW, many different transactions are possible, among which only one of them actualizes. But in the present argumentation, the central point is that possibility is conceived as physically real: "OW and CW are interpreted ontologically in PTI as *physically real possibilities*. In this context, «real» means physically efficacious but not necessarily *actualized*" (Kastner 2013, 68; emphasis in the original). In other words, "the dynamical possibilities referred to by state vectors in PTI are Heisenbergian potentia" (Kastner 2013, 68; see also Kastner, Kauffman, and Epperson 2018). In turn, the actualization of one of the possible transactions is an irreducible stochastic phenomenon, which "is not compatible with any causal process within the confines of ordinary spacetime." (Kastner 2013, 68).

Despite their differences, the MHI and the PTI (and also the Spontaneous Collapse Theory, depending on how probabilities are interpreted in this context, see Frigg and Hoefer 2007) agree in their possibilist, non-actualist interpretation of possibility. According to the MHI, the wavefunction (or, better, the vector state) represents objective probabilities, which measure the propensity or potentiality to actualization of possible events. Although those possibilities and the corresponding probabilities are not reducible to actuality, they are real. In other words, reality unfolds in two realms: possibility and actuality. The Aristotelian *dictum* about being can be applied to this case: being can be said in different ways, as possible being or as actual being, neither of which can be defined or reduced in terms of the other. When the prejudice that reality is exhausted by actual reality is dispelled, it can be admitted that, strictly speaking, the MHI advocates a "wavefunction realism" or, better, a "*quantum state realism*": the quantum state describes reality in its possibility domain.

The fact that propensities belong to the realm of possibility does not mean that they do not have physical consequences in the domain of actuality. Indeed, they produce definite effects on actual reality even if they never become actual. An interesting manifestation of such effectiveness is the case of the so-called *interaction-free measurements*, in which the presence of a quantum system can be detected without any interaction between it and the measuring device. A particular case of such experiments is the Elitzur-Vaidman bomb tester (Elitzur and Vaidman 1993): by finding its roots in the double-slit experiment and by resorting to a Mach-Zehnder interferometer, in this case non-actualized possibilities are used to test bombs without exploding them.

On the basis of this interpretation of possibility, the term 'propensity' will be used to refer to the possibilist possibilities corresponding to the I-case-properties. Therefore, in the framework of the MHI, states encode ontological propensities to actualization (da Costa, Lombardi, and Lastiri 2013):

> **OP12**: Given a bundle $\mathfrak{B}$, its *state e* encodes the *ontological propensity to actualization* for all the I-case-properties of all the I-type-properties belonging to the bundle.

If the bundle $\mathfrak{B}$ is mathematically represented by the space of self-adjoint operators $\mathcal{O} = \mathcal{B}(\mathcal{H})$, the state e is mathematically represented by a functional ρ acting on $\mathcal{O}$. From a physical viewpoint, the state ρ cyphers the probabilities that allow calculating the expectation values for all the observables of the quantum system.

At this point, it is important to emphasize that a probability equal to zero does not mean impossibility. Let us assume that the case-property, say, ω_1 (a definite value of energy) belongs to the spectrum of the type-property H (the Hamiltonian). In this situation, ω_1 is possible even if the system's state is an eigenstate of H different of $|\omega_1\rangle$, which assigns probability zero to ω_1. The fact that ω_1 belongs to the spectrum of H is not only theoretically manifested by the mathematical form of H, but is also experimentally confirmed: the system may be in an infinite number of states that do not assign probability zero to ω_1 and in which ω_1 may become actual. Or, put differently: if ω_1 were really impossible, it would not belong to the spectrum of *H*, and this would physically mean that, for this Hamiltonian, in absolutely no state would property ω_1 be observed—become actual. It is worth recalling that the fact that probability zero does not mean impossibility is not a peculiarity of quantum mechanics: for example, in classical statistical mechanics, there are states with Lebesgue measure equal to zero which are not impossible according to the laws of classical mechanics (this has been discussed in the foundations of statistical mechanics at least since the book of Tolman 1938).

10.9 Two levels of possibility

One concern of the current metaphysics of science is the "*locus*" of modality. In some cases, this problem is closely linked to which ontological category is considered to be fundamental: if the primary ontological category is single, modality is single-located. For example, the substantialist admits the ontological categories of substance and property, but gives ontological priority to substance. In turn, "if we further conceive of properties as powers, or dispositions, the fundamental substances come to be seen as the 'seats' of such powers" (French 2018, 702). By contrast, in general the ontic structural realist shifts the "seat" of modality from objects to structures: structures display or possess primitive modality (Ladyman 2001). From a different perspective, the fundamental ontological category plays no role in the seat of possibility, because modality is located in natural laws (Dretske 1977, Armstrong 1983, Tooley 1987): modality is what endows laws with counterfactual force.

All these positions, although very different, have in common a unifying assumption: the idea that modality acquires a single form and manifests itself in a single way, for example, in dispositions, or in structures, or in laws, or... I will argue here that, even in the ontological domain referred to by a single scientific theory, modality can be located in different "places" and can appear under different forms (see Borge and Lombardi 2024).

As explained in the previous section, in the MHI ontology, propensities constitute an ontological domain as real as that of actuality and are encoded in the quantum state. These propensities to actualization apply to all the I-case-properties of all the I-type-properties belonging to the bundle-system. With different expressions and conceptual details, this type of possibility is, in general, what underlies the ideas of latencies (Margenau 1954), potentialities (Heisenberg 1958), propensities (Popper 1959), propensitons (Maxwell 1988), and selective propensities (Suarez 2007; see also his 2004a, 2004b), among others.

However, the Kochen-Specker theorem confronts us with another level of possibility, logically prior to the previous one as far as actualization is concerned. In fact, the theorem says that not all the observables of a quantum system can acquire a definite value simultaneously. In ontological terms, given a bundle, not all its I-type-properties actualize, i.e., acquire an AI-case-property among all its I-case-properties. This means that, given a bundle-system, all I-type-properties have the possibility to actualize, but only some of them do so. In the framework of the MHI, this possibility, like the propensities just mentioned in **OP12**, is also conceived as a possibilist possibility because it does not reduce to actuality. However, unlike propensities, this is a *deterministic possibility*, since actualization is determined by the MHI actualization rule it is the I-type-property physically represented by the Hamiltonian of the system that *univocally fixes* the preferred context, i.e., the set of the I-type-properties that actualize in that bundle.

Someone might reply that this level of possibility is superfluous since speaking of ontological possibility in a deterministic context makes no sense. However, there is no reason to accept this opinion. For example, the phase space of a classical system represents all the possible classical states of the system, although at each time only one of them is the actual one. In turn, several authors explicitly use the term "deterministic propensity" for a sure-fire disposition (see, e.g., Suárez 2007, Bigaj 2014).

However, someone could counter-argue by means of the following questions. If the Hamiltonian determines the I-type-properties that actualize, what is the ontological role of the remaining I-type-properties of the bundle, that is, those that do not actualize? Why not consider that, ontologically, the quantum system is nothing but the bundle composed only of the I-type-properties that actualize? If we were dealing with a system that remains independent forever, the critic's suggestion could be adopted by removing from the bundle the I-type-properties that will never acquire an AI-case-property. But, in general, quantum systems do not remain independent eternally. By contrast, systems interact with each other giving rise to new systems: in ontological terms, the interaction between two bundles gives rise to a new bundle. And that new bundle resulting from the interaction includes all the I-type-properties of the original bundles and also their combinations, *regardless of whether they actualized or not* before the interaction. The strategy of "trimming" the bundles by removing what does not actualize could not account for the structure of the bundle resulting from an interaction, in which, for example, some I-type-properties that did not actualize in the original bundles may acquire AI-case-properties in the new bundle.

In summary, in the ontological framework of the MHI, modality is not single-located but bi-located. Possibility appears at two levels, one deterministic and the other indeterministic. At the deterministic level, the actualization rule based on the Hamiltonian univocally fixes the preferred context, that is, the set of the I-type-properties of the bundle that actualize. Once the preferred context is fixed, for each I-type-property of that context, only one of its I-case-properties enters the realm of actuality, and it does so in a completely indeterministic way. Modality in the quantum ontology is structured in the interplay of these two levels of possibility.

10.10 Quantum indeterminacy

As mentioned at the end of the previous chapter, quantum mechanics seems to provide a source of examples of genuine metaphysical indeterminacy, which thus becomes quantum indeterminacy. After analyzing quantum metaphysical indeterminacy (QMI) in Relational Quantum Mechanics (Calosi and Mariani 2020), Claudio Calosi (2022) presents the first investigation of QMI in the MHI, concluding that it provides a significant example of a realist interpretation of SQM that involves QMI.

Although there are several accounts of metaphysical indeterminacy, Calosi focuses on one of them, the so-called *determinable-based metaphysical indeterminacy* (DBMI) as originally proposed by Jessica Wilson (2013) and developed for the quantum domain by several authors (e.g., Bokulich 2014, Wolff 2015, Calosi and Wilson 2019, 2021, Mariani and Torrengo 2021, Schroeren 2021, Mariani 2022). According to the determinable-based approach, there is metaphysical indeterminacy when an object instantiates a determinable property but no unique determinate of that determinable property at some level of determination. On the basis of this formulation, metaphysical indeterminacy can arise for two reasons: because the considered determinable has no determinate, or because it has more than one determinate. Wilson (2013) calls the first case "*gappy indeterminacy*" and the second case "*glutty indeterminacy*."

The determinable-based approach has been applied to quantum mechanics to offer an ontological account of metaphysical indeterminacy in the case of superpositions, incompatible observables, and entanglement. For example, Calosi and Wilson (2021) propose a glutty implementation of the approach to provide the basis for an intelligible explanation of the distinctive interference patterns characteristic of the double-slit experiment. Alisa Bokulich (2014) endorses a gappy implementation of metaphysical indeterminacy in order to accommodate the incompatibility between position and momentum observables, and also to account for entanglement. Johanna Wolff (2015) also describes incompatibility, in this case between spin in different directions, from a gappy reading of metaphysical indeterminacy. In turn, the discussions about metaphysical indeterminacy have also focused on particular interpretations and theories: Schroeren (2021) on Standard Quantum Mechanics, Mariani (2022) on Spontaneous Collapse Theory, Wilson (2020) and Calosi and Wilson (2021) on the Many Worlds Interpretation. Although the application of the determinable-based approach to metaphysical indeterminacy to quantum mechanics remains an area of active research, I will not enter this general field. Here, I will only consider Calosi's contribution to bringing the MHI into this metaphysical discussion.

In order to apply Wilson's approach to the MHI, two considerations must be taken into account from the very beginning. First, since in the MHI ontology there are no objects, the determinable-based account of metaphysical indeterminacy must be reformulated accordingly, that is, without referring to objects. Second, whereas Wilson assumes that determinables admit different levels of determination, in the MHI case this assumption can be left aside because each determinable property is represented by a single operator with its eigenvalues, which represent the determinates of that determinable: as discussed in Section 9.2 of the present chapter, in physical terms, each observable is defined by its eigenprojectors and eigenvalues, so no different sets of eigenprojectors and eigenvalues can correspond to a same observable. When these two points are considered, Wilson's DBMI can be reformulated as follows:

> **DBMI**: There is *metaphysical indeterminacy* when a determinable property has no unique determinate of that determinable property.

Calosi's (2022) first step is to furnish the translation between the language used in the DBMI approach and the ontological language of the MHI. Using the terminology of the present work, (i) 'determinable' corresponds to 'I-type-property'—corresponding to 'observable' in physics and to 'self-adjoint operator' in mathematics—, and (ii) 'determinate' corresponds to 'I-case property'—corresponding to 'value of the observable' in physics and to 'eigenvalue of the operator' in mathematics. I already advanced this correspondence in Section 2 of this chapter; therefore, the translation provided by Calosi is correct and reflects the ontological meaning of the MHI proposal.

In Section 10.3 of the present chapter, I explained why I preferred the terminology 'type-property' and 'case-property' over the traditional terminology 'determinable' and 'determinate': whereas the determinable-determinate distinction seems to confine possibility to the determinable side, it is necessary to stress that determinate properties are possible properties, which may or may not become actual properties. This distinction between possible determinates and actual determinates applies in a classical domain: the determinable "color" has many possible determinates, "red," "blue," "yellow," etc.; nevertheless, the distinction may be overlooked in most discussions because, according to the principle of omnimode determination, one and only one of those determinates become actual in a given object. However, in the quantum domain the difference between possibility and actuality should not be ignored, because it may be the case that no possible determinate becomes actual. I consider that current discussions in metaphysics of science about the determinable/determinate distinction would gain much in depth if the modal dimension of the distinction were taken into account.

After the step devoted to translation, Calosi focuses on the MHI actualization rule which can be expressed in ontological terms as follows: the I-type-properties that actualize, i.e., that acquire an actual AI-case-property among all the I-case-properties, are those corresponding to the Hamiltonian and to all the observables that commute with the Hamiltonian and do not break its symmetries. As a consequence, in any bundle-system there are always certain I-type-properties with no corresponding AI-case-property: in the determinable/determinate language of the DBMI, in any bundle there are always certain determinable properties with no determinate property, which leads to a clear case of *gappy indeterminacy*. It is worth recalling that the MHI actualization rule only fixes the preferred context of the I-type-properties that actualize, but the fact that not all the I-type-properties may acquire an AI-case-property simultaneously is a theoretical result derived from the Kochen-Specker theorem.

Finally, Calosi correctly stresses that the—gappy—indeterminacy in the MHI ontology is *not eliminable*. According to the author, this is an important conclusion because it provides a challenge to an argument due to David Glick (2017, 2022), according to which realist interpretations of quantum mechanics do not feature fundamental indeterminacy, and that derivative indeterminacy is eliminable. Glick focuses only on Bohmian Mechanics, Spontaneous Collapse Theory, and Many Worlds Interpretation. However, "there is no reason to limit oneself to those realist interpretations. Indeed, Modal Interpretations are explicitly realistic interpretations [...] that seem to challenge Glick's argument. They offer an example (at least some of them) of a realist interpretation of quantum mechanics where indeterminacy is fundamental." (Calosi 2022, 183).

Summing up, Calosi argues that there is an ineliminable metaphysical indeterminacy according to the MHI, on the basis of the structure of its ontology and its rule of actualization. He stresses that this is crucial for, at least, two reasons: "First, it advances the debate on quantum ontology by showing precisely how and why indeterminacy arises in a quantum world, as described by the MHI. Second, it offers a naturalistic example of genuine metaphysical indeterminacy, an example coming from our best physics." (Calosi 2022, 183).

11
Facing the Ontological Challenges of Quantum Mechanics

In the previous chapter, the structure of the ontology proposed by the MHI was introduced. In this ontology, properties are the fundamental items and quantum systems are non-objectual bundles of properties. However, as is well known, metaphysics is underdetermined by physics; in particular, quantum mechanics is compatible with different ontological pictures. Thus, arguing in favor of a certain quantum ontology over others requires showing how fruitful it is in the task of offering reasonable solutions to interpretative problems. This chapter is devoted to show the advantages of the proposed picture for dealing with the ontological challenges of quantum mechanics: contextuality, non-separability, and indistinguishability. Finally, the top-down approach presented in Chapter 8 will be revisited, showing how it acquires an ontological significance from the perspective of the ontology of properties proposed by the MHI.

11.1 The ontological challenges of quantum mechanics

Although all embodied in the formalism of the theory, the ontological challenges posed by quantum mechanics are of different nature and come into tension with different ontological principles:

- *Contextuality* prevents the simultaneous assignment of determinate values to all the properties of a quantum system. Thus, it is in conflict with the *principle of omnimode determination.*
- *Non-separability* affects the independent existence of non-interacting systems. Thus, it is in tension with the *principle of separability.*
- *Indistinguishability* is related to the statistical behavior of quantum systems. Thus, it challenges the *principle of identity of indiscernibles.*

I decided to present the three problems in this order because I believe it is a natural order with respect to the systems involved: (i) contextuality is a feature of a single system, (ii) non-separability commonly arises from the interaction of at least two systems, and (iii) indistinguishability appears with the statistics of many systems.

Most interpretations of quantum mechanics are designed to cope with the quantum measurement problem: in general, ontological problems do not receive the same

The Modal-Hamiltonian Interpretation of Quantum Mechanics. Olimpia Lombardi, Oxford University Press.
© Olimpia Lombardi (2026). DOI: 10.1093/9780198951407.003.0011

attention. But when these problems are analyzed, a common strategy is to focus on just one of them and to tailor an interpretative view to solve it, leaving aside the remaining difficulties. The ontology of properties proposed by the MHI aspires to formulate a "global" solution, according to which all the ontological problems can be adequately tackled in terms of a single ontology: this is the aim of the following sections of this chapter.

11.2 The problem of contextuality

One of the first reactions to the probabilistic character of quantum mechanics was the attempt to interpret it as a statistical theory, in the style of classical statistical mechanics, so that the probabilities could be explained as frequencies in ensembles of systems with definite but "hidden" values of their observables. The coup de grace for such classical-style statistical interpretations was the Kochen-Specker theorem (Kochen and Specker 1967), which proves the impossibility of ascribing precise values to *all* the observables of a quantum system simultaneously while preserving the functional relations between commuting observables. It follows that the selection of observables to which precise values can be attributed must be contextual.

The distinction between determinable and determinate properties, which allows the principle of omnimode determination to be formulated, is invoked by metaphysicians when discussing metaphysical indeterminacy. However, this distinction was scarcely taken into account in the traditional debates on the interpretation of quantum mechanics. This is surprising because quantum contextuality defies precisely that traditional and intuitive principle: whereas in all classical objects all determinables are determinate, in the quantum realm non-commuting observables correspond to determinables that are not simultaneously determinate. Furthermore, according to the Kochen-Specker theorem (Kochen and Specker 1967), a quantum system always has determinables (observables in the physical language) that are not determinate (that do not have precise values). It is interesting to note that the breach of the principle of omnimode determination is counterintuitive for any kind of object, not only for individuals, but also for stuff. In fact, one expects that not only in the case of a billiard ball but also in the case of soup, the determinable property, say, "color," is determinate, say, "yellow."

Different approaches have been proposed to accommodate quantum contextuality. One of them is based on the adaptation of the logic used in the quantum framework: starting from the fact that contextuality is related to the non-Boolean structure of elementary quantum propositions, a non-classical propositional logic can be formulated in terms of the non-distributive, orthocomplemented lattice of the theory (see, e.g., Jauch and Piron 1969, Piron 1976, Beltrametti and Cassinelli 1981). From a more physical perspective, other authors have dealt with quantum contextuality by selecting a context, via an interpretative assumption (see, e.g., Bub and Clifton 1996, Dieks 1998, 2005), or via a physical process such as decoherence (see, e.g., Zurek 1982, 2003).

However, the general problem of which ontological category a quantum system belongs to and what structure the quantum realm should have for contextuality to be ontologically natural has not been answered in a systematic way, as far as I know.

In the MHI ontological language, quantum contextuality is expressed by saying that, given a bundle-system, not all of its I-type-properties actualize, that is, acquire an *actual* AI-case-property among all their *possible* I-case-properties: which I-type-properties acquire an AI-case-property is a "contextual" fact. Of course, in each context one could insist on the classical idea of I-type-properties with their definite AI-case-properties with no contradiction. In other words, the picture of a bundle of actual case-properties that stands for a classical object could be retained in each context. But as soon as one tries to extend this ontological picture to all the contexts by conceiving the object as a "bundle of bundles," the Kochen-Specker theorem imposes an insurmountable barrier: in a quantum system, it is not possible to actually ascribe AI-case-properties corresponding to all the I-type-properties in a non-contradictory manner. Therefore, the classical idea of a bundle of *actual* case-properties does not work in the quantum ontology.

The Kochen-Specker theorem introduces a constraint with respect to AI-case-properties, more precisely, with respect to which I-case-properties of a bundle can enter actuality. But this restriction does not affect the concept of quantum system as proposed by the MHI ontology, because it is defined not as a bundle of *actual* case-properties, as in the traditional bundle theory, but as a bundle of I-type-properties, each with its corresponding *possible* I-case-properties. Precisely for this reason, these quantum bundles do not constitute objects in the traditional sense: they are non-objectual bundles. Since the basic items of the ontology are properties, which are grouped into bundles that do not constitute objects, *the principle of omnimode determination*, only valid for objects, is not false but simply *does not apply*. As a consequence, this ontology devoid of objects is immune to the ontological challenge represented by the Kochen-Specker theorem.

11.3 The problem of non-separability

Unlike the classical world, the quantum domain seems to admit surprising correlations between the properties of distant non-interacting systems, such as those of the famous Einstein-Podolsky-Rosen (EPR) experiment (Einstein, Podolsky, and Rosen 1935). Taken at face value, EPR-correlations strongly suggest *non-locality*, that is, non-local influences between spatially distant systems, i.e., systems between which no light signal can travel. However, since this idea is incompatible with special relativity, the exact nature of those quantum correlations is a subject of ongoing controversy (see Berkovitz 2016). For instance, according to collapse interpretations, EPR-correlations imply a certain action-at-a-distance which, nevertheless, does not allow sending information at a superluminal velocity. In the case of Bohmian mechanics, it is the quantum field that possesses the necessary non-local features to induce EPR-correlations. From another perspective, those correlations are consequences of the holistic nature of quantum systems, understanding holism as the opposite of separability. According to Richard Healey and Henrique Gomes (2022), *separability* implies that, if a physical object is constructed by assembling its physical parts, then its physical properties are completely determined by the properties of the parts and their relations; *holism*, by contrast, is the characteristic of some physical objects that are not composed of physical parts, but are indivisible wholes, so that EPR-correlations are correlations between properties of a single holistic object. From the holistic perspective, it has been claimed that entanglement is evidence that the whole universe has ontological priority with respect to its parts (Schaffer 2010).

Since the terms "locality" and "separability" have been used in a variety of different ways, Don Howard (1989) clearly distinguishes between the corresponding concepts. The *locality principle* asserts that the state of a system is unaffected by events in regions of the universe so removed from the given system that no signal could connect them. The aim of the locality principle is to rule out objectionable kinds of action-at-a-distance. The *separability principle*, by contrast, is a fundamental ontological principle governing the individuation of physical systems: it asserts that the presence of a non-vanishing space-time interval is a sufficient condition for the individuation of physical systems—and their associated states. As Howard stresses, Einstein's dissatisfaction with quantum mechanics is closely related to the violation of the separability principle: "it is characteristic of [...] physical things that they are conceived of as being arranged in a space-time continuum. Further, it appears to be essential for this arrangement of the things introduced in physics that, at a specific time, these things claim an existence independent of one another insofar as these things 'lie in different parts of space.' Without such an assumption of the mutually independent existence (the 'being-thus') of spatially distant things, an assumption which originates in everyday thought, physical thought in the sense familiar to us would not be possible." (Einstein 1948, 321–322; Howard's translation); "if one renounces the assumption that what is present in different parts of space has an independent, real existence, then I do not at all see what physics is supposed to describe." (Einstein in Born 1971, 223–224; Howard's translation). According to Howard, it is possible to retain locality by understanding non-separability as "a kind of radical ontological holism in which the whole of the forward light cone of any event is regarded as one nonseparable whole because of the pervasiveness of physical interactions" (Howard 1989, 252).

Despite disagreements about how to explain EPR-correlations, in general the arguments about non-separability are based on the assumption that quantum systems are individual objects, and subsystems are also individuals. Consequently, the problem is to explain how the properties of those individual subsystems are instantaneously correlated even though they are not in interaction and they are located in different spatial positions. However, from an ontology of properties, the problem appears in a new light, which is in resonance with Howard's understanding of non-separability.

Recall that the ontological category of individual requires some "principle of individuality" that, regardless of its specific form, identifies a particular individual as distinct from another and as the same over time (see discussion in French and Krause 2006). In turn, individuals can form aggregates, in which they can be counted (see Section 9.2 of Chapter 9): "countability" depends on the possibility of ontologically distinguishing each individual from the others. Therefore, if two individual systems interact as to yield a composite system in an entangled state, they should retain their identity as individual parts of the new whole. For this reason, EPR-correlations are conceived as the correlations linking the properties of those individual subsystems, so that when the subsystems are distant in space and do not interact, such correlations become puzzling.

By contrast, in the MHI ontology of properties, quantum systems are not individuals, not even objects: they are non-objectual bundles of properties. Therefore, there is no principle of individuality that allows them to retain their individual identity when they merge into a new bundle-system. As a consequence, the issue of interpreting EPR-correlations acquires a new formulation from the outset. The problem is no longer to explain the correlations between the properties of distant non-interacting objects. Since the EPR-bundle is a single whole, non-analyzable in component bundles, the EPR-correlations are *correlations between properties of a single item*, precisely, a non-objectual bundle. Thus, the mystery of the original formulation, which seems to require a certain unexplainable harmony between distant objects, vanishes from the new perspective, since correlations between properties of a single system are natural even in the classical ontological domain. For example, it is not surprising that the area of a table is correlated to its length, or that the kinetic energy of a car is correlated to its mass and to its velocity: there is no need of an enigmatic harmony to explain these correlations.

In a sense, this view implies a kind of holism. In a traditional ontology of objects and properties, the indivisible whole is also an individual. But, according to that traditional ontology, if an individual is not "atomic," it can be split into individual parts that are different from the original individual (see Section 9.2 of Chapter 9). This implies that, from a holistic perspective framed in a traditional ontology, the challenge is to account for the fact that the individual composite system is a whole that cannot be decomposed into individual parts. This problem also disappears in an ontology deprived of the category of individual: since the holistic item is not an individual, the fact that it lacks individual parts turns out to be an expected consequence.

Sharing Einstein's concerns, someone might argue that radical ontological holism would prevent any physical theory from being formulated and tested. However, this is not the case. In fact, despite holism, it is always possible to specify two sets of co-measurable observables and to define a factorizable state description that provides all the needed information. Of course, this so-defined factorizable state is not the quantum state of the whole system because, in general, it gives wrong predictions for observables not belonging to the specified sets: this is precisely the manifestation of non-separability. Nevertheless, in the specified measuring context, the factorizable state gives all the right predictions: "to put the idea differently, the universe is 'really' one, but once we put a specific question to it, it falls apart quite naturally into apparent parts." (Howard 1989, 253). In short, ontological holism is not an obstacle to develop physical knowledge.

11.4 The problem of indistinguishability

Most discussions about the ontological commitments of quantum mechanics focus on the challenge posed by the indistinguishability of so-called *identical particles*—particles with the same state-independent properties—to the ontological category of individual. The usual story begins by counting how many distributions or complexions of two identical particles over two states are possible. The classical answer is given by the Maxwell-Boltzmann statistics, according to which there are four possible distributions of two individuals over two states. By contrast, in quantum statistics (Bose-Einstein and Fermi-Dirac), a permutation of the identical particles does not lead to a different complexion since they are "indistinguishable." Although the theory has formal resources to deal with quantum statistics, from a conceptual viewpoint the problem is to explain why a permutation of individual particles does not lead to a different complexion in the quantum case.

As explained in the previous chapter, according to the "Received View" about quantum indistinguishability, fundamental particles cannot be regarded as individuals (see French and Bigaj 2024). From this perspective, indistinguishability is considered a feature that leads to the violation of the weakest version of the *principle of identity of indiscernibles*.

Several perspectives confront the Received View in order to restore the category of individual. For example, Bas van Fraassen (1985) recovers quantum statistics by renouncing the equiprobability of the different distributions of quantum particles in quantum states. Another position contrary to the Received View comes from Steven French (1989), who claims that states of indistinguishable particles that are not symmetric or anti-symmetric are ontologically possible and only physically inaccessible: indistinguishability is a physical situation, not an ontological condition from which non-individuality can be inferred. Another alternative proposal to the Received View is based on the idea of "weak discernibility" (Saunders 2003, Muller and Saunders 2008): in the case of two fermions in a singlet state, the relation "having the opposite direction of each spin component with respect to..." that each fermion has with respect to the other is sufficient to establish numerical distinction between the objects, even if they are indistinguishable with respect to their monadic and relational properties (for an analogous argument for bosons, see Muller and Seevinck 2009). French and Décio Krause (2006) have rejected this strategy by claiming that it entails circularity: in order to appeal to such relations, one has had to discriminate beforehand between the related objects; therefore, the numerical difference between the objects has been presupposed by the relation itself.

Although the problem of indistinguishability in quantum mechanics has generated extensive discussions, a common ground underlies all of them. In fact, despite the

Received View and its critics differ with respect to conceiving quantum systems as individuals, they nevertheless agree with respect to subsuming them under the category of object. However, by dispensing with the category of object, the problem acquires a completely different formulation.

Whereas the Received View claims that quantum systems are non-individuals, it has also been suggested that they are not even objects at all (Quine 1976). Nevertheless, these stances give no metaphysical characterization of those items beyond their non-individuality or non-objectuality. By contrast, in the proposed ontology of properties, non-objectual quantum systems are not only negatively characterized: they are positively and precisely defined in metaphysical terms as bundles of I-type-properties. Moreover, I-type-properties, with their corresponding I-case-properties, are physically/formally represented with precision by observables/operators of an algebra. As explained in Section 10.6 of the previous chapter, indistinguishability is primarily a relation between two I-case-properties of the same U-case-property, and derivatively it is a relation between I-type-properties when their I-case-properties are indistinguishable. From this meaning applied to properties, indistinguishability acquires a derived meaning when applied to bundles: two bundle-systems are indistinguishable when their respective I-type-properties are indistinguishable. Indistinguishable I-case-properties, indistinguishable I-type-properties, and indistinguishable bundle-systems are only numerically different. Nevertheless, this does not imply that the principle of identity of indiscernibles is false for them: whereas the principle refers to the identity of indiscernible objects, in the MHI ontology indistinguishability is a relation between items belonging to the ontological category of property and, as a consequence, *the principle of identity of indiscernibles does not apply* to them. It is precisely this positive characterization of quantum systems that makes this non-applicability of the principle conceptually meaningful.

Now it is important to show the formal consequences of this ontological picture. Let us consider two bundles $\mathfrak{B}^1$ and $\mathfrak{B}^2$, which combine to lead to an aggregate $\mathfrak{B}^{agg} = \mathfrak{B}^1 * \mathfrak{B}^2$ (see **OP11** in Section 10.7 of the previous chapter). If the bundles $\mathfrak{B}^1 = \{[A_i^1]\}$ and $\mathfrak{B}^2 = \left\{[A_j^2]\right\}$ are physically/mathematically represented by the spaces of observables/self-adjoint operators $\mathcal{O}^1$ and $\mathcal{O}^2$, such that $A_i^1 \in \mathcal{O}^1$ and $A_j^2 \in \mathcal{O}^2$, then the I-type-properties $[C^{agg}]$ belonging to $\mathfrak{B}^{agg}$ are represented by the operators of the form

$$C^{agg} = \sum_{ij} k_{ij} \left(A_i^1 \otimes A_j^2\right) \in \mathcal{O}^{agg} = \mathcal{O}^1 \otimes \mathcal{O}^2. \qquad (11\text{-}1)$$

Let us now consider that the two bundles are atomic, $\mathfrak{B}_{at}^1$ and $\mathfrak{B}_{at}^2$, and indistinguishable, $\mathfrak{B}_{at}^1 \triangleq \mathfrak{B}_{at}^2$ (see **OP8** in Section 10.6 of the previous chapter). In this case, it is natural to expect that the I-type-properties belonging to the new bundle $\mathfrak{B}^{agg} = \mathfrak{B}_{at}^1 * \mathfrak{B}_{at}^2$ do not distinguish between the original bundles. Simply phrased, when two indistinguishable atomic bundles merge into a single whole, which component bundle is taken first and which second does not matter at all. In ontological terms, this means that *the operation of aggregation of indistinguishable atomic bundles is commutative*:

$\mathfrak{B}^{agg} = \mathfrak{B}^1_{at} * \mathfrak{B}^2_{at} = \mathfrak{B}^2_{at} * \mathfrak{B}^1_{at}$ (see da Costa, Lombardi, and Lastiri 2013). This commutativity adds a restriction to the I-type properties $[C^{agg}]$ belonging to the bundle $\mathfrak{B}^{agg}$ resulting from the aggregation. It requires the $[C^{agg}]$ to be mathematically represented by operators C^{agg} (of eq. (11-1)) such that

$$A_i^1 \otimes A_j^2 = A_i^2 \otimes A_j^1. \tag{11-2}$$

In other words, the instances $[C^{agg}] \in \mathfrak{B}^{agg}$ are represented by observables that are *symmetric* with respect to the permutation between $\mathfrak{B}^1_{at}$ and $\mathfrak{B}^2_{at}$. Taking the simpler case $C^{agg} = A^1 \otimes B^2$, if the components of A^1 and B^2 are a_{ij} and b_{mn} respectively, the components of C^{agg} are $c_{ijmn} = a_{ij}b_{mn} = a_{mn}b_{ij} = c_{mnij}$.

In summary, the indistinguishability between properties leads to the indistinguishability between bundles. Due to their indistinguishability, when two indistinguishable atomic bundles combine in an aggregate, the operation of aggregation is commutative. And it is this commutativity that leads to the permutation symmetry of the bundle resulting from the aggregation: the operators representing the instances belonging to the new bundle are symmetric with respect to the permutation of the indices corresponding to the two component operators.

11.5 The indistinguishability principle in an ontology of properties

According to the traditional story about indistinguishability, since quantum particles of the same kind obey either Bose-Einstein or Fermi-Dirac statistics, then, under the assumption of equiprobability, they must be conceived as indistinguishable. As a consequence, all quantum states differing only by a permutation of indistinguishable particles must be observationally indistinguishable. This idea is expressed by the so-called *indistinguishability principle* (IP), which can be expressed as (for this formulation, see Butterfield 1993):

> **IP**: If the vector $|\psi\rangle$ represents the state of a composite system whose components are indistinguishable particles, then the expectation value of any observable represented by an operator O must be the same for $|\psi\rangle$ and for any permutation $|\psi'\rangle = P|\psi\rangle$: $\langle O\rangle_{|\psi\rangle} = \langle O\rangle_{|\psi'\rangle}$.

Two well-known ways to satisfy IP are to restrict states to be symmetric (bosons) or anti-symmetric (fermions), which obey Bose-Einstein or Fermi-Dirac statistics, respectively; so, the symmetrization postulate (SP) is introduced.

Although this is, roughly, the traditional story, a closer inspection brings to light different positions. Some think that IP rules out both non-symmetric states and non-symmetric observables (Readhead and Teller 1992), whereas others maintain that it poses a restriction only on observables (Huggett and Imbo 2009). It has also been claimed that IP is not a postulate because it is the definition of indistinguishability (Belousek 2000), or because it follows from SP (Muller and Leegwater 2022). In turn, some consider that SP and IP are independent because SP applies to states and IP applies to observables (Belousek 2000), and they search for a justification of SP independent of IP (Bigaj 2022). Furthermore, this debate is seasoned with the discussions about whether the ruled-out vectors/operators only represent empirically inaccessible states/observables or do not represent states/properties at all.

Here, I will not delve into these discussions but I will focus on the two formulations of IP. The above formulation introduces a restriction on states:

$$\mathbf{IP}_{st}: \ \langle O \rangle_{|\psi\rangle} = \langle O \rangle_{|\psi'\rangle}, \text{with } |\psi'\rangle = P\,|\psi\rangle \qquad (11\text{-}3)$$

However, since the permutation operator P acts on a generic observable O as $O' = P^{\dagger} O P$, it is easy to see that

$$\langle O \rangle_{|\psi'\rangle} = \langle \psi' |\, O \,| \psi' \rangle = \langle \psi |\, P^{\dagger} O P \,| \psi \rangle = \langle \psi |\, O' \,| \psi \rangle = \langle O' \rangle_{|\psi\rangle}. \qquad (11\text{-}4)$$

Then, the IP restriction can be expressed by saying that, for any observable of a system whose components are indistinguishable subsystems, its expectation value in any state must be the same for that observable and for the permutation of that observable (Messiah and Greenberg 1964):

$$\mathbf{IP}_{obs}: \ \langle O \rangle_{|\psi\rangle} = \langle O' \rangle_{|\psi\rangle}, \text{with } O' = P^{\dagger} O P \qquad (11\text{-}5)$$

The two formulations are mathematically equivalent. However, as van Fraassen (1991) emphasizes, they are not equivalent from an interpretative standpoint—once again, mathematical equivalence does not imply interpretative equivalence. IP_{st} establishes a restriction on the states of the whole system, allowing just symmetric-bosonic states or anti-symmetric-fermionic states. By contrast, IP_{obs} imposes a restriction on the observables of the whole system, allowing just symmetric observables. From the algebraically expressed viewpoint of an ontology of properties, IP_{obs} is the formulation that makes sense, since it requires that all the I-type-properties of a bundle-system, which is an aggregate of indistinguishable atomic bundle-systems, are represented by operators that are symmetric with respect to the permutation of the component bundles. But the point to stress here is that, from this ontological perspective, IP_{obs} is not an ad hoc postulate of the theory but a consequence of the *ontologically motivated permutation symmetry* of the observables of the whole system.

Another advantage of this approach to indistinguishability is that the symmetry of the Hamiltonian does not need to be added as a supplementary requirement to that of the symmetrization or antisymmetrization of states. Given that bosons cannot be transformed into fermions and vice versa, no transition must occur between bosonic symmetric states and fermionic anti-symmetric states during the time-evolution of the whole system. Then, the permutation operator must be invariant under the evolution: $U^{\dagger}(t) P U(t) = P \Leftrightarrow [P, U(t)] = 0$. Since the evolution operator is $U(t) = e^{-iHt/\hbar}$, the commutation between P and $U(t)$ implies that $[P, H] = 0$ and, as a consequence, that H is a parity symmetric observable. Therefore, the preservation of the states' symmetry type requires the symmetry of H. If the IP is interpreted as IP_{st}, that is, as a restriction on states, the symmetry of the Hamiltonian must be posited as an additional condition. By contrast, since the Hamiltonian of the whole system is one of its observables, its symmetry follows directly from IP_{obs} with no need for further requirements, and the preservation of the symmetry type of the states turns out to be a consequence of such a symmetry.

But, what about the symmetry or anti-symmetry of states? It is not difficult to see that, when the quantum state is conceived as an expectation value functional acting on a space of observables $\mathcal{O}$, the symmetry of the observables renders the symmetrization or antisymmetrization of states unnecessary. In fact, the state of the entire system can be mathematically represented by an operator ρ, which can always be decomposed as a sum of two operators, one symmetric and one anti-symmetric with respect to the permutation of indices corresponding to the component systems: $\rho = \rho^S + \rho^A$. As is well known, if A is a symmetric operator and B is an anti-symmetric operator, then $Tr(AB) = 0$. Therefore, if the observable O of the whole system is permutation symmetric, its expectation value in the generic state represented by ρ is computed as

$$\langle O \rangle_\rho = Tr(O\rho) = Tr(O\rho^S + O\rho^A) = Tr(O\rho^S) + 0 = \langle O \rangle_{\rho^S}. \qquad (11\text{-}6)$$

This means that the non-symmetric part of the representation of the state has no effect in its application onto symmetric observables and, as a consequence, is superfluous. In turn, in the particular case of pure states, the symmetric $\rho^S = |\varphi\rangle\langle\varphi|$ may be expressed in terms of a symmetric state vector or in terms of an anti-symmetric state vector. In any case, the symmetrization and the antisymmetrization of states vectors lose their ad hoc flavor since, strictly speaking, they are not necessary. Of course, nothing prevents physicists from still using symmetric and anti-symmetric states in practice for simplicity reasons. However, the requirement of the IP is fulfilled due to the symmetry of the observables of the whole system which, in turn, is a consequence of the ontological picture supplied by the MHI.

Finally, it is worth stressing what the word 'permutation' means in this conceptual framework. It is quite clear that it does not refer to the operation of "exchanging" particles, because there are not particles in the ontology. Permutation is a mathematical operation, represented by the permutation operator P, that applies on states and observables. In the case of states, it allows expressing the symmetry or anti-symmetry of certain pure states. Nevertheless, when the restriction of symmetry is applied on observables, symmetrization and antisymmetrization of states is no longer needed. Therefore, the only remaining question is what permutation means when applied to observables. Mathematically, the operation of permutation is the formal means of expressing the permutation symmetry of certain observables. But such a formal symmetry is not an ad hoc requirement, but the mathematical way of representing ontological indistinguishability—only numerical difference—between non-objectual items—bundles—by means of a formal language that inevitably objectifies those items with the use of labels. In other words, the permutation invariance of observables is nothing but an artifact of a mathematical formalism that has no formal resources to represent genuinely indistinguishable items.

11.6 Indistinguishability and structure

One of the main questions in the philosophy of mathematics is that referred to the nature of mathematical entities. According to Platonism, mathematical entities are objects of a non-empirical but real domain, whose properties must be discovered; conventionalists, on the contrary, conceive mathematics as the activity of developing formal systems and, as a consequence, the question about the nature of mathematical entities makes sense only in this conceptual framework. Under the influence of Dedekind's idea that mathematics characterizes its objects up to isomorphism, mathematical structuralism presents itself as an alternative to both positions (Benacerraf 1965). According to this view, structures are the primary items of mathematics. In particular, in the *ante rem* version of Stewart Shapiro (1997), although structures have ontological priority over the places in the structures, those places are treated as genuine, bona fide objects, which constitute the referents of certain terms of the language of the theory under consideration.

In spite of its appeal, Shapiro's proposal has been challenged by arguing that places in structures may violate the principle of identity of indiscernibles. In fact, it can be shown (Keränen 2001) that, in the case of structures having non-trivial automorphisms, some places cannot be distinguished from each other. It has been claimed that this problem in the philosophy of mathematics is analogous to the problem of indistinguishability in quantum mechanics (Narvaja, Córdoba, and Lombardi 2012): although the difficulties are apparently unrelated since coming from different disciplines, they are manifestations of a single ontological problem. On this basis, it has been suggested that the solution proposed in one of the domains may be extrapolated and adapted to the other.

The idea of linking indistinguishability with structure symmetry is not new. In particular, Décio Krause and Antonio Coelho (2005) characterize indistinguishability in terms of invariance under automorphisms. The authors begin by defining a *structure* as:

$$\mathcal{A} = \langle D, \{R_i\}_{i \in I} \rangle, \tag{11-7}$$

where D is the *domain* of the structure, and the R_i are *relations* defined on the elements of D, with $I = \{1, 2, 3, \ldots, n\}$. In this structure, a *property* is a sub-collection $X \subseteq D$ such that it is invariant under all the automorphisms of $\mathcal{A}$: for all the automorphisms g of $\mathcal{A}$, $g(X) = X$. On this basis, the concept of distinguishability in a structure can be defined as follows:

> For any two elements $a, b \in D$, a and b are *$\mathcal{A}$-distinguishable* (distinguishable in the structure $\mathcal{A}$) if there exists a property X such that $a \in X$ if and only if $b \notin X$; otherwise, they are *$\mathcal{A}$-indistinguishable*.

The rough idea behind this definition is that two elements of a structure are indistinguishable when they share all the properties definable in that structure. On this basis, it is said that a structure is *rigid* if and only if its only automorphism is the identity. It can be proved that, in a rigid structure, indistinguishability and identity coincide: a and b are $\mathcal{A}$-indistinguishable if and only if $a = b$ (see Krause and Coelho 2005).

In this theoretical framework, a new concept is relevant to the discussion about indistinguishability. Given a structure $\mathcal{A} = \langle D, \{R_i\}_{i \in I} \rangle$, another structure $\mathcal{B}$ is an *expansion of* $\mathcal{A}$ if and only if $\mathcal{B} = \langle D, \{R_i\}_{i \in I \cup J} \rangle$, where $J = \{n+1, n+2, \ldots\}$ and $I \cap J = \emptyset$. Roughly speaking, $\mathcal{B}$ is an expansion of $\mathcal{A}$ when $\mathcal{B}$ is obtained by adding new relations to $\mathcal{A}$. Moreover, when $\mathcal{B}$ is rigid, it is called a *rigid expansion of* $\mathcal{A}$. Nevertheless, two cases must be distinguished:

- $\mathcal{B}$ is a *trivial rigid expansion* of $\mathcal{A}$ if and only if $\mathcal{B}$ is a rigid expansion of $\mathcal{A}$ and $\mathcal{B}' = \langle D, \{R_i\}_{i \in J} \rangle$ is rigid. This means that the new relations added to $\mathcal{A}$ are what make $\mathcal{B}$ rigid, independently of the original relations of $\mathcal{A}$.
- $\mathcal{B}$ is a *non-trivial rigid expansion* of $\mathcal{A}$ if and only if $\mathcal{B}$ is a rigid expansion of $\mathcal{A}$ and $\mathcal{B}' = \langle D, \{R_i\}_{i \in J} \rangle$ is not rigid. In this case, the new relations added to $\mathcal{A}$ are not, alone, sufficient to make $\mathcal{B}$ rigid: the original relations of $\mathcal{A}$ in combination with the new relations added in the expansion are responsible for the rigidity of $\mathcal{B}$.

On this basis, it is easy to see that *every structure has a trivial rigid expansion.* The trivial way to obtain it is by adding all the singletons of the elements of the domain to the original structure.

This formal framework paves the way to approach the problem of quantum indistinguishability from a different perspective. Although indistinguishability and identity do not coincide in the quantum domain, one can still ask if indistinguishability can be avoided, that is, if there is some way to distinguish elementary particles without violating the principles of the theory. Of course, there is always a trivial rigid expansion that distinguishes particle a from other particles by means of a property like "being identical to a." But the relevant question is whether quantum mechanics admits a non-trivial rigid expansion that makes it possible to distinguish elementary particles. As Krause and Coelho point out, the hypothesis that elementary particles of the same kind are indistinguishable in an ontological sense "amounts to sustain that a mathematical structure of quantum theory cannot have a non-trivial rigid expansion, which intuitively means that the rigidity of a structure for quantum theory can be achieved only by considering new relations which, by themselves, regardless the quantum nature of the elements of the domain, guarantee such a rigidity." (Krause and Coelho 2005, 206).

When discussing weak discernibility—a discernibility that only depends on irreflexive qualitative relations (see Saunders 2003, Muller and Saunders 2008, Muller and Seevinck 2009)—in situations with perfect symmetry, Dennis Dieks and Marijn Versteegh (2008) emphasize the difference between the classical case and the quantum case. In the classical domain, the symmetry of a given situation, such as that of Black's spheres, can be broken by introducing a point of reference or a gauge that distinguishes between the entities involved in the situation: it can be said that the two Black's spheres are *weakly distinguishable.* Although in a different context, two entities related by a symmetry transformation that leaves invariant the whole situation, as the two semi-cones of a light cone or the two spin directions, have been called *formally identical*; when we assign different names to formally identical entities, we introduce a *conventional difference* between them (Castagnino, Lombardi, and Lara 2003, Castagnino and Lombardi 2009). In the formal framework supplied by Krause and Coelho, the introduction of a conventional difference between formally identical entities amounts to introducing a trivial rigid expansion to the original structure. Nevertheless, this does not mean that the formally identical entities collapse into a single entity.

However, Dieks and Versteegh point out that the quantum case is different because symmetrization is a postulate: symmetry cannot be broken without violating one of the principles of the theory, a violation that would lead to a wrong statistics. In a similar vein, Dieks and Andrea Lubberdink also stress that "[i]n classical physics perfect symmetry of particle configurations, if it occurs at all, is something contingent; but in quantum mechanics it is a law-like feature that all indices must always occur, in any expression and in whatever situation, in a fully symmetrical way. [. . .] This is very much different from the case of Black's spheres. In quantum mechanics it is a matter of principle that we can never associate different physical characteristics with different indices in the formalism." (Dieks and Lubberdink 2011, 1063). Furthermore, from a theoretical viewpoint, perfect symmetry in a classical domain is not only contingent, but also extremely improbable: perfectly symmetric configurations have measure zero in the space of all theoretically possible configurations (for a proof in a particular case, see Castagnino, Lombardi, and Lara 2003). This stands in stark contrast to the quantum case, where symmetry is necessary since imposed by one of the principles of the theory.

But the constraints introduced by indistinguishability are even more serious. In fact, those who appeal to weak discernibility to retain the principle of identity of indiscernibles in the quantum domain need to label entities in order to formulate the argument, in contrast to the fact that quantum elementary particles cannot be labelled. In Steven French and Krause terms: "Doesn't the appeal to irreflexive relations in order to ground the individuality of the objects which bear such relations involve a circularity? In other words, the worry is that in order to appeal to such relations, one has already had to individuate the particles which are so related and the numerical diversity of the particles has been presupposed by the relation which hence cannot account for it." (French and Krause 2006, 170–171). In other words, in quantum mechanics we cannot even introduce a conventional difference between indistinguishable particles by assigning them different labels. This seems to mean that quantum mechanics does not even admit a trivial rigid expansion. But, how to make this statement compatible with the undoubtedly true claim that every structure has a trivial rigid expansion? (see da Costa and Lombardi 2014).

The concept of structure as characterized by Krause and Coelho includes a domain defined as a set of the standard set theory. But, in this theory, the elements of a set are individual entities that can be distinguished at least by assigning them different names. This situation opens two alternatives. As already mentioned, some authors consider that the standard set theory is not adequate to deal with the indistinguishable particles of quantum mechanics. The semi-extensional *quasi-set theory*, proposed by Krause (1992; see also da Costa and Krause 1994, 1997, 1999), and the intensional *quaset theory*, developed by Maria Luisa dalla Chiara and Giuliano Toraldo di Francia (1993, 1995), describe collections of *objects* having cardinality but not order type, that is, objects to which the concept of individual of classical logic does not apply. The other alternative is to deny that the elements of the domain are objects at all, as in the MHI.

From the MHI viewpoint, quantum indistinguishability has nothing to do with the principle of identity of indiscernibles, since the principle applies to objects whereas here there are no objects involved. This means that indistinguishability is not a relation between individual particles, but a relation between properties, which leads to a symmetry that is internal to a non-objectual and indivisible whole. In other words, the MHI ontology of properties offers a solution to the problem of quantum indistinguishability that is based on two factors, holism and symmetry: the system (i) is a whole where the components cannot be re-identified precisely due to their non-objectual nature, and (ii) has an internal symmetry manifested by the invariance under permutation of indices, which is an ontological consequence of the indistinguishability of the components' properties. The first point agrees with the idea expressed by Dieks and Versteegh: "one possible stance in such discussions is to argue that there is no multiplicity at all: that there is only one undivided physical system." (2008, 926). The second point, in turn, implies to deny the usual conception of indices as denoting quantum particles; as Dieks and Lubberdink claim, "indices in the quantum mechanical formalism of 'identical particles' refer to the individual factor spaces from which the total Hilbert space in the formalism is constructed—they are merely mathematical quantities." (2011, 1063). According to the authors, the individual particles encountered in experiments pertain to classical limit situations, and do not correspond to the indices in the quantum formalism (see also Lombardi and Dieks 2016). When the problem is formulated in these terms, many of the ontological perplexities derived from considering objects without individuality suddenly vanish.

Up to this point, the replacement of the standard set theory seems to be unnecessary, since there are no objects to which the concept of individual of classical logic does not apply. However, how to talk about an ontology of properties? By means of what logic can bundles of properties be treated? It is at this point that the quasi-set theory can play a relevant role in the MHI framework.

Although Krause's *quasi-set theory* $\mathfrak{Q}$ is a formal theory, it was devised by its author with an intended interpretation: it includes two kinds of *Urelemente*, M-atoms—classical elements—and m-atoms—indiscernible elements. Here, *m-atoms* are conceived as the correlates of elementary particles, which ontologically are non-individuals (see Arenhart, Bueno, and Krause 2019), so that the formal relation of indiscernibility between m-atoms represents the physical relation of indistinguishability between elementary particles of the same kind. Nevertheless, precisely due to its formal nature, quasi-set theory can be detached from that intended interpretation and can be applied to items belonging to any ontological category, whenever the logical "behavior" of those items agrees with the structure of the theory. In particular, quasi-set theory can be applied to a universe of properties, in particular, of I-type-properties, which under certain precise circumstances can be considered as legitimately indistinguishable due to their own ontological nature. In turn, the indistinguishability between atomic bundles is inherited from the indistinguishability of the bundles' components. It can be shown that collections of indistinguishable I-type-properties can be formally characterized as *quasi-sets* or *qsets*, and bundles can be formed by applying the Axiom of Choice to those qsets: a bundle is a qset built with only one element of each one of them. Moreover, qsets of bundle-qsets, with definite quasi-cardinality, can also be defined in order to treat collections of indistinguishable bundles with a definite number of members, in spite of the fact that those members can neither be counted nor labeled (for formal details, see Holik, Jorge, Krause, and Lombardi 2022).

More recently, Krause and Juan Pablo Jorge (2025) realized that the theory of quasi-sets can be further detached from its original intended interpretation, tied to the existence of *Urelemente*. Indeed, the postulation of M-atoms and m-atoms reveals the persistence of the commitment to items that, although non-individuals, still belong to the ontological category of object. Therefore, in order to speak of items not belonging to the category of object, it is necessary to remove the postulation of atomic elements. In the light of this idea, the authors have proposed the *quasi-set theory* $\mathfrak{Q}^-$, which is not committed to the existence of *Urelemente*: the new theory is "pure" in the sense that all that exists are qsets. Thus, the I-type-properties, which in the previous version (Holik, Jorge, Krause, and Lombardi 2022) were represented by m-atoms, are represented by qsets in $\mathfrak{Q}^-$. It is quite clear that a set theory of these characteristics fits in a natural way to an ontology of properties devoid of objects. However, questions remain to be answered. For example: What is the q-extension of the qset associated with an I-type-property?; What role do the I-case-properties of an I-type-property play in determining the q-extension of its associated qset? If the quasi-set theory is to be the appropriate metalanguage for the MHI ontology, the I-case-properties should also be represented by qsets, whose indistinguishability, although ontologically grounded, should be undefined in the logical context and should allow defining the indistinguishability between the qsets associated to I-type-properties. But these are still general ideas in the framework of a project under construction (see Krause, Jorge, and Lombardi 2025).

11.7 Entanglement of indistinguishable particles

Entanglement in many-body systems of "identical particles" has received great attention during the last decades due to its applications in quantum information and condensed matter. These works have led to revise the very concept of entanglement in the case of indistinguishability, which poses a specific challenge: according to the traditional definition of entanglement in pure states as non-factorizability, symmetrized, and anti-symmetrized states should be entangled (except in some cases involving bosons). But there are good reasons to think that this is not the case: not always should non-factorizable states of bosonic or fermionic systems be conceived as legitimate cases of entanglement. In this section, I will consider the problem in detail.

Although the issue of defining entanglement for mixed states is still an open question (see Earman 2015), there is a general consensus regarding pure states: a state $|\psi_{12}\rangle$ of a composite system is entangled if and only if it is *non-factorizable* in terms of the tensor product of the states $|\psi_1\rangle$ and $|\psi_2\rangle$ of the component systems (see, e.g., Myrvold 2022a):

$$|\psi_{12}\rangle \neq |\psi_1\rangle \otimes |\psi_2\rangle . \tag{11-8}$$

This characterization of entanglement works fine when the component systems are distinguishable, but in the case of indistinguishability difficulties appear (for a detailed discussion, see Fortin and Lombardi 2022).

Let us consider a system $\mathcal{S}$, whose component systems are $\mathcal{S}^1$ and $\mathcal{S}^2$, corresponding to the Hilbert spaces $\mathcal{H}^1$ and $\mathcal{H}^2$, respectively. As is well known, if $\mathcal{S}^1$ and $\mathcal{S}^2$ are distinguishable, the Hilbert space of $\mathcal{S}$ is $\mathcal{H} = \mathcal{H}^1 \otimes \mathcal{H}^2$. However, it is usually accepted that, if $\mathcal{S}^1$ and $\mathcal{S}^2$ are indistinguishable particles, according to the SP, the Hilbert space of $\mathcal{S}$ is the symmetric subspace $\mathcal{H}^B$ of $\mathcal{H}$ in the case of bosons, and the anti-symmetric subspace $\mathcal{H}^F$ of $\mathcal{H}$ in the case of fermions. The vectors belonging to $\mathcal{H}^B$ and to $\mathcal{H}^F$ are eigenvectors of the *permutation operator* P, which is defined as follows: given a basis $\{|\alpha_i\rangle_1 \otimes |\beta_j\rangle_2 \equiv |\alpha_i\rangle_1 |\beta_j\rangle_2\}$ of $\mathcal{H} = \mathcal{H}^1 \otimes \mathcal{H}^2$,

$$P\left(|\alpha_i\rangle_1 |\beta_j\rangle_2\right) = |\beta_j\rangle_1 |\alpha_i\rangle_2, \tag{11-9}$$

with $|\alpha_i\rangle_1, |\beta_j\rangle_1 \in \mathcal{H}^1$ and $|\beta_j\rangle_2, |\alpha_i\rangle_2 \in \mathcal{H}^2$. This operator satisfies the following properties: $P^2 = I$ and, therefore, $P = P^{-1}$; P is unitary, that is, $P P^\dagger = P^\dagger P = I$; and, therefore, P is self-adjoint, $P = P^\dagger$.

Now let us consider the possible states $|\uparrow\rangle_1$, $|\downarrow\rangle_1$, $|\uparrow\rangle_2$, and $|\downarrow\rangle_2$ of two particles (the labels refer to the Hilbert spaces to which the states belong). If the particles are distinguishable, the definition of entanglement in terms of factorizability works well: the state of the whole system is entangled if and only if it is non-factorizable, e.g., (the symbol $\otimes$ is omitted),

$$|\Theta_{ent}\rangle = 1/\sqrt{2}\,(|\uparrow\rangle_1 |\downarrow\rangle_2 - |\downarrow\rangle_1 |\uparrow\rangle_2), \tag{11-10}$$

and it is non-entangled if and only if it is factorizable, e.g.,

$$|\Theta_{non\text{-}ent}\rangle = |\uparrow\rangle_1 |\downarrow\rangle_2 . \tag{11-11}$$

By contrast, if the two particles are indistinguishable, the situation is not so clear. Let us focus on the case of two fermions. There are non-factorizable states that are uncontroversially accepted as entangled since they were obtained by the anti-symmetrization of a non-factorizable state. For example,

$$|\Psi_{ent}\rangle = {}^{1}\!\big/\!\sqrt{4}\,(|L\rangle_1|\uparrow\rangle_1|R\rangle_2|\downarrow\rangle_2 - |L\rangle_1|\downarrow\rangle_1|R\rangle_2|\uparrow\rangle_2 - |R\rangle_1|\downarrow\rangle_1|L\rangle_2|\uparrow\rangle_2 + |R\rangle_1|\uparrow\rangle_1|L\rangle_2|\downarrow\rangle_2), \tag{11-12}$$

where $|L\rangle$ and $|R\rangle$ represent spatial locations left and right. This is a legitimate anti-symmetric state since $P|\Psi_{ent}\rangle = -|\Psi_{ent}\rangle$ and it can be obtained from the anti-symmetrization of the non-anti-symmetric non-factorizable state $|\psi_{12}\rangle$:

$$|\Psi_{ent}\rangle = A|\psi_{12}\rangle = A\left[\frac{1}{\sqrt{2}}(|L\rangle_1|\uparrow\rangle_1|R\rangle_2|\downarrow\rangle_2 - |L\rangle_1|\downarrow\rangle_1|R\rangle_2|\uparrow\rangle_2)\right], \tag{11-13}$$

where A is the anti-symmetrizer projector. However, factorizable states of the form of eq. (11-11) are not legitimate in the case of indistinguishable particles. In the case of two fermions, a factorizable vector as $|\Psi'_{12}\rangle = |L\rangle_1|\uparrow\rangle_1|R\rangle_2|\downarrow\rangle_2$ must be anti-symmetrized,

$$|\Psi_{tang}\rangle = A|\Psi'_{12}\rangle = {}^{1}\!\big/\!\sqrt{2}\,(|L\rangle_1|\uparrow\rangle_1|R\rangle_2|\downarrow\rangle_2 - |R\rangle_1|\downarrow\rangle_1|L\rangle_2|\uparrow\rangle_2), \tag{11-14}$$

which is a legitimate anti-symmetric state since $P|\Psi_{tang}\rangle = -|\Psi_{tang}\rangle$. Fred Muller and Gijs Leegwater (2022) call states of this kind, whose non-factorizability is the result of symmetrization or antisymmetrization, *tangled*. The question is, then, whether tangled states are entangled or not.

The seemingly simplest answer to the problem is to retain the definition of entanglement as non-factorizability even in the indistinguishable case (see, e.g., Jean-Marc Lévi-Leblond quoted in Ghirardi 2005, 340). However, there are good reasons to reject this view: tangled states do not give rise to correlations that violate Bell's inequalities, the probability distribution associated with them factorizes, they are fully compatible with local hidden-variable models, and they cannot be used as resources for quantum information processing as legitimate entangled states. This difficulty leads to two different but related problems: the need for a differentiated treatment of distinguishability and indistinguishability, and the need to make a decision about tangled states.

Let us consider the first problem. In a detailed and repeatedly referenced paper on entanglement of composite quantum systems, GianCarlo Ghirardi, Luca Marinatto, and Tullio Weber (2002) (see also Ghirardi and Marinatto 2003, 2004, 2005) explicitly criticize the definition of entanglement as non-factorizability, and offer a new characterization that intends to be valid for both the distinguishable and the indistinguishable case. Inspired by the idea of "element of physical reality" as proposed in the EPR paper (Einstein, Podolsky, and Rosen 1935), the authors state that two systems are non-entangled when they possess a complete set of properties. In the distinguishable case, this definition is equivalent to the factorizability of the state. In the indistinguishable case, although factorizability cannot be the criterion of non-entanglement due to symmetrization, this fact "does not forbid to attribute a complete set of physical properties to the subsystems: the only claim that one cannot make is to attribute the possessed property to one rather than to the other constituent." (Ghirardi and Marinatto 2003, 383). In other words, even if one cannot say that "particle 1 has a complete set of properties and particle 2 also has a complete set of properties," one can meaningfully say that "there are two particles, each one with its complete set of properties." On the basis of their definition of entanglement, Ghirardi and collaborators prove that what we called tangled states are always non-entangled (see Theorems 2.1 and 2.2 in Ghirardi and Marinatto 2004, 2–3). They also show that these states cannot violate Bell's inequalities (see Ghirardi and Marinatto 2004, 3).

Despite proposing a generic criterion for non-entanglement—possessing a complete set of properties—, Ghirardi, Marinatto, and Weber (2002) analyze entanglement of distinguishable particles and of indistinguishable particles in different ways, in particular, in separate parts of their article, Part II and Part III, respectively. Regarding this point, Cord Friebe (2014) argues that the authors appeal to a kind of double standard: they use different individuation criteria in the distinguishable and the indistinguishable cases. According to them, two distinguishable particles, say an electron and a proton, are individuated by their different state-independent properties. So, their individuality is independent of the state: they can share their state-dependent properties. But the authors cannot use that individuation criterion for indistinguishable particles, since they share all their state-independent properties: "The talk of «distinguishing» particles by telling «which is which» is ambiguous, because GMW use two different sorts of labels, namely labels according to the description theory of proper names, in the case of non-similar particles, and labels according to the direct reference theory of proper names in the case of similar particles." (Friebe 2014, 93).

The word "*factorism*," coined by Adam Caulton (2011), refers to the traditional assumption that, given a factorization of the Hilbert space of a system, the labels of the factor Hilbert spaces correspond to the component particles. As Dieks and Lubberdink (2022) clearly explain, factorism is not valid in the case of indistinguishable particles due to the requirement of symmetrization or antisymmetrization of the states. In fact, whereas the formalism of Hilbert spaces is adequate for the distinguishability case, in the indistinguishability case "a different formalism is required to characterize the subunits between which entanglement is considered." (Tichy, Mintert, and Buchleitner 2011, 9). In this sense, some authors claim that the Hilbert space formalism includes a "surplus structure" (Redhead 1980) that must be eliminated by formal means. For example, in their highly-cited paper, Jon Magne Leinaas and Jan Myrheim (1977) remove what they call "redundancy in notation" by redefining the configuration space. In turn, James Ladyman, Øystein Linnebo, and Tomasz Bigaj (2013) propose to replace the tensor product between Hilbert spaces by the wedge product, which identifies states related by particle permutation. As already commented, some authors even consider that dealing with indistinguishability requires a different logical formalism; this is the case of Krause (1992) and his quasi-set theory.

Of course, this need for a differentiated treatment of the cases of distinguishability and indistinguishability has not been an insurmountable obstacle to exploit the peculiarities of indistinguishability in information processes and quantum computation. However, from a conceptual point of view, a unified treatment is desirable, not only for practical reasons, but also because it would cancel the need to explain the peculiar nature of indistinguishable *particles*. In fact, the above aspects depend essentially on adopting a particle-centered perspective, that is, on accepting from the very beginning the existence of entities conceived as particles but which, at the same time and paradoxically, have a completely different nature from that of classical particles. In this respect it is interesting to mention that, in their comprehensive study of entanglement in indistinguishable particle systems, Fabio Benatti, Roberto Floreanini, Fabio Franchini, and Ugo Marzolino (2020) propose three reasonable consistence requirements that the definition of entanglement for quantum systems made of indistinguishable particles should satisfy: "i) entanglement should emerge as a manifestation of non-local correlations between commuting subalgebras of operators; ii) the entanglement of effectively distinguished identical particles [through measurement] should agree with the standard distinguishable particle entanglement; iii) in absence of other quantum resources, entanglement should be identifiable as the only means to overcome classical performances." (2020, 18). On the basis of the analysis of four approaches to the issue based on particle-partitioning of quantum systems, the authors conclude that all of them violate at least one of the above requirements, which are necessary to guarantee the consistency of a concept of entanglement applicable to both distinguishable and indistinguishable particles.

Regarding the second problem, that referred to the nature of tangled states, Muller and Leegwater (2022) argue against what they call "∀-Factorism," that is, the standard assumption according to which factorism applies unrestrictedly. The authors claim that the correct position is "∃-Factorism": only in certain factorizations of certain Hilbert spaces the labels do refer. In order to show the confusions ∀-Factorism leads to, the authors consider a product state of the form

$$|\Xi_{prod}\rangle = {}^1\!/\!\sqrt{2}\big(|L\rangle_1|\uparrow\rangle_1 + |R\rangle_1|\uparrow\rangle_1\big){}^1\!/\!\sqrt{2}\big(|L\rangle_2|\downarrow\rangle_2 + |R\rangle_2|\downarrow\rangle_2\big). \tag{11-15}$$

Antisymmetrization turns this state into

$$\begin{aligned}|\Xi_{tang}\rangle = {}^1\!/\!\sqrt{2}\,(&|L\rangle_1|L\rangle_2 + |R\rangle_1|R\rangle_2 + |L\rangle_1|R\rangle_2 + \\ &+ |R\rangle_1|L\rangle_2)\ {}^1\!/\!\sqrt{2}\,(|\uparrow\rangle_1|\downarrow\rangle_2 - |\downarrow\rangle_1|\uparrow\rangle_2).\end{aligned} \tag{11-16}$$

On the basis of Ghirardi and collaborators' criterion for non-entanglement, state $|\Xi_{tang}\rangle$, since tangled, should be non-entangled and, as a consequence, should not violate Bell's inequalities. Nevertheless, Muller, and Leegwater point out that $|\Xi_{tang}\rangle$ seems to violate Bell's inequalities. So, they close their article with the rhetorical question: "Were Ghirardi et al. mistaken to propound that, *contra communis opinio*, tangled states do not deserve to be called 'entangled' *because* they do not lead to a violation of some Bell inequality?" (Muller and Leegwater 2022, 15). In other words, is the state of eq. (11-16) entangled even though it is tangled, so the link tangled-non-entangled is not valid? Or is it non-entangled but can nevertheless violate Bell's inequalities, so the link entangled-violation of Bell's inequalities is not valid?

In a detailed study of the theoretical and experimental aspects of entanglement, Malte Tichy, Florian Mintert, and Andreas Buchleitner (2011) distinguish between two forms of entanglement of indistinguishable particles: (i) traditional entanglement, as defined by Ghirardi and collaborators, which applies when the systems are measured by observables corresponding to each single particle, and (ii) *measurement-induced entanglement*, which arises when the particles are measured by observables of the whole system but that do not correspond to each single particle. For instance, let us consider two fermions in the following state:

$$|\Psi_{12}\rangle = 1\big/\sqrt{2}\,(|L\rangle_1|\uparrow\rangle_1|R\rangle_2|\downarrow\rangle_2 - |R\rangle_1|\downarrow\rangle_1|L\rangle_2|\uparrow\rangle_2). \tag{11-17}$$

This is a tangled state because it is non-factorizable exclusively due to the antisymmetrization of the state $|\Psi\rangle = |L\rangle_1|\uparrow\rangle_1|R\rangle_2|\downarrow\rangle_2$. If the state of eq. (11-17) is measured in the directions L and R, that is, by the observables $|L\rangle\langle L|$ and $|R\rangle\langle R|$, then Bell's inequalities are not violated, as expected according to Ghirardi and collaborators' criterion. But let us consider that the whole system is measured in directions L' and R', different than L and R, that is, by the observables $|L'\rangle\langle L'|$ and $|R'\rangle\langle R'|$ such that (see Tichy, Mintert, and Buchleitner 2011, 14)

$$|L'\rangle = 1\big/\sqrt{2}\,(|L\rangle + |R\rangle) \text{ and } |R'\rangle = 1\big/\sqrt{2}\,(|L\rangle - |R\rangle). \tag{11-18}$$

In this case, quantum correlations between the measurements in the two new directions are found, and Bell's inequalities are violated (for a detailed discussion, see Dieks 2020). This case shows that tangled states may violate Bell's inequalities in the case of measurement-induced entanglement. This kind of entanglement is not merely formal, but can be exploited as a resource for quantum information and quantum computing processing (see, e.g., Roch et al. 2014, Hensen et al. 2015, Lo Franco and Compagno 2018, Nosrati, Castellini, Compagno, and Lo Franco 2020; for an updated presentation of the operational approach to entanglement in identical particle systems, see Morris et al. 2020). The state of eq. (11-16), which worries Muller and Leegwater, is a particular case of this situation. Therefore, if one accepts that the violation of Bell's inequalities is a distinctive mark of entanglement, tangled states seem to be peculiar in the sense that the question of whether they are entangled or not does not have a univocal answer.

The two problems presented in the previous paragraphs, which derive from a particle-view, appear under a new light in an ontology of properties. As explained in the previous chapter, in the MHI ontology, indistinguishability is not a relation between individual particles, but is manifested as a symmetry internal to a non-objectual and indivisible bundle-system; this symmetry derives from the indistinguishability of the properties that compose the bundle. And, as argued in Section 11.5 of this chapter, symmetrization and antisymmetrization of states are not conceptually necessary, because the right value of the expectation values is given by symmetric observables (see Fortin and Lombardi 2022). On this basis, the particle-view and the property-view can be compared. From the traditional perspective, the need to symmetrize or anti-symmetrize the states of indistinguishable particles (i) prevents the factorization of the Hilbert space of the whole system into factor spaces corresponding to particles, and (ii) goes against the traditional definition of entanglement in terms of factorization. From the new perspective, by contrast, (i) the algebra of observables can always be decomposed into subalgebras and, since there are no particles, there is no privileged decomposition (see Sections 8.1 and 8.2 of Chapter 8), and (ii) entanglement can be defined in a generic way in the formal algebraic framework. In fact, A. P. Balachandran, T. R. Govindarajan, Amilcar R. de Queiroz, and A. F. Reyes-Lega (2013a, 2013b) offer a completely algebraic approach to entanglement. The guiding idea underlying their work is that, given a system $\mathcal{S}$ with component systems $\mathcal{S}^1$ and $\mathcal{S}^2$, a pure state $|\Psi\rangle$ of $\mathcal{S}$ is non-entangled if and only if the corresponding reduced states of $\mathcal{S}^1$ and $\mathcal{S}^2$ are also pure (see Ghirardi, Marinatto, and Weber 2002). However, when the tensor product structure $\mathcal{H}^1 \otimes \mathcal{H}^2$ cannot be obtained, as in general in the bosonic and fermionic cases, the operation of partial trace is ill-defined and, as a consequence, reduced states cannot be computed as usual. Nevertheless, the definition of reduced state can be retained (see Section 8.3 of Chapter 8). What Balachandran and collaborators show is that the partial trace operation has a natural generalization in the operation of *restriction to a subalgebra*. Let us consider the algebra of operators $\mathcal{O}$, a state ω on $\mathcal{O}$, and a partition of subalgebras $\mathcal{O}^1 \vee \mathcal{O}^2$. A state ω_r^1 is said to be the *restriction* of ω *to* $\mathcal{O}^1$, $\omega_r^1 = \omega|_{\mathcal{O}^1} : \mathcal{O}^1 \to \mathbb{C}$, when

$$\forall O^1 \in \mathcal{O}^1, \quad \left\langle O^1 \otimes I^2 \right\rangle_\omega = \left\langle O^1 \right\rangle_{\omega_r^1}, \tag{11-19}$$

where I^2 is the identity operator of $\mathcal{O}^2$. In the authors' terms: "when $\mathcal{H}$ is not of the form of a «simple tensor product,» partial trace is not a suitable operation. In contrast, if the system is described in terms of a state ω_ρ on an algebra $\mathcal{A}$, it is still sensible to describe a subsystem in terms of a corresponding subalgebra $\mathcal{A}_0$ and of the restriction $\omega_{\rho,0}$ of ω_ρ to $\mathcal{A}_0$." (Balachandran et al. 2013b: 2). Entanglement for pure states can be characterized on this basis: a pure state ω on $\mathcal{O}$ is non-entangled with respect to the partition $\mathcal{O}^1 \vee \mathcal{O}^2$ if the state restrictions $\omega_r^1 = \omega|_{\mathcal{O}^1}$ and $\omega_r^2 = \omega|_{\mathcal{O}^2}$ are also pure. It is easy to see that this characterization of entanglement agrees with Ghirardi's characterization—in terms of reduced states computed as partial traces in the Hilbert space formalism—when the system can be represented as a tensor product structure. This algebraic approach to entanglement is general enough to be applied not only to bosons and fermions, but also to the case of parastatistics and in any other physical situation in which a relevant tensor product structure is not available: "It allows us to meaningfully treat the entanglement of identical and nonidentical particles on an equal footing, without the need to resort to different criteria according to the case under study." (Balachandran et al. 2013b: 1). In a series of papers, Howard Barnum, Lorenza Viola, and collaborators (Barnum et al. 2003, 2004, Viola and Barnum 2010) give a new twist to entanglement from an algebraic perspective. These authors propose a further generalization of concept of entanglement, according to which the relativity of entanglement (see Section 8.2 of Chapter 8) no longer needs to be referred to a partitioning of the system into subsystems: this generalized concept is relative to a preferred or "distinguished" subspace of observables, which can be selected for operational reasons but has not ontological significance.

Once the particle concept is abandoned and the algebraic formalism is adopted, dealing with indistinguishability does not need a differentiated treatment as in the traditional approaches (see Lombardi 2023a). To the extent that there are no particles corresponding to Hilbert spaces, the problem of factorism disappears: one can always speak of subalgebras representing the subsystems or the parts of a whole system, but these components are not related to individual particles belonging to the category of object and located in some spatial position. From this perspective, there is no need for a new formalism to deal with entanglement in the indistinguishability case: the algebraic approach is a tool adequate for a generic treatment. In turn, the "surplus structure" of the Hilbert space formalism becomes innocuous: obtaining the correct mean values is guaranteed by the symmetry of the observables, without the need to symmetrize or anti-symmetrize the states. In other words, "[s]ince states are devices designed exclusively to compute expectation values, if two vectors yield the same expectation values for all the observables of the algebra representing the system, they are two mathematical representations of the same physical situation: the two states are physically equivalent." (Fortin and Lombardi 2022, 240–241).

After criticizing the approaches based on the notion of particle, Benatti and collaborators (2020) analyze the entanglement between orthogonal modes based on the second-quantization approach, showing that this type of entanglement is free from the limitations of particle-based approaches. In a sense, the proposal of an ontology of properties without particles points in the same direction, but with no need to conceptually appeal to second quantization. In fact, the present proposal agrees with Benatti and collaborators' idea of focusing "no longer on the particle aspect, rather on generic degrees of freedom thus accommodating properties more general than those of particle type." (Benatti, Floreanini, Franchini, and Marzolino 2020, 16).

Regarding the perplexity of tangled states, the new way of conceiving the quantum ontology allows us to think about the conceptual relations between tangled, entangled, violation of Bell's inequalities, and quantum correlations from a different perspective (see Lombardi 2023a). From the perspective of an ontology of properties that takes entanglement relativity seriously, the conceptual perplexities introduced in the papers of Muller and Leegwater (2022) and of Tichy and collaborators (2011) disappear. A given state of a system may be tangled relative to the "traditional" partition of the system into what are usually conceived of as particles; with respect to the observables of these components, Bell's inequalities are not violated and EPR-correlations do not obtain. But the same state may be entangled relative to a different decomposition, in terms of observables that do not correspond to those supposed "particles"; with respect to this decomposition, Bell's inequalities are violated and EPR-correlations are obtained. Nevertheless, since the quantum ontology is inhabited by properties and not by objectual particles, all decompositions are equally legitimate: there is no difference between a "true" entanglement and an entanglement that seems to be merely apparent since "induced" by measurement. In other words, once it is accepted that there are no longer particles in the ontology, distinguishing between a "real" entanglement and a "measurement-induced entanglement" loses all its meaning. Of course, implementing certain observables in a laboratory may not be an easy task, particularly when they are "non-local" in the sense that they do not refer to properties confined to a localized region of space. Nevertheless, as mentioned above, experimental physicists are creative enough to devise means of performing such non-local measurements. And when those non-local observables are implemented, the entanglement, the Bell's inequalities violation, and the EPR-correlations thus obtained are not less objective or legitimate than those obtained in the traditional local case.

In summary, Ghirardi and collaborators are right when they claim that the association between

entanglement – violation of Bell's inequalities – EPR - correlations

is not broken by tangled states. But what must not be forgotten is that these associated concepts are always *relative to the partition of the whole system*. And when, from the perspective of an ontology of properties, it is accepted that there is no privileged decomposition, any possible confusion and perplexity is simply dissolved.

11.8 Revisiting the top-down view

In Chapter 8, a top-down approach to quantum mechanics was introduced from a physical viewpoint. According to this approach, the theory applies primarily to closed systems and only derivatively to their constituents. In fact, strictly speaking only closed systems are quantum systems, since described by the dynamical postulate of the theory. When certain degrees of freedom are cut out from the closed system, the non-unitarily evolving result is a mere part of the whole, defined only in relation to the remaining parts.

After the journey through the ontology of properties proposed by the MHI, that top-down approach acquires an ontological significance. Quantum mechanics applies primarily to closed systems due to the ontological priority of the whole bundle over the bundle's components, which lack the identity necessary to be reidentified when they merge into the whole. From this perspective, the different aspects of the top-down approach prove to be endowed with an ontological meaning.

Reduced states or, more generally, restriction states are not the quantum states of the open "subsystems" because the bundles that merge into a new bundle are not subsystems: they lose their existence once combined. Thus, reduced states and restriction states are formal devices that make it possible to compute the expectation values of certain observables, which nevertheless could be computed by means of the quantum state of the whole quantum system. In the MHI ontology of properties, the relevant concept is the probability of actualization of each I-case property of the I-type properties of the whole bundle, and not some putative state of an open individual system.

It makes no sense to say that a quantum state is entangled or not without specifying the partition of the closed bundle-system to which the state corresponds. In ontological terms, entanglement is relative to the partition of the whole bundle into collections of I-type-properties. In turn, entanglement embodies quantum correlations. But in an ontology without objects, they are not correlations between properties of different objects, but correlations between properties of a single bundle. This way of conceiving correlations is in perfect conceptual harmony with the radical holistic interpretation of non-separability: any bundle is a single whole, ontologically non-decomposable into component bundles. In other words, the relativity of entanglement with respect to the multiple partitions of the system is nothing but the manifestation of the multiple correlations between the properties of a single bundle, which have nothing to do with non-local influences between spatially separated individual subsystems.

This view about entanglement has relevant consequences with respect to how decoherence must be understood. In the proposed ontological context, it makes no sense to say that a system, conceived as an object, decoheres when interacting with an environment, because there are no objects in the ontology. Decoherence is a relative phenomenon that depends on the I-type-properties of the whole closed bundle that are selected to be studied: certain collections of I-type-properties decohere and certain others do not. Nevertheless, the observable viewpoint, rooted in the ontological priority of properties, makes classicality a property of observables with no ambiguity or relativity. Therefore, classicality is not a feature of objectual systems: in an ontology of properties without objects, classicality is an objective and non-relative feature of certain I-type-properties of the whole bundle. This makes it possible to understand that, in the same system, certain properties may behave as if they were classical and others may preserve their quantum behavior.

The observable viewpoint can be naturally justified from a physical perspective: we have empirical access to the quantum domain only through the expectation values of observables. Moreover, this viewpoint finds an adequate mathematical expression in the algebraic formalism. But it is in the MHI ontology that it acquires its philosophical justification. In fact, if the fundamental items of the ontology are the I-type-properties with its corresponding I-case-properties, the perspective that takes observables, physical counterparts of those I-type-properties, as the primary elements of the theory turns out to be a natural consequence of the ontological picture.

11.9 Ontology: A summary

In Chapter 4, where the basic elements of the MHI were introduced, the importance of distinguishing between the mathematical language, the physical language, and the ontological language was pointed out. However, in many cases the three languages have overlapped in a strictly imprecise manner in favor of readability; for example, when talking about an algebra of observables or the eigenvalues of an observable. It is therefore worthwhile to close these chapters devoted to the ontological facet of the MHI by recalling the distinction in a summarized form (Table 1).

A final remark is in order. The MHI has been under development for years with the aim of providing a coherent and comprehensive answer to the interpretative challenges of quantum mechanics, both those of a more technical nature—such as the measurement problem and the interpretation of decoherence—and those of an ontological nature. In this presentation, an effort has been made to show how the different aspects of the interpretation fit harmoniously with each other. However, it is necessary to recognize that the proposed ontology of properties could be adopted for metaphysical reasons but independently of the MHI actualization rule. Indeed, a realist reading of quantum mechanics, interested in the most appropriate ontological picture for the theory, could embrace a domain devoid of objects, with its peculiar structure, but without committing itself to the solution provided by the MHI to the measurement problem in its different aspects and without considering the application of the interpretation to physical models. Or, on the contrary, somebody who did not feel interested in ontological questions could leave aside the ontological facet of MHI and retain the actualization rule and its applications. These are certainly not the paths followed in this presentation, but the possibility of making such decisions remains undeniably open.

Table 1 Correspondence between the mathematical, the physical, and the ontological languages.

Mathematics	**Physics**	**Ontology**
	Physical magnitude A	U-type-property $\langle A \rangle$
Self-adjoint operator A	Observable A	I-type-property $[A]$ of the U-type-property $\langle A \rangle$
Eigenvalues a_i of the self-adjoint operator A	Values a_i of the observable A	I-case properties $[a_i]$ of the I-type-property $[A]$
—–––––	—–––––	AI-case property $[\boldsymbol{a}_k]$ of the I-type-property $[A]$
Space $\mathcal{O}$ of the self-adjoint operators of the algebra $\mathcal{A}$	Quantum system $\mathcal{S}$	Bundle $\mathfrak{B}$ of I-type-properties
Probability function	Probability	Propensity
Linear functional ρ on $\mathcal{A}$	Quantum state ρ	Encoding of propensities

12

Revisiting the Measurement Problem

Chapter 6 was devoted to argue that the MHI offers a solution to the measurement problem for both ideal and non-ideal measurements, and to show how it provides a distinction between reliable and non-reliable measurements in the non-ideal case. However, in the last decades more complex measurement scenarios have been considered, some as old as the so-called *Wigner's friend experiment*, and others recently designed to test the theoretical predictions of quantum mechanics. For this reason, in the present chapter, I will revisit the measurement problem from the perspective of the challenges that those scenarios pose to interpretation.

12.1 Maudlin's characterization of the measurement problem

According to the standard von Neumann model, a quantum measurement is an interaction between a system and a measuring apparatus. The problem consists in explaining why, being the post-interaction state a superposition of the eigenstates of the apparatus' pointer, the pointer acquires a definite value. This formulation of the problem rests on three implicit assumptions: that the state of a system specifies its properties, that the state evolution is always linear, and that the apparatus' pointer acquires a single definite value. On this basis, Tim Maudlin (1995) presents what he calls "the problem of outcomes" as the incompatibility between the following three claims:

(a) The quantum state of a system is complete, i.e., the quantum state specifies (directly or indirectly) all the physical properties of a system.
(b) The wavefunction always evolves in accordance with a linear dynamical equation (the Schrödinger equation).
(c) Measurements always have unique and determinate outcomes: e.g., the measurement of the spin of an electron always (or at least usually) has a determinate outcome, i.e., at the end of the measurement, the measuring device is either in a state which indicates spin up (and not down) or spin down (and not up).

In the light of this presentation of the measurement problem, it is often claimed that there are only three interpretative strategies to deal with it: denying completeness (a), which leads to hidden-variable theories, in particular, to Bohmian Mechanics, denying linearity (b), which leads to some version of the Spontaneous Collapse Theory, or denying uniqueness (c), which leads to the Many Worlds Interpretation or to some other version of the Everett-Like Interpretations. In general, Modal Interpretations are

The Modal-Hamiltonian Interpretation of Quantum Mechanics. Olimpia Lombardi, Oxford University Press.
© Olimpia Lombardi (2026). DOI: 10.1093/9780198951407.003.0012

not included in the list. The few times they are discussed in the context of Maudlin's presentation, they are subsumed in the first strategy, as a case of hidden-variable interpretations. In fact, Modal Interpretations accept both (b) and (c), so that it follows that they must violate (a). However, according to Modal Interpretations, the quantum state does not fix all the actual physical properties of a system (i.e., the actual values of all its definite-valued physical quantities), which is the criterion of incompleteness presupposed in (a). This tension shows that the issue deserves further examination.

As Laura Ruetsche (2003) argues, it is important to distinguish whether the Modal Interpretations are viewed as hidden-variables theories, in which value states are added as hidden variables to the original formalism in order to obtain a full description of the physical situation, or rather as an interpretative move that equips the original formalism with a new semantics. For modal theorists, the second alternative is the correct one.

The fact that the quantum state does not specify all actual properties is not appropriately expressed by saying that quantum mechanics, modally conceived, is "incomplete." The incompleteness terminology suggests the idea of a shortcoming of the theory to be solved by completing it (with hidden variables). But modal theorists maintain that quantum mechanics is not in need of an addition, but of an interpretation, and Modal Interpretations already provide everything that can reasonably be expected from theories describing a fundamentally *indeterministic* world in which modality is objective. First, modal schemes completely specify which observables are definite-valued. Second, they fix the possible values that may be taken by these observables and they provide the probabilities of actualization of these values. Since it is an essential modal principle that the actualization of properties in the quantum world is a fundamentally indeterministic process, nothing beyond this probabilistic description is possible. From this standpoint, insisting that modally interpreted quantum mechanics is necessarily incomplete is tantamount to stipulating that any probabilistic theory is incomplete by definition, even if it applies to a fundamentally probabilistic world.

In other words, Maudlin's proposition (a) can be denied in two ways. One is by considering that the quantum state of a system is incomplete and, consequently, certain additional (perhaps hidden) variables must be added to complete it, that is, to specify all the physical properties of the system. But proposition (a) can be denied from a more fundamental point of view, precisely, by rejecting that the role of the quantum state is to specify in some way the properties of the system. In the particular case of the MHI, which properties acquire definite values is determined, not by the state, but by the Hamiltonian of the system; and they acquire their definite values in a completely indeterministic way. The role played by the quantum state is to provide a specification of the propensities to acquire those definite values, measured by the corresponding probabilities. The quantum state specifies propensities/probabilities, not properties. Therefore, the fact that it does not fix all the properties of the system is not a manifestation of incompleteness, because fixing properties is not the function of the quantum state.

12.2 Wigner's friend

The "Wigner's friend" experiment is a *Gedankenexperiment* originally devised by Eugene Wigner (1961) for the purpose of arguing that consciousness affects quantum processes. Wigner considers the superposition state of a particle in a sealed laboratory where an observer *F* (his friend) is confined. A "super-observer" *W* (Wigner) is placed outside the laboratory. When Wigner's friend measures the system, they get one of the possible outcomes and the system collapses into the corresponding state. By contrast, as long as Wigner does not perform a measurement, for him, the whole system, measured system+friend, remains in superposition. Since there is no privileged perspective or ultimate observer, the two different quantum descriptions must be considered on a par: "When Wigner's friend observes an outcome, does the state collapse for Wigner as well? If not, how can we reconcile their different accounts of the process?" (Brukner 2018, 1). In simple terms, the problem involved in Wigner's friend experiment consists in the fact that Wigner and his friend give not only different, but also incompatible descriptions of the same quantum situation.

The thought experiment of Wigner's friend was widely discussed and was approached from different interpretative perspectives. In the Many Worlds Interpretation, the issue is solved by positing that there are many copies of Wigner's friend in different worlds, each one of which observes a different outcome, so the superposition dissolves. From the perspective of Bohmian Mechanics, the perplexity is the result of forgetting that the wavefunction alone does not supply the complete description of the quantum situation. According to QBism, since the quantum state does not represent a physical situation but the state of knowledge of an observer, there is nothing paradoxical in the fact that Wigner and his friend assign different states to the same system. In turn, Relational Quantum Mechanics was specifically designed to cope with Wigner's friend scenarios: the descriptions of Wigner and his friend are not incompatible because quantum facts are always relative to the observer.

It is important to notice that the troubles with Wigner's friend *Gedankenexperiment* appear under the assumption of collapse. For this reason, non-collapse approaches tend to dissolve the perplexities. By contrast, the solutions provided by some of the interpretations that incorporate collapse, whether objective or subjective, introduce relativity with respect to the observer. Now the question is how to account for Wigner's friend experiment in the context of the MHI.

The first step in that direction is to purge the problem of subjective or mind-dependent aspects. Since Wigner proposed his thought experiment in the framework of the discussion about if and how mind influences the physical world, the situation is commonly described in terms of Wigner's and his friend's descriptions, or of their perceptions or mental contents. But once one takes distance from that framework, the quantum arrangement must be described in terms of systems and measuring devices. This is the path I will follow in this section.

The experiment involves a system $\mathcal{S}$ to be measured, and two measuring apparatuses $\mathcal{M}^W$ and $\mathcal{M}^F$ corresponding to Wigner and his friend, respectively. The respective Hilbert spaces are $\mathcal{H}^S$, $\mathcal{H}^W$, and $\mathcal{H}^F$. Let us consider that $\mathcal{S}$ is in a superposition state

$$|\psi_S\rangle = \sum_i c_i\,|a_i\rangle \in \mathcal{H}^S, \tag{12-1}$$

where the $|a_i\rangle$ are the eigenvectors of the observable A of $\mathcal{S}$. The measurement of $\mathcal{S}$ by $\mathcal{M}^F$ requires an interaction that introduces the correlation between the $|a_i\rangle$ and the eigenstates $|p_i^F\rangle$ of the pointer P^F of $\mathcal{M}^F$. As explained in Section 6.2 of Chapter 6, if the interaction Hamiltonian $H_{\text{int}}^{(S,M^F)}$ is

$$H_{\text{int}}^{(S,M^F)} = -\frac{\lambda\hbar}{\Delta t_1}\left(A \otimes R^{P^F}\right), \tag{12-2}$$

where λ is a constant, Δt_1 is the time the interaction lasts, and R^{P^F} is the observable conjugate to P^F, $[P^F, R^{P^F}] = i\hbar I^F$, where I^F is the identity on $\mathcal{H}^F$, then the state of the system $\mathcal{S} \cup \mathcal{M}^F$ becomes (see Mittelstaedt 1998)

$$|\psi_{SM^F}\rangle = \sum_i c_i\,|a_i\rangle \otimes |p_i^F\rangle \in \mathcal{H}^S \otimes \mathcal{H}^F. \tag{12-3}$$

Since there is no collapse, the state of $\mathcal{S} \cup \mathcal{M}^F$ state remains in superposition. Nevertheless, according to the MHI, if $\mathcal{M}^F$ is correctly designed, its pointer P^F acquires one of its possible values p_i^F, and this is the value reported by Wigner's friend. Of course, such a value cannot be predicted because actualization is essentially indeterministic.

Now the question is what happens outside the sealed laboratory, where Wigner has access to his measurement apparatus $\mathcal{M}^W$. As long as the laboratory remains sealed, Wigner's apparatus $\mathcal{M}^W$ records nothing and, therefore, Wigner cannot make any physically grounded statement about what is going on inside the laboratory. Wigner can give a physical description of what happened with his friend's measurement only when the lab is opened and his measuring apparatus physically interacts with the lab's content. In this case, the interesting question is: if Wigner measures the pointer P^F of $\mathcal{M}^F$(strictly speaking, the observable $I^S \otimes P^F$ of $\mathcal{S} \cup \mathcal{M}^F$, where I^S is the identity on $\mathcal{H}^S$) by means of the pointer P^W of $\mathcal{M}^W$, will his measurement result match his friend's measurement result? In order to answer this question, a second interaction must be introduced, such that it correlates the eigenstates $|p_i^F\rangle$ of the pointer P^F of $\mathcal{M}^F$ with the eigenstates $|p_i^W\rangle$ of the pointer P^W of $\mathcal{M}^W$. In this case, the interaction Hamiltonian $H_{\text{int}}^{(M^F,M^W)}$ is

$$H_{\text{int}}^{(M^F,M^W)} = -\frac{\lambda\hbar}{\Delta t_2}\left(P^F \otimes R^{P^W}\right), \tag{12-4}$$

where λ is a constant, Δt_2 is the time this second interaction lasts, and R^{P^W} is the observable conjugate to P^W, $[P^W, R^{P^W}] = i\hbar I^W$, where I^W is the identity on $\mathcal{H}^W$. As a consequence of this interaction, the state of the system $\mathcal{S} \cup \mathcal{M}^F \cup \mathcal{M}^W$ becomes

$$|\psi_{SM^FM^W}\rangle = \sum_i c_i \, |a_i\rangle \otimes |p_i^F\rangle \otimes |p_i^W\rangle \in \mathcal{H}^S \otimes \mathcal{H}^F \otimes \mathcal{H}^W. \tag{12-5}$$

It is interesting to notice that this reasoning is similar to that used to explain consecutive measurements in Section 6.8 of Chapter 6. The difference is that, in that case, the second measurement re-measures the system on the same or in another basis, whereas, in this case, the second measurement measures the first measurement apparatus in the basis of its pointer. Since in this case there is no change of basis, the situation is much simpler and the correlation between the values of the two pointers is direct. In fact, the probability that P^W acquires the value p_i^W given that P^F acquired the value p_i^F can be computed as

$$pr\left(p_i^W | p_i^F\right) = \frac{pr\left(p_i^W \wedge p_i^F\right)}{pr\left(p_i^F\right)}. \tag{12-6}$$

As explained when discussing consecutive measurements, $pr\left(p_i^W \wedge p_i^F\right)$ is a legitimate probability since (i) the pointers P^W and P^F commute because they belong to different measuring apparatuses which, as a consequence, are represented by different Hilbert spaces, and (ii) the two measurements are not simultaneous. The probabilities involved in eq. (12-6) can be directly computed from eq. (12-5), with the following result: $pr\left(p_i^W \wedge p_i^F\right) = pr\left(p_i^F\right) = |c_i|^2$. Therefore,

$$pr\left(p_i^W | p_i^F\right) = 1. \tag{12-7}$$

In other words, when Wigner performs a measurement by means of his measuring apparatus, from the result he obtains he can infer the result obtained by his friend when they measured the system by means of their measurement apparatus.

In summary, when the Wigner's friend thought experiment is described strictly in quantum terms without collapse, not only is there no paradox derived from the descriptions of different observers, but also the agreement between the results obtained by the two friends in their measurements can be theoretically explained.

12.3 The Frauchiger-Renner argument

In April 2016, Daniela Frauchiger and Renato Renner published online an article entitled "Single-world interpretations of quantum theory cannot be self-consistent," in which they introduced a *Gedankenexperiment* that led them to conclude that, if "quantum theory is applied to model an experimenter who herself uses quantum theory," then "no single-world interpretation can be logically consistent." (Frauchiger and

Renner 2016, 1). Their argument intended to support the many-worlds interpretation of quantum mechanics, to the extent that it would force us "to give up the view that there is one single reality." (Frauchiger and Renner 2016, 22). In a new version of the paper, now entitled "Quantum theory cannot consistently describe the use of itself" and published in *Nature Communications* in September 2018, the authors moderated their original claim. This new version proposes the same thought experiment, but now to "investigate the question whether quantum theory can, in principle, have universal validity," and the conclusion is "that quantum theory cannot be extrapolated to complex systems, at least not in a straightforward manner." (Frauchiger and Renner 2018, 1); on this basis, the authors consider how the different interpretations of Standard Quantum Mechanics and the different quantum theories should face their result.

Since its first online publication, the Frauchiger and Renner's (F-R) argument has caused quite a splash in the field of quantum foundations. In general, it has been considered as a new no-go result for quantum mechanics. In many cases, even more extreme reactions can be found, based on conceiving the F-R argument as a kind of proof of the internal inconsistency of quantum mechanics. The immense impact of Frauchiger and Renner's work is due to the fact that their argument is neutral regarding interpretation: on the basis of three very generic and apparently reasonable assumptions that do not include interpretative premises, the argument seems to lead to a contradiction. This fact has been regarded as pointing to a deep shortcoming of quantum mechanics itself, which contrasts with the extraordinary success of the theory. The purpose of the present section is to offer a careful reconstruction of the F-R argument in order to clarify its conceptual scope.

The F-R *Gedankenexperiment* relies on duplicating Wigner's setup. Let us consider two friends F_1 and F_2 located in separate and isolated labs L_1 and L_2, respectively, where the labs are represented by the Hilbert spaces $\mathcal{H}^1$ and $\mathcal{H}^2$, respectively. F_1 measures the observable C of a biased "quantum coin" in the state $\left(1/\sqrt{3}\right)|h\rangle + \left(\sqrt{2/3}\right)|t\rangle$, where $|h\rangle$ and $|t\rangle$ are the eigenstates of C, and h and t are its respective eigenvalues. F_1 prepares a qubit in the state $|\downarrow\rangle$ if the outcome is h, or in the state $|\rightarrow\rangle$ if the outcome is t, and sends it to F_2. When F_2 receives the qubit, they measure its observable S_z, with eigenstates $|\uparrow\rangle$ and $|\downarrow\rangle$ and respective eigenvalues $+1/2$ and $-1/2$. After these two measurements, the state of the whole system composed of the two labs is

$$|\Psi\rangle = \frac{1}{\sqrt{3}}|H\rangle|\Downarrow\rangle + \sqrt{\frac{2}{3}}|T\rangle|\Rightarrow\rangle \quad \in \quad \mathcal{H}_1 \otimes \mathcal{H}_2, \qquad \text{(12-8)}$$

where

- $|H\rangle$ and $|T\rangle$, eigenstates of an observable A with respective eigenvalues H and T, are the states of the entire lab L_1 when the outcome of F_1's measurement is h and t, respectively,

- $|\Uparrow\rangle$ and $|\Downarrow\rangle$, eigenstates of an observable B with respective eigenvalues $\Uparrow$ and $\Downarrow$, are the states of the entire lab L_2 when the outcome of F_2's measurement is $+1/2$ and $-1/2$, respectively,
- and $|\Rightarrow\rangle = 1/\sqrt{2}(|\Uparrow\rangle + |\Downarrow\rangle)$.

The *Gedankenexperiment* continues by considering two "Wigner" observers, W_1 and W_2, located outside the labs, who will respectively measure the observables X and Y of labs L_1 and L_2:

- X has eigenvalues fail_X and ok_X, with respective eigenstates $|\text{fail}_X\rangle$ and $|\text{ok}_X\rangle$ such that:

$$|\text{fail}_X\rangle = \frac{1}{\sqrt{2}}(|H\rangle + |T\rangle) \qquad |\text{ok}_X\rangle = \frac{1}{\sqrt{2}}(|H\rangle - |T\rangle). \tag{12-9}$$

- Y has eigenvalues fail_Y and ok_Y, with respective eigenstates $|\text{fail}_Y\rangle$ and $|\text{ok}_Y\rangle$ such that:

$$|\text{fail}_Y\rangle = \frac{1}{\sqrt{2}}(|\Downarrow\rangle + |\Uparrow\rangle) \qquad |\text{ok}_Y\rangle = \frac{1}{\sqrt{2}}(|\Downarrow\rangle - |\Uparrow\rangle). \tag{12-10}$$

Before analyzing the consequences of the thought experiment, Frauchiger and Renner point out that their argument can be conceived as a no-go theorem that proves that three "natural-sounding" assumptions, (Q), (C), and (S), cannot all be valid (2018, 2):

(Q) *Compliance with quantum theory*: Quantum mechanics is universally valid, that is, it applies to systems of any complexity, including observers. Moreover, an agent knows or is certain that a given proposition is true whenever the Born rule assigns probability 1 to it.
(C) *Self-consistency*: Different agents' predictions are not contradictory.
(S) *Single-world*: From the viewpoint of an agent who carries out a particular measurement, this measurement has one single outcome.

In the 2016 paper, Frauchiger and Renner implicitly consider (Q) and (C) as unavoidable: as a consequence, they claim that their argument shows that "no single-world interpretation can be logically consistent" (2016, 1) and, therefore, "we are forced to give up the view that there is one single reality" (2016, 22). By contrast, in the 2018 paper, they stress that "[t]he theorem itself is neutral in the sense that it does not tell us which of these three assumptions is wrong" (2018, 2); as a consequence, they admit the possibility of different theoretical and interpretative perspectives regarding their result, and include a table showing which of the three assumptions each interpretation or quantum theory violates (2018, 8).

On the basis of the above elements—experimental setup and assumptions—, the F-R argument proceeds as follows. First, in order to compute the probability that the measurements of X and Y yield the results ok_X and ok_Y, respectively, the state described by eq. (12-8) must be expressed as

$$|\Psi\rangle = \frac{1}{\sqrt{12}}\,|\text{ok}_X\rangle\,|\text{ok}_Y\rangle - \frac{1}{\sqrt{12}}\,|\text{ok}_X\rangle\,|\text{fail}_Y\rangle + \frac{1}{\sqrt{12}}\,|\text{fail}_X\rangle\,|\text{ok}_Y\rangle + \sqrt{\frac{3}{4}}\,|\text{fail}_X\rangle\,|\text{fail}_Y\rangle\,. \tag{12-11}$$

From this eq. (12-11) it is clear that the probability of obtaining ok_X and ok_Y is 1/12.

The second part of the argument consists in showing that the observers involved in the experiment can draw a different conclusion from the above one on the basis of the following reasoning. Let us consider the probability that F_2 obtains $-1/2$ in their S_z measurement and W_1 obtains ok_X in their X measurement; in order to compute this probability, the state described by eq. (12-8) must be expressed as

$$|\Psi\rangle = \sqrt{\frac{2}{3}}\,|\text{fail}_X\rangle\,|\Downarrow\rangle + \frac{1}{\sqrt{6}}\,|\text{fail}_X\rangle\,|\Uparrow\rangle - \frac{1}{\sqrt{6}}\,|\text{ok}_X\rangle\,|\Uparrow\rangle\,. \tag{12-12}$$

From this eq. (12-12) it is easy to see that the considered probability is zero. Then, if W_1 obtains ok_X in their X measurement on lab L_1, they can infer that the outcome of F_2's S_z measurement on the qubit was $+1/2$. In turn, if F_2 obtains $+1/2$ in their S_z measurement on the qubit, they can infer that the outcome of F_1's C measurement on the quantum coin was t, because otherwise F_1 would send F_2 the qubit in state $|\downarrow\rangle$. And if F_1 obtains t in their C measurement on the quantum coin, they can infer that the outcome of W_2's Y measurement on lab L_2 will be fail_Y, because the outcome t is perfectly correlated with the state $|\Rightarrow\rangle$ of lab L_2, and $|\Rightarrow\rangle = |\text{fail}_Y\rangle$ (see eqs. (12-10)). Therefore, from a nested reasoning it can be concluded that, when W_1 gets ok_X, they can infer that W_2 certainly gets fail_Y. But this conclusion contradicts what was inferred from eq. (12-11), that is, that there is a non-zero probability that W_1 gets ok_X and W_2 gets ok_Y.

The reactions to the F-R argument have been multiple and varied. Some authors consider that the conclusion of the argument follows from the collapse hypothesis (see, e.g., Laloë 2018, Araújo 2018). These opinions are not completely unfounded, since the entire explanation of the "Wigner's friend paradox" included in the Introduction of the 2018 article presupposes collapse, and the new *Gedankenexperiment* is presented as an extension of Wigner's one. However, the single-world assumption (S) does not amount to nor imply collapse. Whereas the hypothesis of collapse imposes the non-unitary modification of the system's state due to measurements, (S) says nothing about the system's state: it only establishes that any measurement has a single outcome, and this may happen even if the system persists in its unitary evolution, as in the case of the Modal Interpretations. If the conclusion of the F-R argument depended on collapse, it would lose much of its appealing: it not only would depend on an interpretative assumption, but also would offer not much novelty when compared with the original Wigner's friend thought experiment. By contrast, what has shocked most of the physics community is that the argument seems to show an internal inconsistency of quantum mechanics *at the level of probabilities*, independently of any interpretative addition.

An interesting response emphasizes an implicit assumption of the argument: the non-relational view of quantum mechanics as an indispensable premise of the derivation. This is the stance of Časlav Brukner, who considers, from an operational perspective, that the self-consistency condition (C) is too restrictive, since "the states referring to outcomes of different observers in a Wigner-friend type of experiment cannot be defined without referring to the specific experimental arrangements of the observers, in agreement with Bohr's idea of contextuality" (Brukner 2018, 8). From a non-operational standpoint, Dennis Dieks (2019; see also Bene and Dieks 2002) advocates, in the line of Carlo Rovelli's relational view (1996), for a perspectivalist interpretation of quantum mechanics: when the perspectival nature of quantum states is included as a premise, no contradiction can be inferred from the F-R argument. According to Richard Healey (2018), the F-R argument implicitly depends on an inconclusive additional assumption, "intervention insensitivity," which guarantees that the truth-value of the outcome of a counterfactual measurement is insensitive to the occurrence of a physically isolated intervening event.

From a completely different perspective, the conclusion of the F-R argument was rejected on the basis of Bohmian mechanics, the paradigmatic one-world non-collapse quantum theory. For instance, Anthony Sudbery (2017) offers a Bell-Bohmian reconstruction of the argument, claiming that it supplies a counter-example to the conclusion obtained by Frauchiger and Renner. With a similar reasoning, Dustin Lazarovici and Mario Hubert (2019) assert that any Bohm-type theory provides a logically consistent description of F-R *Gedankenexperiment* if the state of the entire system and the effects of all measurements are taken into account.

After supplying his clear and elegant reconstruction of the F-R argument, Jeffrey Bub (2020a) claims that what he calls the "Frauchiger-Renner contradiction" shows that quantum mechanics should be understood probabilistically, as a new sort of non-Boolean probability theory, rather than representationally, as a theory about the elementary constituents of the physical world. Bub (2020b) also considers that the soundness of the F-R argument relies on the fact that it does not speak of the values of observables nor of the agent's knowledge, but only involves certainty. However, the term "certainty" does not have any standard meaning, but is a technical term entirely defined by the assumptions of the F-R argument: "As used in the Frauchiger-Renner argument, 'being certain' is a technical term entirely specified by the three assumptions. In other words, 'being certain' can mean whatever you like, provided that the only inferences involving the term are those sanctioned by the three assumptions." (Bub 2020b, 210). The problem with this strategy is that it makes the F-R argument circular (for a detailed criticism, see Fortin and Lombardi 2020).

In order to analyze the F-R argument, the first step is to clean the discourse of any reference to observers and to what they know, or see, or are certain of, since quantum mechanics, as a physical theory, does not speak of the mental or visual states of agents. For this purpose, I will reformulate the F-R argument in terms of quantum propositions of the form "the property P has the value p," which will be represented as "$P : p$" for conciseness. From now on, I will use the symbols "$\neg$," "$\wedge$," "$\vee$," "$\rightarrow$," and "$\leftrightarrow$" for negation, conjunction, disjunction, conditional, and biconditional respectively, as usual. On the basis of these clarifications, the first part of the F-R argument leads to the conclusion that, with probability 1/12, the following proposition can be asserted:

$$X : \text{ok}_X \wedge Y : \text{ok}_Y. \tag{12-13}$$

The second part of the F-R argument is a reasoning that reads:

(a) $X : \text{ok}_X \rightarrow S_z : +1/2,$
(b) $S_z : +1/2 \rightarrow C : t,$
(c) $C : t \rightarrow Y : \text{fail}_Y,$
(concl) $X : \text{ok}_X \rightarrow Y : \text{fail}_Y.$

By defining $\rightarrow$ in terms of $\wedge$, and by considering that fail_Y and ok_Y are the only two eigenvalues of Y, the conclusion (concl) can be expressed as:

$$X : \text{ok}_X \rightarrow Y : \text{fail}_Y \equiv \neg(X : \text{ok}_X \wedge \neg Y : \text{fail}_Y) \equiv \neg(X : \text{ok}_X \wedge Y : \text{ok}_Y). \tag{12-14}$$

The contradiction of the F-R argument results from the conjunction of eqs. (12-13) and (12-14).

Now the F-R argument is formulated clearly enough to formally show that it does not require collapse, but does require assumption (Q): a quantum proposition can be asserted/denied when the Born rule assigns probability 1/0 to it. It is also necessary to remember that the eigenstates/eigenvalues of the observable A of lab L_1 are correlated with the eigenstates/eigenvalues of the observable C of the coin, and the eigenstates/eigenvalues of the observable B of lab L_2 are correlated with the eigenstates/ eigenvalues of the observable S_z of the qubit:

$$C : h \leftrightarrow A : H \qquad C : t \leftrightarrow A : T, \tag{12-15}$$

$$S_z : +1/2 \leftrightarrow B : \Uparrow \qquad S_z : -1/2 \leftrightarrow B : \Downarrow . \tag{12-16}$$

- Proposition (12-13) can be asserted in some situations because $\Pr(X : \text{ok}_X \wedge Y : \text{ok}_Y) = 1/12$, which is implied by eq. (12-11).
- Proposition (a) can be transformed by taking into account eqs. (12-16), the definition of $\rightarrow$ in terms of $\wedge$, and the fact that $\Uparrow$ and $\Downarrow$ are the only two eigenvalues of B:

$$X:\mathrm{ok}_X \rightarrow S_z: +1/2 \equiv X:\mathrm{ok}_X \rightarrow B:\Uparrow \equiv \neg(X:\mathrm{ok}_X \wedge \neg B:\Uparrow) \equiv$$
$$\neg(X:\mathrm{ok}_X \wedge B:\Downarrow). \tag{12-17}$$

In order to assert $\neg(X:\mathrm{ok}_X \wedge B:\Downarrow)$, it must be proved that the probability $\Pr(X:\mathrm{ok}_X \wedge B:\Downarrow) = 0$, which in turn requires to express the state $|\Psi\rangle$ *in the basis* X-B of $\mathcal{H}_1 \otimes \mathcal{H}_2$ as follows:

$$|\Psi\rangle = \sqrt{\frac{2}{3}}\,|\mathrm{fail}_X\rangle\,|\Downarrow\rangle + \frac{1}{\sqrt{6}}\,|\mathrm{fail}_X\rangle\,|\Uparrow\rangle - \frac{1}{\sqrt{6}}\,|\mathrm{ok}_X\rangle\,|\Uparrow\rangle. \tag{12-18}$$

- Proposition (b) can be transformed by taking into account eqs. (2-15) and (2-16), again the definition of $\rightarrow$ in terms of $\wedge$, and the fact that T and H are the only two eigenvalues of A:

$$S_z: +1/2 \rightarrow C:t \equiv B:\Uparrow \rightarrow A:T \equiv \neg(B:\Uparrow \wedge \neg A:T) \equiv \neg(B:\Uparrow \wedge A:H). \tag{12-19}$$

In order to assert $\neg(B:\Uparrow \wedge A:H)$, it must be proved that the probability $\Pr(B:\Uparrow \wedge A:H) = 0$, which in turn requires to express the state $|\Psi\rangle$ *in the basis* B-A of $\mathcal{H}_1 \otimes \mathcal{H}_2$ as follows:

$$|\Psi\rangle = \frac{1}{\sqrt{3}}\,|H\rangle\,|\Downarrow\rangle + \frac{1}{\sqrt{3}}\,|T\rangle\,|\Uparrow\rangle + \frac{1}{\sqrt{3}}\,|T\rangle\,|\Downarrow\rangle. \tag{12-20}$$

- Proposition (c) can be transformed by taking into account eqs. (12-15), again the definition of $\rightarrow$ in terms of $\wedge$, and the fact that fail_Y and ok_Y are the only two eigenvalues of Y:

$$C:t \rightarrow Y:\mathrm{fail}_Y \equiv A:T \rightarrow Y:\mathrm{fail}_Y \equiv \neg(A:T \wedge \neg Y:\mathrm{fail}_Y) \equiv$$
$$\neg(A:T \wedge Y:\mathrm{ok}_Y). \tag{12-21}$$

In order to assert $\neg(A:T \wedge Y:\mathrm{ok}_Y)$, it must be proved that the probability $\Pr(A:T \wedge Y:\mathrm{ok}_Y) = 0$, which, in turn, requires to express the state $|\Psi\rangle$ *in the basis* A-Y of $\mathcal{H}_1 \otimes \mathcal{H}_2$ as follows:

$$|\Psi\rangle = \frac{1}{\sqrt{6}}\,|H\rangle\,|\mathrm{fail}_Y\rangle + \frac{1}{\sqrt{6}}\,|H\rangle\,|\mathrm{ok}_Y\rangle + \sqrt{\frac{2}{3}}\,|T\rangle\,|\mathrm{fail}_Y\rangle. \tag{12-22}$$

This detailed reconstruction of the F-R argument makes easy to see that, in the reasoning of the second part of the argument, the conclusion (concl) is obtained by applying the classical inference rule of the transitivity of the conditional, according to which, from the conjunction $(p \rightarrow q) \wedge (q \rightarrow r)$, the proposition $(p \rightarrow r)$ can be inferred. Therefore, the conjunction between the conditional propositions (a), (b), and (c) is necessary in order to infer (concl) by transitivity. Each one of those propositions was obtained from a probabilistic assertion that was computed by expressing the state $|\Psi\rangle$ in a different basis of the Hilbert space $\mathcal{H}_1 \otimes \mathcal{H}_2$ of the whole system, formed by the two labs L_1 and L_2: the bases *X-B*, *B-A*, and *A-Y*, for (a), (b), and (c), respectively (see eqs. (12-18), (12-20), and (12-22), respectively). But they are three *different bases*, rotated with respect to each other. In other words, arriving at the contradiction by means of the F-R argument requires *making classical conjunctions between propositions corresponding to different contexts*, something that the non-Boolean structure of the quantum propositions forbids, as is well-known since 1967, when Simon Kochen and Ernst Specker demonstrated their famous theorem.

According to Frauchiger and Renner (2018, 5, caption of Figure 3, my emphasis), "If a theory T (such as quantum theory) enables consistent reasoning (C) then it must allow any agent A to promote the conclusions drawn by another agent A' to his own conclusions, provided that A' has the same initial knowledge about the experiment and *reasons within the same theory* T." This means that, if the shared initial knowledge is warranted, (C) requires that two agents reason within the same theory T and use the same system of logic to obtain their conclusions. But, what is the relation between a theory T—in this case, a physical theory as quantum mechanics—and the system of logic by means of which the agents draw their own conclusions? Answering this question confronts us with a dilemma (for a detailed argumentation, see Fortin and Lombardi 2020).

The first horn of the dilemma consists in considering that the logic by means of which the agents make inferences is classical, defined on a Boolean structure of propositions, independently of the theory T. This is the perspective adopted by Frauchiger and Renner, since they allow the agents to use classical logic to make their inferences. On this basis, they prove that accepting the assumptions (Q)—compliance with T = quantum theory—, (S)—single outcomes of measurements or single truth values of propositions—, and (C)—agents making inferences on the basis of T = quantum theory with classical logic—leads to a contradiction. This view sounds pretty reasonable, since in general human agents seem to reason with classical logic. Moreover, since the assumptions are statements about the theory T, they are not constrained by T. Therefore, the first horn of the dilemma admits the contradictory conclusion. However, in this case the argument would not offer a novel result, since it was already proved by the Kochen-Specker theorem: classical conjunctions of propositions corresponding to different contexts lead to contradiction. In other words, according to this first horn, the F-R argument is an original and interesting way to get a result already obtained by other means: the argument is a no-go theorem, but what does not go was already well known and provides no new insights about quantum mechanics—and much less does it offer evidence of an internal inconsistency of the theory.

According to the second horn of the dilemma, it is considered that a physical theory T constrains the range of systems of logic that can be used to make inferences with its propositions. In other words, the mathematical structure of the theory T embodies the algebra of T-propositions, which restricts the logic that is appropriate for operating with these propositions. Therefore, in the case of quantum mechanics, making inferences with classical logic on quantum propositions is not legitimate, because it conflicts with the non-Boolean algebra of those propositions. From this viewpoint, the F-R argument is illegitimate, to the extent that it assumes compliance with quantum theory by (Q), but according to (C) allows the agents to make inferences with classical logic. This position might sound quite extreme: asking agents to drop out classical logic and to use inferences based on a "quantum logic" seems to be a too demanding requirement. Nevertheless, the situation is not as serious as it seems. The only condition that agents must satisfy is not to construct conjunctions between propositions corresponding to different contexts (bases). And this is not a very strong condition; in fact, quantum physicists in their daily work not only are aware that certain conjunctions are not admissible, but also know exactly which those conjunctions are. Or, from another perspective: any professor of quantum mechanics will certainly give a low mark to a student's exam that includes a conjunction of propositions corresponding to the values of non-commuting observables.

Recently, it has been suggested that the fact that inferences in the F-R argument are all carefully timed would circumvent any objection based on the contextuality of quantum mechanics (Bub, personal contact). However, if timing really matters in the F-R argument, then it must be reconstructed by means of a formalism that allows dealing with quantum properties at different times, something that is not formally represented in the standard formalism. Such a timed formalism is given by the Theory of Consistent Histories, developed by the contributions of various quantum physicists since the 1980s (Griffiths 1984, 2002, 2013, Omnès 1987, 1994, 1999, Gell-Mann and Hartle 1990, 1993). The Theory of Consistent Histories extends the formalism of quantum mechanics with the introduction of the notion of history, which generalizes the notion of event: an elementary history is defined as a sequence of events at different times, where an event is the occurrence of a property. But since probabilities cannot be assigned to the set of all histories, the theory selects the subsets of histories that satisfy the condition of consistency. By applying the Theory of Consistent Histories, it can be shown that the contradiction resulting from the F-R argument is inferred by computing probabilities in a *family of histories* that is *not consistent* (for the formal proof, see Losada, Laura, and Lombardi 2019).

12.4 Encapsulated measurements

The Frauchiger-Renner argument belongs to a class of works dealing with measurement situations that have come to be called "extended Wigner's friend" scenarios. These scenarios involve "*encapsulated*" measurements, in which a super-observer measures any arbitrary observable of a complex quantum system that includes another observer. The common strategy of these works is to propose certain apparently reasonable assumptions, and to prove the inconsistency of those assumptions by means of encapsulated measurements (see, e.g., Brukner 2018, 2020, Bong et al. 2020, Allard-Guérin et al. 2021).

In this context, Veronika Baumann and Stefan Wolf (2018) published a very interesting article, in which the aim is not to prove the inconsistency of certain assumptions, but to show that different interpretations of quantum mechanics, traditionally considered empirically equivalent, may lead to different probabilistic predictions. Their argument also appeals to encapsulated measurements: "Different levels of observation: Observer O performs a measurement on a quantum system S in state $|\phi\rangle$. A super-observer SO then performs a measurement on the joint system including the observer O. For this type of consecutive measurements, the standard quantum formalism and the relative-state formalism yield different conditional probabilities for the measurement results." (Bauman and Wolf 2018, 5, caption of Figure 2). Although the authors' explicit purpose is to compare the "standard quantum formalism" with the "relative state quantum formalism," actually the scope of their argument is more general, since it allows the comparison between the two versions of Standard Quantum Mechanics (SQM), SQM_1, in which the projection postulate is included—the collapse version—and SQM_2, in which it is not included—the non-collapse version—(see Section 2.2 of Chapter 2), as well as between SQM_1 and any non-collapse interpretation.

In the present section, I will discuss Baumann and Wolf's argument but, in agreement with my remark in the previous section, I will not consider observers' perceptions or mental states, but only probabilities of measurement results. I will use a terminology similar to that of the authors, but with slight modifications that, I hope, will clarify the presentation. The detailed development of the argument will allow me to raise certain questions that can be inferred from but are not explicit in Baumann and Wolf's work.

The first step is *stating the problem*. Let us consider a system $\mathcal{S}$, a measuring apparatus $\mathcal{M}^1$ that measures system $\mathcal{S}$, and a measuring apparatus $\mathcal{M}^2$ that measures the joint system $\mathcal{S} \cup \mathcal{M}^1$. The Hilbert spaces of these three systems are $\mathcal{H}^S$, $\mathcal{H}^{M_1}$, and $\mathcal{H}^{M_2}$, respectively, and the Hilbert spaces of $\mathcal{S} \cup \mathcal{M}^1$ is $\mathcal{H}^{SM_1} = \mathcal{H}^S \otimes \mathcal{H}^{M_1}$ and of $\mathcal{S} \cup \mathcal{M}^1 \cup \mathcal{M}^2$ is $\mathcal{H}^{SM_1M_2} = \mathcal{H}^S \otimes \mathcal{H}^{M_1} \otimes \mathcal{H}^{M_2}$. The observables involved in the experiment are:

- $a \in \mathcal{H}^S \otimes \mathcal{H}^S$, with eigenvalues a_i and eigenstates $|a_i\rangle \in \mathcal{H}^S$.
- $A \in \mathcal{H}^{M_1} \otimes \mathcal{H}^{M_1}$, with eigenvalues A_i and eigenstates $|A_i\rangle \in \mathcal{H}^{M_1}$.
- $B \in \mathcal{H}^{M_2} \otimes \mathcal{H}^{M_2}$, with eigenvalues B_i and eigenstates $|B_i\rangle \in \mathcal{H}^{M_2}$.
- $b \in \mathcal{H}^{SM_1} \otimes \mathcal{H}^{SM_1}$, with eigenvalues b_i and eigenstates $|b_i\rangle \in \mathcal{H}^{SM_1}$.

Let us suppose that system $\mathcal{S}$ is initially in the state $|\phi\rangle \in \mathcal{H}^S$, such that

$$|\phi\rangle = \sum_i \alpha_i \, |a_i\rangle, \text{ with } \alpha_i = \langle a_i| \, \phi\rangle . \tag{12-23}$$

The first measurement measures the observable a of the system $\mathcal{S}$ by means of the pointer A of the apparatus $\mathcal{M}^1$. Before the interaction between $\mathcal{S}$ and $\mathcal{M}^1$, the state of the joint system $\mathcal{S} \cup \mathcal{M}^1$ is (the tensor product symbols are omitted in this section)

$$|\phi_{SM^1}\rangle = \sum_i \alpha_i \,|a_i\rangle\, |A_0\rangle \in \mathcal{H}^{SM^1}, \tag{12-24}$$

where $|A_0\rangle$ is the ready-to-measure eigenstate of A. After the interaction, the state of $\mathcal{S} \cup \mathcal{M}^1$ becomes

$$|\phi'_{SM^1}\rangle = \sum_i \alpha_i \,|a_i\rangle\, |A_i\rangle \in \mathcal{H}^{SM^1}. \tag{12-25}$$

The second measurement measures the observable b of the system $\mathcal{S} \cup \mathcal{M}^1$ by means of the pointer B of the apparatus $\mathcal{M}^2$. In order to express the correlation resulting from the second interaction, the state of $\mathcal{S} \cup \mathcal{M}^1$ must be expressed in the basis of b. The relation between the $|b_i\rangle$ and the $|a_j\rangle\,|A_j\rangle$ is given by

$$|b_i\rangle = \sum_j \beta_{ij} \,|a_j\rangle\, |A_j\rangle, \text{ with } \beta_{ij} = \left(\langle a_j|\,\langle A_j|\right)|b_i\rangle, \tag{12-26}$$

or, in the other direction,

$$|a_i\rangle\,|A_i\rangle = \sum_j \chi_{ij}\,|b_j\rangle, \text{ with } \chi_{ij} = \langle b_j|\left(|a_i\rangle\,|A_i\rangle\right). \tag{12-27}$$

After the second measurement, two conditional probabilities can be defined:

- The probability that the observable b of the joint system $\mathcal{S} \cup \mathcal{M}^1$ acquires the value b_l in the second measurement, given that the observable a of system $\mathcal{S}$ acquired the value a_k in the first measurement, represented as $pr(b : b_l \,|\, a : a_k)$.
- The probability that the pointer B of the apparatus $\mathcal{M}^2$ acquires the value B_l in the second measurement, given that the pointer A of the apparatus $\mathcal{M}^1$ acquired the value A_k in the first measurement, represented as $pr(B : B_l \,|\, A : A_k)$.

Since the eigenstates of a are perfectly correlated with the eigenstates of A, and the eigenstates of b are perfectly correlated with the eigenstates of B, it is reasonable to expect that the two conditional probabilities have the same value. However, there is a relevant conceptual difference between them: whereas $pr(B : B_l \,|\, A : A_k)$ involves compatible observables, $[B, A] = 0$, since belonging to different systems, $pr(b : b_l \,|\, a : a_k)$ involves incompatible observables, $[b, a] \neq 0$. In what sense is it possible to speak of the conditional probability between non-commuting observables? The advocate of physical collapse might argue that the two measurements are not simultaneous, nor can they be, because between them there is a collapse understood as a physical phenomenon. Whoever appeals to the projection postulate exclusively as a rule to compute probabilities might say that the conditional probability $pr(b : b_l \,|\, a : a_k)$ is meaningless because it involves non-commuting observables and because in fact it cannot be tested independently of the actual measurements made with the apparatuses $\mathcal{M}^1$ and $\mathcal{M}^2$. Precisely because of the conflicting status of probability $pr(b : b_l \,|\, a : a_k)$, here I will focus only on probability $pr(B : B_l \,|A : A_k)$, which is assumed, at least in theory, to be empirically corroborated from the records of the apparatuses $\mathcal{M}^1$ and $\mathcal{M}^2$. Then, the task will be to compute this conditional probability in the collapse and the non-collapse cases in order to compare the two results.

Let us first consider the *collapse case*. If in the first measurement the apparatus $\mathcal{M}^1$ measures the value A_k, the state $|\phi'_{SM^1}\rangle$ of eq. (12-25) collapses to $|\phi'^{c}_{SM^1}\rangle$:

$$|\phi'_{SM^1}\rangle \longrightarrow |\phi'^{c}_{SM^1}\rangle = |a_k\rangle\,|A_k\rangle\,. \tag{12-28}$$

In order to describe the second measurement, the state $|\phi'^{c}_{SM^1}\rangle$ must be expressed in the basis of b by means of eq. (12-27):

$$|\phi'^{c}_{SM^1}\rangle = |a_k\rangle\,|A_k\rangle = \sum_j \chi_{kj}\,|b_j\rangle. \tag{12-29}$$

Now, the second measurement introduces the correlation between the $|b_j\rangle$ and the $|B_j\rangle$:

$$|\phi''^{c}_{SM^1M^2}\rangle = \sum_j \chi_{kj}\,|b_j\rangle\,|B_j\rangle\,. \tag{12-30}$$

By means of this equation, the probability that the observable B acquires the value B_l when the system $\mathcal{S} \cup \mathcal{M}^1$ is in the state $|\phi''^{c}_{SM^1M^2}\rangle$ can be computed as:

$$pr_{\,|\phi''^{c}_{SM^1M^2}\rangle}(B : B_l) = Tr\left(I^S \otimes I^{M^1} \otimes |B_l\rangle\langle B_l|\,|\phi''^{c}_{SM^1M^2}\rangle\langle\phi''^{c}_{SM^1M^2}|\right) = |\chi_{kl}|^2. \tag{12-31}$$

Since this probability was computed by assuming the collapse to $|a_k\rangle\,|A_k\rangle$ in the first measurement, it can be interpreted as the conditional probability that the pointer B of $\mathcal{M}^2$ acquires the value B_l in the second measurement, given that the pointer A of system $\mathcal{M}^1$ acquired the value A_k in the first measurement:

$$pr_{\,|\phi''^{c}_{SM^1M^2}\rangle}(B : B_l/A : A_k) = pr_{\,|\phi''^{c}_{SM^1M^2}\rangle}(B : B_l) = |\chi_{kl}|^2. \tag{12-32}$$

Let us now consider the *non-collapse* case. In order to describe the second measurement, the state $|\phi'_{SM^1}\rangle$ of eq. (12-25) must be expressed in the basis of b by means of eq. (12-27):

$$|\phi'_{SM^1}\rangle = \sum_i \alpha_i |a_i\rangle |A_i\rangle = \sum_i \alpha_i \left(\sum_j \chi_{ij} |b_j\rangle \right) = \sum_j \gamma_j |b_j\rangle, \text{ with } \gamma_j = \left(\sum_i \alpha_i \chi_{ij} \right). \tag{12-33}$$

Now the second measurement introduces the correlation between the $|b_j\rangle$ and the $|B_j\rangle$:

$$|\phi''_{SM^1M^2}\rangle = \sum_j \gamma_j |b_j\rangle |B_j\rangle. \tag{12-34}$$

Now it is necessary to make a new change of basis by means of eq. (12-26):

$$|\phi''_{SM^1M^2}\rangle = \sum_j \gamma_j \left(\sum_i \beta_{ji} |a_i\rangle |A_i\rangle \right) |B_j\rangle = \sum_j \sum_i \gamma_j \beta_{ji} |a_i\rangle |A_i\rangle |B_j\rangle. \tag{12-35}$$

In this case, the conditional probability must be obtained in terms of the total state $|\phi''_{SM^1M^2}\rangle$ of the whole system $\mathcal{S} \cup \mathcal{M}^1 \cup \mathcal{M}^2$. Since the observables $I^S \otimes I^{M^1} \otimes B$ and $I^S \otimes A \otimes I^{M^2}$ commute, the conditional probability $pr(B : B_l \,|\, A : A_k)$ can be computed with the standard definition:

$$pr_{\,|\phi''_{SM^1M^2}\rangle}(B : B_l \,|\, A : A_k) = \frac{pr_{\,|\phi''_{SM^1M^2}\rangle}(B : B_l \wedge A : A_k)}{pr_{\,|\phi''_{SM^1M^2}\rangle}(A : A_k)}. \tag{12-36}$$

So, with a little bit of patience, it can be proved that

$$pr_{\,|\phi''_{SM^1M^2}\rangle}(B : B_l \,|\, A : A_k) = \frac{|\gamma_l|^2 |\beta_{lk}|^2}{\sum_j |\gamma_j|^2 |\beta_{jk}|^2}. \tag{12-37}$$

When the collapse case and the non-collapse case are compared, the following interesting conclusion can be drawn. The conditional probability $pr(B : B_l \,|\, A : A_k)$ is certainly meaningful and, in principle, can be tested by checking the outcomes of the apparatuses $\mathcal{M}^1$ and $\mathcal{M}^2$. Moreover, it can be computed with and without collapse. By comparing eqs. (12-32) and (12-37) it can be seen that the two cases do not yield the same value. It is in this sense that Baumann and Wolf are right when claiming that Standard Quantum Mechanics with collapse and without collapse are not merely different interpretations but are different theories because based on inequivalent formalisms. This conclusion agrees with my view as introduced in Chapter 1, Section 1.2: two versions of Standard Quantum Mechanics (SQM) must be distinguished according to whether the projection postulate is included in the formalism (SQM_1) or not (SQM_2).

A traditionally minded physicist would be happy to use this predictive difference as a proof of the correctness of collapse and to discredit non-collapse perspectives. However, this is a biased position, since only in encapsulated measurements is the difference obtained and, as is generally accepted, the effective experimental realization of this type of measurement is not a simple task. Therefore, for the time being, the two perspectives, collapse and non-collapse, are on a par.

Perhaps the issue can be approached by appealing to some theoretical resource such as the Lüders Rule. Nevertheless, this strategy depends on how the rule is interpreted. Authors who accept the collapse hypothesis tend to consider the Lüders Rule as a generalization of the projection postulate to the case of measuring degenerate observables (see, e.g., Busch and Lahti 2009). From this standpoint, the application of the Lüders Rule involves a non-unitary change undergone by the state of the system as a result of measurement. Therefore, if the rule is applied to encapsulated measurements under this reading, it is not surprising that the result so obtained agrees with that computed in the case of collapse. However, from a more general perspective, several authors not tied to the idea of collapse interpret the Lüders Rule as a generalization of the classical concept of conditional probability to the quantum case (see, e.g., Bub 1979, Cassinelli and Zanghí 1983; see also many textbooks in the philosophy of quantum mechanics). On the basis of this interpretation, the Lüders Rule can be applied both to the first measurement to obtain $pr(b : b_l \,|\, a : a_k)$ and to the second measurement to obtain $pr(B : B_l \,|\, A : A_k)$. As above, I will focus on this second conditional probability.

If the Lüders Rule is applied to the second measurement, the relevant Hilbert space is $\mathcal{H}^{SM^1M^2} = \mathcal{H}^S \otimes \mathcal{H}^{M^1} \otimes \mathcal{H}^{M^2}$ corresponding to the joint system $\mathcal{S} \cup \mathcal{M}^1 \cup \mathcal{M}^2$. The conditional probability $pr(B : B_l \,|\, A : A_k)$ computed by the Lüders Rule in the state $|\phi''_{SM^1M^2}\rangle \in \mathcal{H}^{SM^1M^2}$ of eq. (12-35) is

$$pr_{\,|\phi''_{SM^1M^2}\rangle}(B : B_l \,|\, A : A_k) = \frac{Tr\left(I^S \otimes I^{M^1} \otimes \Pi_{B_l} \cdot I^S \otimes \Pi_{A_k} \otimes I^{M^2} \,|\phi''_{SM^1M^2}\rangle\langle\phi''_{SM^1M^2}|\, I^S \otimes \Pi_{A_k} \otimes I^{M^2}\right)}{Tr\left(I^S \otimes \Pi_{A_k} \otimes I^{M^2} \,|\phi''_{SM^1M^2}\rangle\langle\phi''_{SM^1M^2}|\, I^S \otimes \Pi_{A_k} \otimes I^{M^2}\right)}. \tag{12-38}$$

With a bit more patience, it can be proved that

$$pr_{\,|\phi''_{SM^1M^2}\rangle}(B : B_l \,|\, A : A_k) = \frac{|\gamma_l|^2 |\beta_{lk}|^2}{\sum_j |\gamma_j|^2 |\beta_{jk}|^2}. \tag{12-39}$$

The result is that the conditional probability so obtained agrees with the conditional probability computed without collapse in eq. (12-37) and not with that computed with collapse in eq. (12-32). In other words, when the Lüders Rule is applied to obtain the conditional probability $pr(B : B_l \,|\, A : A_k)$, which is certainly meaningful and in principle can be tested by checking the outcomes of the apparatuses $\mathcal{M}^1$ and $\mathcal{M}^2$ in repeated measurements, the result does not endorse collapse but supports non-collapse.

In their analysis of the Lüders Rule, Isabel Guerra Bobo (2013) and John Earman (2019) consider that the rule cannot be legitimately interpreted as a generalization of the classical concept of conditional probability to the quantum case. The reason is that the Lüders Rule "does not express a quantum analog of the core idea of classical conditional probability; namely, the idea that the probability of E conditional on F is the probability of what is «common» to E and F, normalized by dividing by the probability of F. [...] the numerator [in the Lüders Rule] is not the probability of what is common to E and F since it is not the probability of any quantum proposition when E and F do not commute." (Earman 2019, 7). Nevertheless, somebody could reply that the Lüders Rule is not a generalization of the *classical* concept of conditional probability but a completely *new* definition of conditionalization adequate to the quantum domain. I do not intend to delve into this issue here, but it is worth noting that the objection formulated by Guerra Bobo and Earman applies when the rule is used to obtain $pr(b : b_l \,|\, a : a_k)$ since, as pointed out above, it involves incompatible observables: $[b, a] \neq 0$. However, when the Lüders Rule is used to obtain the conditional probability $pr(B : B_l \,|\, A : A_k)$, the objection does not apply because, in this case, the involved observables commute with each other: $[B, A] = 0$.

Let us summarize the argument of this section. In agreement with Baumann and Wolf (2018) (but hopefully with a much clearer presentation) it was shown that conditional probabilities in encapsulated measurements do not yield the same result when computed with and without collapse. Guerra Bobo (2013) and Earman (2019) reasonably argue that conditional probabilities involving non-commuting observables are not physically meaningful. Nevertheless, the disagreement between the collapse and the non-collapse perspectives also persists in the case of the conditional probability between the pointers of the measuring apparatuses, which are commuting observables and whose values can be checked by direct observation. The Lüders Rule, when conceived as a meaningful quantum conditionalization rule, supports the non-collapse view. Of course, this is not a definitive argument, but it should at least be sufficient to not rule out non-collapse approaches just because they do not match the collapse account. As should always be the case in physics, the last word rests on experimental results, if physicists will be able to perform encapsulated measurements in a totally reliable way and without loopholes, something that, as far as I know, has not yet been achieved.

Although it is necessary to await the verdict of experiments, I do not want to close this section without considering what would become of the MHI if such encapsulated measurements were effectively implemented and the verdict were favorable to the collapse option. Would this mean that the MHI should be entirely discarded? I think not, because its main features do not depend on rejecting collapse:

- The MHI actualization rule could be maintained, and its application to the problem of measurement in its different manifestations, as shown in Chapter 6, would continue to be valid. Furthermore, the account of consecutive measurements would be simpler than that provided from the non-collapse perspective. The incorporation of collapse would neither affect the application of the MHI actualization rule to the physical models treated in Chapter 5.
- As for the ontological picture provided by the interpretation, the structure of the ontology would not undergo an essential modification by accepting collapse. The only difference would affect the evolution of propensities, which would no longer always evolve unitarily: the collapse of the state would represent the ontological

phenomenon by which propensities change instantaneously and indeterministically when systems interact and give rise to a new system. In that new system, the collapsed state would still encode the propensities that assign certainty to the actualized determinate properties (AI-case-properties) of all the determinables (I-type-properties) selected by the MHI actualization rule as those belonging to the preferred context.

Someone might argue that a hypothetical MHI with collapse would not be a modal interpretation anymore. Of course, this is true to the extent that the characterization of Modal Interpretations includes being non-collapse approaches. Nevertheless, the specificity of Modal Interpretations does not lie in their position with respect to collapse, but precisely in their modal character; therefore, by adopting collapse they would not lose their essential nature. In any case, these are only labels that do not affect conceptual and foundational issues.

In short, encapsulated measurements bring to light fundamental questions for the interpretation of quantum mechanics. But the reasonable strategy for now is to wait and see.

13
Miscellanea

Up to this point, each one of the chapters has dealt with some topic in an argumentatively ordered way. In this last chapter, on the contrary, I want to present some issues that I consider should not be left out, but which are not closely related to each other. Therefore, the first three sections of this miscellanea chapter can be read in any order. The last one, by contrast, makes sense as a closing section since it introduces my general philosophical position, which explains my interpretative strategy with regard to quantum mechanics.

13.1 Relations to other approaches

After the long tour through the different aspects of the MHI, the presentation would not be complete without devoting a section to a brief comparison of the present proposal with other older and better-known approaches. The content of this section will be an exploratory work in this direction, which deserves to be expanded in future research.

Perhaps the furthest approach from MHI is QBism, according to which quantum mechanics does not describe the world, but is concerned with our knowledge and beliefs about the world. This purely epistemic conception of the theory is at odds with the realist stance of the MHI: the difference between the two interpretations is so deep that there is little point in exploring possible agreements on matters of detail.

Another approach with which the MHI has little in common is the Spontaneous Collapse Theory. Not only does it incorporate the collapse of the wavefunction as a spontaneous random process, but it also modifies the formalism by adding stochastic and nonlinear terms to the Schrödinger equation. Also, from the ontological viewpoint the difference is significant: in the matter-density ontology, the only item in the universe is a stuff that "fills" the entire space, and in the flash ontology, the quantum domain is a realm of events; those of stuff and event are ontological categories totally different than the category of property that underlies the MHI ontology.

The MHI shares with the Many Worlds Interpretation two important aspects: the realist conception of quantum mechanics and the non-collapse view. It is precisely due to the non-collapse perspective that both interpretations must take charge of the results of the encapsulated measurements described in Section 12.4 of Chapter 12, where it was explained how the MHI could face the eventual case of a confirmation of the collapse prediction. However, the two interpretations are clearly different regarding the ontological dimension. Whereas the MHI splits reality into the possible and

The Modal-Hamiltonian Interpretation of Quantum Mechanics. Olimpia Lombardi, Oxford University Press.
© Olimpia Lombardi (2026). DOI: 10.1093/9780198951407.003.0013

the actual, making the indeterministic transition from possibility to actuality intelligible, the Many Worlds Interpretation dissolves possibility in actuality by multiplying worlds and, in this way, ends up being a kind of super-determinism. Another point of disagreement relies on the appeal to decoherence, which in recent versions of the Many Worlds Interpretation plays the role of solving the problem of the preferred basis. This strategy is not admissible from the closed-system perspective of the MHI, from which, as explained in Chapter 8, decoherence is a relative coarse-grained process that depends on the observables selected for the description and, as a consequence, cannot play the fine-grained interpretative role assigned to it by the Many Worlds Interpretation.

A scarcely studied approach is the Transactional Interpretation, perhaps due to the challenge that the idea of advanced wavefunctions propagating backward in time offers to physical intuition. The notion of a transaction between a retarded wavefunction generated by an emitter and an advanced wavefunction generated by an absorber is completely alien to the MHI. However, the MHI agrees with the possibilist version of the Transactional Interpretation precisely in its possibilist ontological stance. In fact, the two interpretations agree in that the wavefunction represents objective probabilities, which measure the propensity or potentiality to actualization of the possible events. Although those possibilities and the corresponding probabilities are not reducible to actuality, they are as real as actual facts.

The MHI agrees with Bohmian Mechanics in rejecting collapse and disagrees with it regarding its deterministic view. As mentioned in Chapter 3, Bohmian Mechanics was conceived by some authors as a modal interpretation, with position as the privileged observable; in this sense, it would be closer to the MHI than other Modal Interpretations, which make the preferred context to depend on the instantaneous state of the system. However, Modal Interpretations are usually characterized as accepting the standard formalism, so in this sense Bohmian Mechanics is not a modal interpretation to the extent that it is based on a different formal background. In any case, this is just a matter of name: the conceptually relevant matters are the agreements and disagreements between the different approaches.

Finally, I find it particularly interesting to compare the MHI with relational approaches, such as the Perspectival Modal Interpretation (PMI) and the Relational Quantum Mechanics (RQM), because they can complement and enrich each other (Lombardi and Ardenghi 2022). The three interpretations agree in their realism, their adoption of the standard formalism, the rejection of a classical cut and of an undefined notion of observer, the conception of measurements as ordinary physical interactions, and the interpretation of the quantum state not as a physical field in space but as encoding probabilities. Moreover, although in the PMI and the RQM it is not completely explicit, I think it can be said that they agree in that the actual acquisition of a definite value by an observable is an event: events are actual, objective, and irreducibly indeterministic phenomena, which occur at the interactions. There are also specific differences and agreements with each of the two interpretations. The MHI shares with the PMI its rejection of collapse, but differs from it in the actualization rule, which in the PMI still depends on the instantaneous state as in the original Modal Interpretations. The MHI agrees with the RQM in starting with the observables of the system, but disagrees with it in that RQM assumes collapse (although not a physical phenomenon, collapse is a change in the state occurring at any interaction).

What concerns me here is to examine the relational aspect of these interpretations, according to which (i) properties of quantum systems have a relational character and are defined with respect to another quantum system, and (ii) the different descriptions, given from different perspectives, are equally objective and all correspond to physical reality. Although the ontology of the MHI includes relations that link instances of universal properties (see Section 10.5 of Chapter 10), providing a structure to the algebra of observables, there is an obvious difference with respect to relational approaches: whereas from the relational perspectives the properties of a system acquire definite values only relative to another system, the MHI endows a closed system with certain definite-valued non-relational properties, in particular those belonging to the preferred context. However, I want to argue that this fact is not an obstacle to the compatibility between the MHI and the relational views, but rather an opportunity for complementarity.

Let us begin by recalling the logical-ontological difference between monadic properties, usually called 'properties,' and dyadic properties, usually called 'relations.' Both cases belong to the logical-ontological category of "property," but they are distinguished by their arity n: $n = 1$ for monadic properties, $n = 2$ for dyadic properties. For example, (i) P_a: "having the value a of the observable A" is a monadic property, a so-called *property*, whereas (ii) R_a: "having the value a of the observable A relative to" is a dyadic property, a so-called *relation*. Let us now consider the interaction between a system $\mathcal{S}$ and an apparatus $\mathcal{M}$, leading to the joint system $\mathcal{S} \cup \mathcal{M}$. According to the MHI, in this interaction stage the Hamiltonian of $\mathcal{S} \cup \mathcal{M}$ has a definite value, but the observable A of $\mathcal{S}$ may not acquire a definite value. This means that the actualization of the possible property P_a may not occur. According to a relational interpretation, $\mathcal{S}$ has a definite value, say a, of the observable A relative to $\mathcal{M}$. Therefore, the actualization of the possible relation R_a occurs. But the non-actualization of P_a and the actualization of R_a are perfectly compatible, because *they are different properties*: one monadic and the other dyadic. If the assignments of monadic and dyadic properties introduced by the MHI and the relational interpretations, respectively, are compatible, then in principle it is not necessary to retain one type of property and discard the other one.

But compatibility is not the whole story: the two interpretative stances can be made complementary if applied at different levels: the MHI gives the perspective of the closed system, whereas the relational interpretations describe the objective and relative values acquired by the properties of the open parts during the interaction. Let us consider a measurement such as the Stern-Gerlach experiment, in which the system $\mathcal{S}$ to be measured and the measuring apparatus $\mathcal{M}$ interact during a period $\Delta t = t_2 - t_1$, leading to the joint system $\mathcal{S} \cup \mathcal{M}$, and at time t_2 they stop interacting, becoming the two closed systems $\mathcal{S}$ and $\mathcal{M}$ as before the interaction. In this case,

- From t_1 to t_2, $\mathcal{S}$ and $\mathcal{M}$ are open and have only relational properties with respect to each other, as stated by the relational interpretations.
- For $t > t_2$, systems $\mathcal{S}$ and $\mathcal{M}$, now closed, acquire non-relational properties according to the actualization rule of the MHI.

Therefore, the assignments of actual monadic and dyadic properties are not only compatible, but can also fruitfully collaborate to make sense of a particular physical situation.

Discussions about the interpretation of quantum mechanics often take the form of confrontations, with each perspective defending a particular stance and emphasizing the difficulties or limitations of the others. The most neutral and balanced presentations try to compare different theoretical and/or interpretative approaches, making their pros and cons explicit. Here, my purpose has been to adopt a more proactive perspective, aimed at seeing how different interpretations can enrich each other through a productive dialogue. In particular, I believe that the complementarity between the MHI and the relational interpretations can be highly fruitful and, thus, deserves further study.

13.2 Time in the Modal-Hamiltonian interpretation

In classical mechanics, the state of a system is defined in terms of the actual properties that the system acquires at each time. In Standard Quantum Mechanics, by contrast, the state only provides probabilities on the possible values of the properties of the system, whereas such properties acquire actual values only at some specific points in time (this view excludes Wavefunction Realism, according to which the wavefunction is a concrete physical field in a high-dimensional space; see, e.g., Albert 2013). Therefore, at least two different notions of time must be distinguished in the quantum realm. In Carlo Rovelli's words: "There are two independent notions of time in ordinary quantum mechanics: the time in which the system evolves, and the «time» that orders the measurements of the observer. These two are not related and may be non-coincident." (Rovelli 1991, 130; see also Busch 2008). More precisely:

- The *parameter-time*, also called "dynamical" or "intrinsic" time by Paul Busch (2008), is the time along which the system unitarily evolves. It is represented by the variable t as it appears in the Schrödinger equation.
- The *event-time*, also called "observable" time by Busch (2008), is the time at which particular events occur. Those events are often thought of as measurement results but, more generally, they are any acquisition of a definite value by a certain observable.

The parameter-time is the time involved in the characterization of the Galilei group; thus, in quantum mechanics it is supposed to be homogeneous and isotropic as in the classical case. The Schrödinger equation rules how the probabilities on the possible values of all the observables of the system change along the parameter-time. In his detailed works on the meaning of the uncertainty principles, Busch (1990a, b, 2008) introduced a further distinction on the basis of how the parameter-time is measured. The *external time* is the time measured by a clock that is not dynamically connected with the system studied in the experiment. This time acts as an external parameter of the system's evolution. The *intrinsic time*, by contrast, is defined through the dynamical behavior of the quantum systems themselves: every *dynamical variable* of a physical system can be used to mark the passage of time. In other words, in principle every non-stationary quantity A defines, for any initial quantum state ρ, a characteristic time $\tau_\rho(A)$ along which the expectation value of A changes significantly.

Although both external and intrinsic times make sense in the context of local non-relativistic quantum mechanics, when the focus is quantum gravity the idea of an external time turns out to be at least controversial. Moreover, as many authors emphasize (e.g., Kuchař 1991, Isham 1993), one of the main obstacles to the formulation of a quantum theory of gravity is the difference between the notions of time in quantum theory and in general relativity, a theory invariant under general coordinate transformations. For these reasons, some authors take the perspective of the intrinsic time and attempt to reconstruct it from correlations. For instance, Don Page and William Wootters (1983; see also Wootters 1984) propose an "evolution without evolution." Their central idea is that the universe is a stationary scenario and that time and evolution emerge in a subsystem of the universe, which is entangled with another subsystem that satisfies the conditions to be considered a clock. In turn, Rovelli (2002, 2006) generalizes the Heisenberg picture by distinguishing between traditional observables and partial observables, which correspond to precise measuring procedures. Since, for Rovelli, all partial observables are on the same footing, dynamics expresses the relation between partial observables, one of which may be selected as a clock.

In contrast to the parameter-time, the event-time has no formal representation in quantum mechanics. Yet it is essential in order to endow the theory with physical meaning: testing the theory is only possible by recording specific events, such as the hit of an electron on a screen or the absorption of a photon by an atom, which occur at particular times. The absence of a formal representation perhaps explains why, despite its central relevance, the nature of the event-time and its relation with the parameter-time has been scarcely discussed in the literature on quantum foundations. An exception to this situation is the study of the event-time in the MHI framework: by appealing to the case of consecutive measurements as a toy model, it can be shown how interactions give rise to a net of events from which the event-time emerges (Fortin, Lombardi, and Pasqualini 2022).

In the framework of the MHI, given that the definite-valued observables of the preferred context always commute with the Hamiltonian, they are constants of motion of the closed system. Thus, actualization events—where each event is the acquisition of a definite value by a definite-valued observable—occur only once in the quantum system, at the constitution of the closed system as such. From the viewpoint of the parameter-time, since then there is no change in the realm of actuality: the definite-valued observables with their corresponding definite values remain unchanged during the entire "parameter-lifetime" of the system. However, in the actuality realm, the event-time can only be constituted by the relations between events: there is no event-time between interaction events. For instance, let us consider two systems, A and B, with Hamiltonians H^A and H^B, respectively, that interact to constitute the system C, with Hamiltonian $H^C = H^A \otimes I^B + I^B \otimes H^A + H^{AB}_{\mathrm{int}}$. Such an interaction leads to the actual event by which the Hamiltonian H^C acquires an actual definite value, say, ω^C_k—and analogously the remaining observables of the preferred context. System C, in turn, interacts with a system D, with Hamiltonian H^D, to constitute the system E, with Hamiltonian $H^E = H^C \otimes I^D + I^D \otimes H^C + H^{CD}_{\mathrm{int}}$, which acquires an actual definite value, say, ω^E_k. But it also may happen that this second interaction ceases, so that C and D become closed systems again with their original Hamiltonians H^C and H^D, which again acquire actual definite values, say, ω^C_l and ω^D_l. In this way, "the event-time arises from the network of interaction relations between the systems that compose the whole closed universe," and "the structure of that event-time is embodied in the internal structure of the Hamiltonian of that universe" (Fortin, Lombardi, and Pasqualini 2022, 19; see a universe toy-model based on consecutive measurements, and Figures 1 and 2 in that paper).

Quantum interactions establish a *strict order relation* in the set of events, that is, a relation that is irreflexive, asymmetric, and transitive. Nevertheless, in general it is not a total order but a *partial order*, since it is possible that, given two instants, neither of them is prior to the other, so that the relation "being earlier than" does not hold for them. This feature can be regarded as an obstacle to accept that the event-time is really a time (see, e.g., Gryb and Thébault 2016). However, in view of the peculiar features of general relativity, the fact that the resulting event-time does not have the structure of classical time can be considered as an advantage for the construction of a truly fundamental quantum gravity (see discussion in Fortin, Lombardi, and Pasqualini 2022).

On this basis, the difference between parameter-time and event-time can be related to the modal dimension of the MHI:

- From the possibilist perspective of the MHI, the quantum state does not refer to actual events, but describes possibilities, measured by objective probabilities: it "inhabits" the realm of possibility. Therefore, its time-evolution also occurs in the realm of possibility: the Schrödinger equation encodes how those probabilities change over time. In turn, as just explained, the time of the unitary evolution of the state is the parameter-time. This means that the parameter-time is the time over which possibilities evolve and, consequently, it is what endows the realm of possibility with temporality. Nevertheless, we have access to potentialities only through actual events. As a consequence, the realm of actuality enters the scene.
- According to the MHI, actual events are the acquisition of actual definite values by the Hamiltonians of closed systems at their constitution through interaction relations. The event-time is, then, the time of the realm of actuality: it emerges

from the network set up by actual events through their inter-relations. As a consequence, its nature is essentially relational. Moreover, since actualization is inherently indeterministic, the event-time is not described by the theory. Therefore, whereas the parameter-time is represented by the variable t of the Schrödinger equation, the event-time has no theoretical representation: it is a physical magnitude resulting from observable facts. Precisely due to its nature, the event-time, which arises from the relations between actual events, is not continuous like the parameter-time, but is discrete: there is no event-time between actual events.

13.3 Extrapolation to quantum field theory

During the late 1990s and early 2000s, a number of papers were concerned with considering the possibility of applying the Modal Interpretations of non-relativistic quantum mechanics to the relativistic domain. In general, the diagnosis was negative: the modal approach, at least in its traditional version, cannot be made Lorentz-invariant in a straightforward way. Perhaps this was, in addition to the problem of non-ideal measurements, one of the obstacles that influenced the decline of interest in Modal Interpretations in the following years.

One of the first works in this direction was that of Michael Dickson and Rob Clifton (1998), who investigated the prospects for fundamental Lorentz-invariance in Modal Interpretations, and found that, in general, they cannot be fundamentally Lorentz-invariant. By taking the Spectral-Decomposition Modal Interpretation (SDMI; see Chapter 3, Section 3.4) proposed by Pieter Vermaas and Dennis Dieks (1995) as representative, Dickson and Clifton concluded that Modal Interpretations need a preferred frame that, nevertheless, cannot be significantly found (see also Arntzenius 1998). By contrast, Wayne Myrvold (2002b) analyzed the case of the modal approach suggested by Jeffrey Bub (1992, 1997), which selects some privileged observable as always-definite. By considering two isolated systems evolving between two hypersurfaces of simultaneity, Myrvold proved that it is impossible to satisfy the Born-rule of joint probabilities for possessed values of the privileged observables of the two systems on all possible spacelike hypersurfaces.

Some years later, Joseph Berkovitz and Meir Hemmo (2005a, 2005b, 2006) proposed a way out of the above challenges: they argued that a relational version of the modal interpretation, in which quantum-mechanical states assign relational rather than intrinsic properties, circumvents all the no-go theorems for relativistic Modal Interpretations. From a similar perspective, Dieks (2021) offered a detailed scheme with the help of which the Perspectival Modal Interpretation (PMI; see Chapter 3, Section 3.4) can be made relativistically covariant. In particular, he provided a relativistically covariant treatment of experiments of the EPR-type.

In the context of algebraic quantum field theory (AQFT), Clifton (2000) generalized the Modal Interpretations whose actualization rule is based on the quantum state to the local algebras of quantum field theory (QFT). By assuming that the algebra of observables of a system is a von Neumann algebra, and by relying on a faithful state—a state that maps no non-zero positive elements of the algebra to zero—, Clifton identified an Abelian subalgebra of the von Neumann algebra over which the state defines a mixture of dispersion free states (for an extension to non-faithful states, see Kitajima 2004). John Earman and Laura Ruetsche (2005) considered Clifton's proposal as the best candidate for a modal interpretation adapted to relativistic QFT. In particular, they claimed that, in Clifton's framework, Myrvold's result does not hold because the operators belonging to algebras corresponding to different regions of space-time always commute with each other.

If the above panorama of the application of Modal Interpretations to the relativistic domain is considered comprehensively, two aspects can be noted. First, all the articles were published before the first presentation of the MHI in 2008 and, as a consequence, they do not analyze it. With the exception of Myrvold's argument, all the results were obtained for traditional Modal Interpretations, whose actualization rule is based on the instantaneous state of a composite system (see Chapter 3). Therefore, they do not apply directly to the MHI. Second, with the exception of Clifton's and Earman and Ruetsche's articles, those results were formulated by taking the traditional Hilbert space formalism of quantum mechanics as a starting point. This second point brings us to the question of what reconciling an interpretation of non-relativistic quantum mechanics with special relativity means.

Non-relativistic quantum mechanics, like classical mechanics, is a theory whose symmetry group is the Galilei group. This particular group of transformations identifies the physical observables of the theory and underlies the results about the invariance of the fundamental laws. Trying to "reconcile" a theory, formulated in terms of its symmetry group, with a different symmetry group does not seem to make much sense: I do not believe that anyone was seriously interested in interpreting classical mechanics so as to make it Lorentz-invariant. But this is precisely what the just-mentioned works, except for the last two, try to do with the Modal Interpretations. In this sense, I agree with what Earman and Ruetsche say about Myrvold's and the other works on the subject, with the exception of Clifton's: "This is not a full-blown manifestly relativistic quantum theory, but a mixture of bits and pieces of ordinary nonrelativistic quantum mechanics and relativity theory. The targets of other extant results indicating a tension between Modal Interpretations and relativity are similarly chimerical. A more fruitful way to investigate the alleged tension is to chart the fortunes of the modal interpretation for well-defined quantum theories that are indisputably relativistic." (Earman and Ruetsche 2005, 560). In other words, the task is not *applying* an interpretation designed for non-relativistic quantum mechanics to relativistic QM or to QFT, but *extrapolating* that interpretation to quantum theories with other symmetries. In the case of Modal Interpretations, it is necessary to adapt the actualization rule to the new theories with their own symmetries. As I will argue below, the MHI seems to be particularly well-suited to follow this strategy.

First, the fact that the MHI takes an algebraic formalism for non-relativistic quantum mechanics as the formal departing point is a first advantage when the task is to extrapolate the interpretation to QFT in its algebraic version. The idea behind the AQFT is that measurements are always performed in a region of the space-time. Then, given an algebra $\mathcal{A}$ of observables, each open-bounded region D of the Minkowski space-time is associated with a *local subalgebra* $\mathcal{A}(D)$: $D \to \mathcal{A}(D) \subset \mathcal{A}$. It is assumed that the self-adjoint elements of the local subalgebra $\mathcal{A}(D)$ represent local operations in D. The set of local subalgebras has a net structure characterized by the following features: (i) *isotony*, that is, if $D_1 \subseteq D_2$, then $\mathcal{A}(D_1) \subseteq \mathcal{A}(D_2)$, (ii) *causality*, that is, if D_1 is spacelike separated from D_2, then $[\mathcal{A}(D_1), \mathcal{A}(D_2)] = 0$, and (iii) the algebra $\mathcal{A}$ is called *quasi-local algebra* and can be obtained as $\cup_D \mathcal{A}(D)$. In turn, the state is represented by a positive, linear, and normalized functional ω on $\mathcal{A}$ such that $\omega(A)$ is the expectation value of the self-adjoint element $A \in \mathcal{A}$.

Second, as discussed in Section 7.10 of Chapter 7, the MHI can be extrapolated to QFT by changing the relevant symmetry group. In fact, the symmetry group of AQFT is the Poincaré group $\mathcal{P}$: each Poincaré transformation $T_\alpha \in \mathcal{P}$ is represented by an automorphism s_α such that, if D_α is the Poincaré transformation of D, then $s_\alpha(\mathcal{A}(D)) = \mathcal{A}(D_\alpha)$. Moreover, the vacuum state ω_0 must be invariant under all the transformations of the Poincaré group. Therefore, if in non-relativistic quantum mechanics the definite-valued observables are those represented by the Casimir operators of the Galilei group (and those commuting with them and having the same symmetries), in AQFT the definite-valued observables are those represented by the Casimir operators of the Poincaré group (and those commuting with them and having the same symmetries). Since mass M and squared spin S^2 are the only Casimir operators of the Poincaré group, they are always definite-valued observables, and this conclusion agrees with the common assumption that elementary particles always have definite values of mass and spin, and that mass and spin are two of the properties that contribute to the classification of elementary particles. Moreover, it can be proved that the definite-valued observables in the non-relativistic and the relativistic case are related through adequate limits, despite that, strictly speaking, the physically meaningful symmetry group of quantum mechanics is not the Galilei group but its central extension (recall Section 2.3 of Chapter 2) and the Poincaré group does not admit non-trivial central extensions (see Ardenghi, Castagnino, and Lombardi 2011).

Since the central idea of the MHI is to consider the observables representing invariances as the definite-valued observables of the system, when this interpretation is extrapolated to the relativistic domain, all the symmetries should be considered. In particular, in relativistic quantum theories, besides the space-time symmetries represented by the Poincaré group, quantum systems have internal gauge-symmetries. Therefore, according to the MHI, the observables corresponding to those gauge-symmetries should also be definite-valued. In the case of relativistic quantum mechanics with gauge U(1) fields, the only Casimir operator of this symmetry group is charge Q and, as a consequence, charge is also a definite-valued observable. Again, this is a physically reasonable result since charge, together with mass and spin, are the properties that any quantum system is supposed to always possess and that can be measured in any physical situation; moreover, their values are precisely what define the different kinds of elementary particles.

From the ontological viewpoint, there do not seem to be serious obstacles to extrapolate the MHI ontology of properties to the case of QFT. The ontological picture of bundles of I-type-properties—determinables—with their corresponding I-case-properties—determinates—could be retained with a slight adjustment. In the QFT case, the I-type-properties, corresponding to the observables of the quasi-local algebra $\mathcal{A}$ and mathematically represented by operator-valued quantum "fields" $\phi(x, t)$ and their conjugate "fields," explicitly depend on space-time points, just like their corresponding I-case-properties. In turn, the quantum state, mathematically represented by an expectation-value functional on $\mathcal{A}$, retains the same ontological meaning as in the non-relativistic case: it embodies the propensities to actualization of all the I-case-properties, now functions of space-time position, of all the I-type-properties. This extrapolation appears straightforward and natural because, with the algebraic formalism being the mathematical framework of QFT in its algebraic version, the MHI ontology mimics the structure of the algebraic formalism but now applied to the non-relativistic case. This quantum ontology of properties for QFT may pave the way to overcome the traditional debate between the particle-interpretation and the field-interpretation of the theory.

Of course, the brief comments of this section do not amount to a fully developed interpretation of QFT. Nevertheless, I believe they are sufficient to show that the MHI admits a significant extrapolation to the relativistic domain that deserves further study.

13.4 Pluralist realism

This book was entirely devoted to present the MHI of non-relativistic quantum mechanics in its different aspects. Although the previous section showed how it can be extrapolated to QFT, someone might question the very meaning of this enterprise: Why devote so much time and effort to understand non-relativistic quantum mechanics, which is already known not to be a "fundamental" theory? I do not know how the many physicists and philosophers who are interested in the foundations of non-relativistic quantum mechanics would answer this question. But I can give my own answer, which is based on a general philosophical position that I have called *Kantian-rooted pluralist realism* (see Lombardi 2023b, 2024).

As traditionally conceived, metaphysical realism is a philosophical position that is committed with a "ready-made" world, inhabited by self-subsistent and self-identifying items. It is a kind of realism that also presupposes the possibility of knowing that world as it is in itself, at least approximately, and which becomes a scientific metaphysical realism when the thesis that science is the privileged means to reach that knowledge is added. As clearly expressed by Roberto Torretti: "«Scientific realists» believe that reality is well-defined, once and for all, independently of human action and human thought, in a way that can be adequately articulated in human discourse. They also believe that the primary aim of science is to develop just the sort of discourse which adequately articulates reality—which, as Plato said, «cuts it at its joints»—, and that modern science is visibly approaching the fulfilment of this aim." (Torretti 2000, 114). However, this kind of realism is disproved by the scientific activity itself. We always access the world through our own frameworks. In Kantian terms, the object of knowledge is always the result of a synthesis between our theoretically founded schemes and the independent noumenal reality. Not only the properties and the behavior of the entities studied by science, but also the very existence of those entities depend on the categorical-conceptual framework implicit in a theory. As framework-dependent is the answer to the question "how is it?" as is the answer to the question "what is it?" As Hilary Putnam claims: "«Objects» do not exist independently of conceptual schemes. We cut up the world into objects when we introduce one or another scheme of description" (Putnam 1981, 52). With his internalist realism, Putnam acknowledges his debt to Kantian philosophy. Does this mean that we should embrace idealism?

We have all learnt that Kantian philosophy is a form of idealism. And this is so because, although the existence of something independent of the subject is conceded, the access to the subject-independent reality is always permeated by the active participation of the subject. The object of knowledge is the result of a synthesis between the subject-independent noumenal reality and the framework introduced by the knowing subject. But, if knowledge essentially requires the two elements, why insist on calling this view "idealism?" In fact, using the term "idealist" to describe Kantian philosophy is a metaphysically realist prejudice. It assumes that, whenever the subject participates in the constitution of knowledge, the "pure essence" of realism breaks down. It seems that the only philosophical position that deserves the name "realism" is metaphysical realism, the "God's Eye" point of view.

I resist that metaphysically realist prejudice and prefer to speak of a *Kantian-rooted realism* for two reasons. First, due to the role played by noumenal reality in Kantian philosophy. Although, against pre-critical realism, Kant insists on the unknowability of noumena, he also stresses that a phenomenon involves in itself the reference to something that is not phenomenon and necessarily takes part in the constitution of knowledge: "the purely phenomenal character of the objects of experience does not exclude, but rather implies a transcendental reality that serves them as a basis, and that, although unknowable, is not for this less effective. [...] phenomenal objects are not mere insubstantial ghosts, [...] the perception in which their presence is manifested reveals an effective existence." (Torretti 2005a, 676–677). The second reason is the possibility of retaining a correspondence theory of truth. Since metaphysical realism is traditionally linked to the adoption of truth as correspondence, it is usually assumed that rejecting metaphysical realism necessarily implies giving up any correspondence definition of truth in favor of some kind of coherentist or pragmatist conception. However, this is not the case: the active role of the subject in the constitution of the ontic domain to which language refers only leads us to reject the link between language and an independent reality, but does not force us to drop the intuition of truth as correspondence with facts, which now are constituted in a Kantian sense (see Lombardi and Pérez Ransanz 2012).

Adopting a Kantian perspective, however, does not imply that Kant's philosophy must be accepted in its entirety. After having witnessed the great revolutions and the enormous ramification of the twentieth-century science, still assuming that our knowledge fits into a single framework is not easy. Torretti clearly stresses the limitation of Kant's philosophy: although Kant "rightly compares the scientist with an investigating judge who directs his questions to nature and fixes the terms in which the answer has to be conceived, he never considers the case that, frustrated because the answers contradict with each other, the inquiries get complicated and stuck and, in general, do not seem to lead anywhere, the investigating judge rethinks his questions, modifies the categories in which the answers must be framed, and even changes the goals of the research or the criteria to evaluate their results." (Torretti 2005b, 15). Therefore, in order to account for this kind of situations, common in scientific research, Kantian thesis about the constitution of the object of knowledge must be complemented by *categorical-conceptual relativity*. According to this thesis, no concept, not even the most basic categories, necessarily need to be included in our categorical-conceptual frameworks: there is no privileged concept of object or of existence. In other words, in their access to reality, human beings develop different frameworks, which are neither convergent nor reducible to a single objective one. Hence, the fundamental criticism of metaphysical realism is directed against its commitment to absolute categories and concepts (the way in which these ideas appear in Jorge Luis Borges's short stories is reviewed in Lombardi 2023b).

Accepting the possibility of different categorical-conceptual frameworks leads to the thesis of *ontological pluralism*: in contrast with Kant's convictions, there are different ontic domains, which are equally objective in different contexts and in function of certain interests and purposes. Ontic pluralism points in the same direction as the underdetermination of theories by evidence: both theses have been used against scientific metaphysical realism, which assumes that science converges to the true description of the independent reality. Since the privileged viewpoint of God's Eye does not exist, there is not a single "true" ontic domain: all the domains have the same metaphysical status if all of them are constituted by equally objective descriptions. It is worth emphasizing that they are not mere "epistemologized" domains as opposed to the "real" world: when there is no metaphysically objective ontic domain, the very expression "epistemologized domain" loses any content. In other words, according to this ontologically pluralist standpoint, the question "What objects does the world consist of?" can be meaningfully posed only in the context of a specific framework. Only when we have adopted a system of categories and concepts, we can assume that certain facts and objects are there to be discovered. In other words, ontological questions only make sense from the perspective of knowledge. Supposing otherwise would amount to putting the cart of ontology before the horse of knowledge.

Although different forms of pluralism have been widely discussed in the context of the problem of scientific change, they have been less addressed in a synchronic sense. However, when we realize that different theories and disciplines, embodying different categorical-conceptual frameworks, are accepted and successfully used at the same historical time, we must also admit that different ontic domains may coexist without priorities or privileges, since all of them are equally categorically conceptually constituted. In other words, the Kantian-rooted ontological pluralism is an anti-reductionist and anti-emergentist stance: there are no good philosophical reasons to assume the ontological dependence of the domain described by certain theories on the domain described by others (for formal, historical, and pragmatic arguments for this claim, see Lombardi 2023b).

The ideas of reduction and of emergence are directed to depicting a unified view of reality. Perhaps for this reason, some authors find in ontological pluralism the threat of a disintegration of reality, resulting from a fragmentary science whose different disciplines, and even different theories, are completely disconnected from each other (see Needham 2006, Hettema 2012). Certainly, the belief that reality is not an incoherent and chaotic plurality has guided scientists throughout the history of science, and the ideal of unification acted as a powerful engine for scientific research. But this does not mean that such an ideal must be conceived in terms of ontological dependence. A Kantian-rooted ontological pluralism retains the aim of unification under a more flexible philosophical approach, which follows the perspective inaugurated by Otto Neurath when asserting that science is not oriented toward a single whole, but proceeds by means of local systematizations and, consequently, preserves a plural and always incomplete character. On this basis, Neurath favors an idea of unification based not on hierarchical links, but rather on a picture of science as an encyclopedia, where the connections between theories adopt very different, stronger or weaker, forms (Neurath 1935). From this perspective, inter-theory relations do not express hierarchical and asymmetric relations between ontological domains: they are *symmetric links* working as "bridges" between theories, which can be "crossed" in the two directions and do not impose relations of priority or dependence between the corresponding ontic domains. Therefore, the theories simultaneously accepted by the scientific community form not a chain but a web, where each theory may be connected with more than two other theories, and through different links with each one of them. For instance, classical mechanics is related to classical statistical mechanics, with special relativity, and with quantum mechanics by means of completely different links. As Gordon Belot and John Earman make the point, "we have a web of independent theories, each of which is thought to be empirically adequate within its own domain of applicability [...] 'Web' rather than 'hierarchy' here because theories often have more than one limit: for instance, special relativity is the curvature $\rightarrow 0$ limit of general relativity, while the curved spacetime formulation of Newtonian gravity is its $c \rightarrow \infty$ limit." (Belot and Earman 1997, 162). This web structure with its symmetric links is what provides a *non-priority-based unification* of science.

It is important to emphasize that talking about categorical-conceptual frameworks does not keep us chained to an analytical and representational style of thinking that ignores the effective practice of science. When the pluralist admits the possible coexistence of different frameworks, such a claim does not mean that all possible frameworks are equivalent in real-life science. On the contrary, along history, and even within a single historical time, some categorical-conceptual frameworks become constitutive of certain scientific disciplines or fields, whereas others pass away or never come to exist. But, since they are not susceptible to be true or false, this fact cannot be explained by appealing to truth values. It is precisely here where the practice of science becomes essential: certain categorical-conceptual frameworks consolidate in the actual scientific activity due to their practical success. The practice of science includes, of course, the *empirical practice* of observation and experimentation, but also the *theoretical practice*, involving formal techniques, general assumptions, paradigmatic models (Kuhn's "exemplars"), among many others aspects, and *financial support-practice*, which is not a mere side or accessory scientific task, but has become a central endeavor in real science that strongly shapes the goal and the content of research (for a more detailed discussion of this point, see Lombardi 2024). Thus, a system of scientific practices is a complex net, with blurred boundaries, of inter-related activities developed by the different agents involved in the social phenomenon called "science." Precisely those practices are what play an essential role in deciding what is successful in science and, with it, in fixing certain categorical-conceptual frameworks over others. Of course, such a practical success does not make those frameworks "true," since categorical-conceptual frameworks, like Hacking's styles of reasoning (Hacking 2002), are not something amenable to have truth value: they are the frameworks in the context of which truth and falsity can be established by empirical means.

Once metaphysical realism has been given up, thus, there is no argument to oppose realism and practice. As Sami Pihlström claims: "pragmatism, as such, is no enemy of (moderate) scientific realism" (Pihlström 2008, 59). Pihlström takes a step further when he characterizes that kind of realism in the context of a pragmatist reading of Kant's philosophy; according to him, the contemporary scientific realist should be prepared to accept a pragmatically naturalized Kantian transcendental perspective on realism. From this "naturalized" Kantism, it is practice that supplies the conditions of possibility of knowledge: "it would seem—at least this rearticulation should be available to the «Kantian pragmatist»—that practices provide transcendental (contextual) conditions for the possibility of there being scientifically representable objects at all—for us." (Pihlström 2012, 88–89). It is interesting to notice that this recovering of Kantian philosophy appears also in some authors strongly committed with the so-called *philosophy of science in practice*. For instance, Hasok Chang (2009) introduces the idea of ontological principles, which, analogously to Kant's categories, do not supply empirical knowledge but allow for the intelligibility of epistemic activities—for example, the epistemic activity of counting involves the ontological principle of discreteness. Even more recently, Mieke Boon (2017) takes Kant's idea of "condition of possibility of knowledge" as a starting point for an epistemology of scientific practices, especially those that aim at knowledge for practical uses. This pragmatic turn brings us back to the second thesis of Marx's *Theses on Feuerbach*: "The question whether objective truth can be attributed to human thinking is not a question of theory but is a practical question. Man must prove the truth—i.e. the reality and power, the this-sidedness of his thinking in practice. The dispute over the reality or non-reality of thinking that is isolated from practice is a purely *scholastic* question." (Marx 1845).

This brief presentation by no means exhausts all aspects of pluralist realism and its applications (see, e.g., Lombardi and Labarca 2005, 2006). Nevertheless, I believe that what has been said is enough to make it clear why I consider continuing to discuss on the interpretation of non-relativistic quantum mechanics to be highly relevant. Indeed, from my perspective there are no "fundamental" theories to which the others reduce: every pragmatically successful theory has its field of application and its limits of applicability, and deserves to be conceptually understood. If we were to believe that only the fundamental theory is worth the effort of interpretation, then we should wait for getting the "theory of everything" before undertaking any interpretative task: in the meantime, we should give up understanding and we would use successful theories in a totally instrumental way, saving our realist beliefs for better times. But, as I argued in Chapter 1, understanding has cognitive and pragmatic advantages that a mere instrumental stance cannot offer.

Of course, admitting the ontological autonomy of the domain of non-relativistic quantum mechanics does not imply renouncing to establish links between it and other theoretical domains in the network of relations that constitutes scientific knowledge. Science is also largely nourished by inter-theoretical relations. But since those links do not involve priority or dependence on other theories, non-relativistic quantum mechanics preserves its intrinsic value, and it is still worth trying to make sense of the world it describes.

Final Remarks

The content of the present book is the result of a work developed over a period of more than fifteen years, to which many students and collaborators, as well as renowned philosophers and physicists, have contributed in various aspects, as the references testify. The aim of this work has been to present the MHI in a systematic and organic way, incorporating recent discussions and making explicit its underlying motivations. I have tried to show that the proposal does not address isolated interpretative topics, but provides a broad, articulated, and consistent view to make sense of the quantum world.

Throughout the chapters, I have argued in favor of the reasonableness and fruitfulness of the proposal. Of course, the reasons put forward are not directly empirical, but this does not make them any less sound. The main arguments presented in the text can be summarized in the following points:

- When applied to well-known models and experimental results, the MHI shows its physical relevance manifested by its agreement with the orthodox practice of physics.
- The MHI offers a solution to the measurement problem for both ideal and non-ideal measurements, and distinguishes between reliable and non-reliable measurements in the non-ideal case. It also adequately accounts for consecutive measurements.
- The MHI approach delivers a new, more physical understanding of the concept of quantum measurement, according to which a quantum measurement is a symmetry-breaking process that renders a symmetry generator of the system's Hamiltonian empirically accessible.
- On the basis of the results of applying the Galilei transformations to the MHI interpretative postulates, the interpretation can be reformulated in an explicitly Galilei-invariant form.
- The MHI ontology of properties, framed in the algebraic formalism of quantum mechanics, provides a comprehensive answer to the ontological challenges of the theory: contextuality, non-separability, and indistinguishability. In this picture, quantum systems turn out to be non-objectual bundles of properties, which do not preserve their identity either synchronically or diachronically.

The Modal-Hamiltonian Interpretation of Quantum Mechanics. Olimpia Lombardi, Oxford University Press.
© Olimpia Lombardi (2026). DOI: 10.1093/9780198951407.003.0014

Three more general aspects can be added to these advantages:

- *Familiarity for science practitioners.* The central role played by energy in physical phenomena is natural for physicists: energy is commonly considered an observable with definite value; moreover, physical interactions are conceived as energy exchanges. Therefore, when the MHI has been presented to an audience of physicists, the reception of the proposal has always been very favorable. The fact that the observables that break the symmetries of the Hamiltonian are not definite-valued seems almost obvious to them, precisely because of the arbitrariness that the definite values of those observables would imply. The leading role of energy is also easy to accept for chemists, whose work is mostly developed in the basis of energy. Several chemists were even astonished by the need to appeal to so many arguments in favor of the MHI, since they conceive the core role of the Hamiltonian as an obvious theoretical assumption.
- *Compatibility with other theories and scientific knowledge.* The general framework of the MHI provides meaningful answers to physical issues that are discussed in the scientific community. For example, the relativity of entanglement with respect to the decomposition of a whole closed system into its components acquires a natural place in this framework. This relativity, in turn, has relevant consequences for the understanding of the phenomenon of decoherence, which has been extensively discussed during the last decades. In turn, the MHI ontology of properties also supplies a natural solution to the problem of entanglement of so-called indistinguishable particles. Beyond the boundaries of non-relativist quantum mechanics, the MHI offers a general interpretative strategy that can be meaningfully extrapolated to relativist quantum theories. In the field of chemistry, the MHI has proved to be particularly fruitful for solving the problem of symmetry, considered as one of the basic conceptual problems of quantum chemistry.
- *Basis for understanding.* The ontological picture underlying the MHI relies on a corpus of metaphysical knowledge that allows for a meaningful description of the structure of the quantum reality: it is no longer a question of what quantum elements are not, but a metaphysically precise characterization of their nature can be offered. The proposed ontology is certainly heterodox and requires considerable philosophical effort to understand its main features. However, once the ontological picture is internalized, it can be meaningfully explained by describing its metaphysical features, and new ontological conclusions can be drawn about the nature of the elements that populate the quantum realm. In other words, the precise structure of the proposed ontology provides the basis for a meaningful understanding of the quantum world.

Although here I have attempted to present a broad and strongly articulated interpretative proposal, much work remains to be done, at least in the following areas:

- The metaphysics of science is a field that has recently developed enormously and that I have addressed here without being a specialist. It would be very positive if scholars specifically engaged in the metaphysics of physics could take up the ideas that I have presented in order to further elaborate them in a deeper way. Among many other issues, a pending task is to study the points of contact between the MHI ontological proposal and Ontic Structural Realism, which seem to be many and significant. I am moving in this direction with the help of Bruno Borge, who has the necessary philosophical background to address these questions.
- In the physical field, there is an interesting work to be developed with respect to the new results that are continuously obtained in novel and sophisticated measurement scenarios. The question is how the MHI conceptualizes these situations, to see if any of them represent a real challenge for the interpretation. I am studying these questions in collaboration with Sebastian Fortin and Juan Sebastian Ardenghi, as a continuation of our work on the Frauchiger-Renner argument and on encapsulated measurements.
- An essential point for the MHI is its extrapolation to quantum field theory. Although in the last chapter I proposed some general ideas in this regard, they must be developed both in the physical and in the ontological directions. I would like to have a better physical background to address this topic. Fortunately, I am working in collaboration with Federico Holik, who palliates my theoretical limitations about this subject.
- From a logical point of view, the challenge of formulating a metalanguage for the MHI ontology still remains. Of course, since I am not a logician, I am completely unable to undertake this work alone. Luckily, on this point I can collaborate with the work of Décio Krause and his student Juan Pablo Jorge: my aim is to offer them an application of their new quasi-set theory, which in turn could become the adequate metalanguage to talk about the MHI ontology.

These possible ramifications of the research work show that, as I said in the Introduction, this is a lively project. I hope that soon other philosophers of physics, metaphysicians of science, and logicians will decide to join us in this endeavor.

References

Acuña, P. (2021). "Charting the landscape of interpretation, theory rivalry, and underdetermination in quantum mechanics." *Synthese*, **198**: 1711–1740.

Adler, S. L. (1995). *Quaternionic Quantum Mechanics and Quantum Fields*. New York: Oxford University Press.

Adler, S. L. (2003). "Why decoherence has not solved the measurement problem: A response to P. W. Anderson." *Studies in History and Philosophy of Modern Physics*, **34**: 135–142.

Akiba, K. (2004). "Vagueness in the world." *Noûs*, **38**: 407–429.

Albert, D. Z. (1996). "Elementary quantum metaphysics." Pp. 277–284 in J. T. Cushing, A. Fine, and S. Goldstein (eds.), *Bohmian Mechanics and Quantum Mechanics: An appraisal*. Dordrecht: Kluwer.

Albert, D. Z. (2013)."Wave function realism." Pp. 58–75 in A. Ney and D. Z. Albert (eds.), *The Wave Function: Essays on the Metaphysics of Quantum Mechanics*. Oxford: Oxford University Press.

Albert, D. Z., and B. Loewer (1988). "Interpreting the Many Worlds Interpretation." *Synthese*, **77**: 195–213.

Albert, D. and B. Loewer (1990). "Wanted dead or alive: two attempts to solve Schrödinger's paradox." Pp. 277–285 in A. Fine, M. Forbes, and L. Wessels (eds.), *Proceedings of the PSA 1990, Vol. 1*. A. East Lansing, Michigan: Philosophy of Science Association.

Albert, D. and B. Loewer (1991). "Some alleged solutions to the measurement problem." *Synthese*, **88**: 87–98.

Albert, D. and B. Loewer (1993). "Non-ideal measurements." *Foundations of Physics Letters*, **6**: 297–305.

Allard-Guérin, P., V. Baumann, F. Del Santo, and Č. Brukner (2021). "A no-go theorem for the persistent reality of Wigner's friend's perception." *Communications Physics*, **4**: #93.

Alvarado, J. T. (2020). "Identity conditions for transcendent universals." Pp 247–259 in J. T. Alvarado (ed.), *A Metaphysics of Platonic Universals and their Instantiations*. Synthese Library, *Vol 428*. Cham: Springer.

Amann, A. (1992). "Must a molecule have a shape?" *South African Journal of Chemistry*, **45**: 29–38.

Anastopoulos, C. (2002). "Frequently asked questions about decoherence." *International Journal of Theoretical Physics*, **41**: 1573–1590.

Anderson, P. W. (1972). "More is different." *Science*, **177**: 393–396.

Anderson, P. W. (2001). "Science: A 'dappled world' or a 'seamless web'?" *Studies in History and Philosophy of Modern Physics*, **34**: 487–494.

Araújo, M. (2018). "The flaw in Frauchiger and Renner's argument." URL: http://mateusaraujo.info/2018/10/24/the-flaw-in-frauchiger-and-renners-argument/

Ardenghi, J. S., M. Castagnino, and O. Lombardi (2009). "Quantum mechanics: Modal interpretation and Galilean transformations." *Foundations of Physics*, **39**: 1023–1045.

Ardenghi, J. S., M. Castagnino, and O. Lombardi (2011). "Modal-Hamiltonian interpretation of quantum mechanics and Casimir operators: the road to quantum field theory." *International Journal of Theoretical Physics*, **50**: 774–791.

Ardenghi, J. S., O. Lombardi, and M. Narvaja (2013). "Modal interpretations and consecutive measurements." Pp. 207–217 in V. Karakostas and D. Dieks (eds.), *EPSA 2011: Perspectives and Foundational Problems in Philosophy of Science*. Berlin: Springer.

Arenhart, J. R. B., O. Bueno, and D. Krause (2019). "Making sense of nonindividuals in quantum mechanics." Pp. 185–204 in O. Lombardi, S. Fortin, C. López, and F. Holik (eds.), *Quantum Worlds. Perspectives on the Ontology of Quantum Mechanics*. Cambridge: Cambridge University Press.

Armstrong, D. M. (1983). *What Is a Law of Nature?* Cambridge: Cambridge University Press.

Armstrong, D. M. (1989). *Universals: An Opinionated Introduction*. Boulder, CO: Westview.

Armstrong, D. M. (1993). "A world of states of affairs." *Philosophical Perspectives*, **7**: 429–440.

Arntzenius, F. (1990). "Kochen's interpretation of quantum mechanics." Pp. 241–249 in A. Fine, M. Forbes, and L. Wessels (eds.), *Proceedings of the PSA 1990, Vol. 1*. East Lansing, Michigan: Philosophy of Science Association.

Arntzenius, F. (1998). "Curiouser and curiouser: A personal evaluation of modal interpretations." Pp. 337–377 in D. Dieks and P. Vermaas (eds.), *The Modal Interpretation of Quantum Mechanics*. Dordrecht: Kluwer.

Auletta, G. (2000). *Foundations and Interpretation of Quantum Mechanics*. Singapore: World Scientific.

Auyang, S. Y. (1995). *How is Quantum Field Theory Possible?* Oxford: Oxford University Press.

Bacciagaluppi, G. (1995). "A Kochen-Specker theorem in the modal interpretation of quantum mechanics." *International Journal of Theoretical Physics*, **34**: 1205–1216.

Bacciagaluppi, G. (1996). *Topics in the Modal Interpretation of Quantum Mechanics. Dissertation.* Cambridge: Cambridge University.

Bacciagaluppi, G. (2020). "The role of decoherence in quantum mechanics." in E. N. Zalta (ed.), *The Stanford Encyclopedia of Philosophy* (Fall 2020 Edition). URL: https://plato.stanford.edu/archives/fall2020/entries/qm-decoherence/

Bacciagaluppi, G. and M. Dickson (1999). "Dynamics for modal interpretations." *Foundations of Physics*, **29**: 1165–1201.

Bacciagaluppi, G. and M. Hemmo (1996). "Modal interpretations, decoherence and measurements." *Studies in History and Philosophy of Modern Physics*, **27**: 239–277.

Bader, R. (1990). *Atoms in Molecules. A Quantum Theory.* Oxford: Oxford University Press.

Bader, R. (2011). "On the non-existence of parallel universes in chemistry." *Foundations of Chemistry*, **13**: 11–37.

Baker, D. (2010). "Symmetry and the metaphysics of physics." *Philosophy Compass*, **5**: 1157–1166.

Balachandran, A. P., T. R. Govindarajan, A. R. de Queiroz, and A. F. Reyes-Lega (2013a). "Algebraic approach to entanglement and entropy." *Physical Review A*, **88**: #022301.

Balachandran, A. P., T. R. Govindarajan, A. R. de Queiroz, and A. F. Reyes-Lega (2013b). "Entanglement and particle identity: A unifying approach." *Physical Review Letters*. **110**: #080503.

Ballentine, L. (1970). "The statistical interpretation of quantum mechanics." *Reviews of Modern Physics*, **42**: 358–381.

Ballentine, L. (1998). *Quantum Mechanics: A Modern Development.* Singapore: World Scientific

Barnes, E. (2010). "Ontic vagueness: A guide for the perplexed." *Noûs*, **44**: 601–627.

Barnes, E. and J. R. G. Williams (2011). "A theory of metaphysical indeterminacy." Pp. 103–148 in K. Bennett and D. W. Zimmerman (eds.), *Oxford Studies in Metaphysics. Vol. 6.* Oxford: Oxford University Press.

Barnum, H., E. Knill, G. Ortiz, R. Somma, and L. Viola (2004). "A subsystem-independent generalization of entanglement." *Physical Review Letters*, **92**: 107902.

Barnum, H., E. Knill, G. Ortiz, and L. Viola (2003). "Generalizations of entanglement based on coherent states and convex sets." *Physical Review A*, **68**: 032308.

Barrett, J. (2023). "Everettian quantum mechanics." in E. N. Zalta and U. Nodelman (eds.), *The Stanford Encyclopedia of Philosophy* (Summer 2023 Edition). URL: https://plato.stanford.edu/archives/sum2023/entries/qm-everett/

Barrett, T. W. (2019). "Equivalent and inequivalent formulations of classical mechanics." *The British Journal for the Philosophy of Science*, **70**:1167–1199.

Barrett, T. W. and H. Halvorson (2016). "Glymour and Quine on theoretical equivalence." *Journal of Philosophical Logic*, **45**: 467–483.

Baumann, V. and S. Wolf (2018). "On formalisms and interpretations." *Quantum*, **2**: #99.

Belot, G. and J. Earman (1997). "Chaos out of order: quantum mechanics, the correspondence principle and chaos." *Studies in History and Philosophy of Modern Physics*, **28**: 147–182.

Belousek, D. W. (2000). "Statistics, symmetry, and the conventionality of indistinguishability in quantum mechanics." *Foundations of Physics*, **30**: 1–34.

Beltrametti, E. and G. Cassinelli (1981). *The Logic of Quantum Mechanics.* Boston: Addison-Wesley.

Benacerraf, P. (1965). "What numbers could not be?" *Philosophical Review*, **74**: 47–73.

Benatti, F., R. Floreanini, F. Franchini, and U. Marzolino (2020). "Entanglement in indistinguishable particle systems." *Physics Reports*, **878**: 1–27.

Ben-Menahem, Y. (2018). *Causation in Science.* Princeton: Princeton University Press.

Bene, G. (1997). "Quantum reference systems: A new framework for quantum mechanics." *Physica A*, **242**: 529–565.

Bene, G. and D. Dieks (2002). "A perspectival version of the modal interpretation of quantum mechanics and the origin of macroscopic behavior." *Foundations of Physics*, **32**: 645–671.

Benovsky, J. (2008). "The bundle theory and the substratum theory: deadly enemies or twin brothers?" *Philosophical Studies*, **141**: 175–190.

Berkovitz, J. (2016). "Action at a distance in quantum mechanics." in E. N. Zalta (ed.), *The Stanford Encyclopedia of Philosophy* (Spring 2016 Edition), URL: https://plato.stanford.edu/archives/spr2016/entries/qm-action-distance/

Berkovitz, J., R. Frigg, and F. Kronz (2006). "The ergodic hierarchy, randomness and Hamiltonian chaos." *Studies in History and Philosophy of Modern Physics*, **37**: 661–691.

Berkovitz, J. and M. Hemmo (2005a). "Can modal interpretations of quantum mechanics be reconciled with relativity?" *Philosophy of Science*, **72**: 789–801.
Berkovitz, J. and M. Hemmo (2005b). "Modal interpretations of quantum mechanics and relativity: a reconsideration." *Foundations of Physics*, **35**: 373–397.
Berkovitz, J. and M. Hemmo (2006). "A new modal interpretation in terms of relational properties." Pp. 1–28 in W. Demopoulos and I. Pitowsky (eds.), *Physical Theory and its Interpretation: Essays in honor of Jeffrey Bub*. New York: Springer.
Berlin, Y. A., A. L. Burin, and V. V. Goldanskii (1996). "The Hund paradox and stabilization of molecular chiral states." *Zeitschrift für Physik D*, **37**: 333–339.
Bernoulli, J. (1713). *Ars Conjectandi, Opus Posthumum. Accedit Tractatus de Seriebus Infinitis, et Epistola Gallice Scripta de Ludo Pilae Reticularis*. Basel: Thurneysen.
Bigaj, T. (2014). "Quantum dispositions and the notion of measurement." *Filozofia Nauki*, **22**: 5–24.
Bigaj, T. (2022). "How to justify the symmetrization postulate in quantum mechanics." *Journal for General Philosophy of Science*, **53**: 239–257.
Black, M. (1952). "The identity of indiscernibles." *Mind*, **LXI**: 153–164.
Boethius (1880). *Commentary on Aristotle's 'On Interpretation,'* 2nd ed., C. Meiser (ed.), Teubner: Leipzig.
Bohm, A. and M. Gadella (1989). *Dirac Kets, Gamow Vectors and Gel'fand Triplets. The Rigged Hilbert Space Formulation of Quantum Mechanics*, Springer Lecture Notes in Physics, *Vol. 348*. Berlin: Springer.
Bohm, D. (1952). "A suggested interpretation of the quantum theory in terms of 'hidden' variables." *Physical Review*, **85**: 166–179,180–193.
Bokulich, A. (2014). "Metaphysical indeterminacy, properties, and quantum theory." *Res Philosophica*, **91**: 449–475.
Bokulich, A. and G. Jaeger (2010). *Philosophy of Quantum Information and Entanglement*. Cambridge: Cambridge University Press.
Bong, K.-W., A. Utreras-Alarcón, F. Ghafari, Y.-C. Liang, N. Tischler, E. G. Cavalcanti, G. J. Pryde, and H. M. Wiseman (2020). "A strong no-go theorem on the Wigner's friend paradox." *Nature Physics*, **16**: 1199–1205.
Bonifacio, R., S. Olivares, P. Tombesi, and D. Vitali (2000). "Model-independent approach to nondissipative decoherence." *Physical Review A*, **61**: #053802.
Boon, M. (2017). "Philosophy of science in practice: A proposal for epistemological constructivism." Pp 289–310 in H. Leitgeb, I. Niiniluoto, P. Seppälä, and E. Sober (eds.), *Logic, Methodology and Philosophy of Science. Proceedings of the 15th International Congress (CLMPS 2015)*. London: College Publications.
Borge, B. and O. Lombardi (2024). "Modalidad en mecánica cuántica." *Revista de Filosofía* (Universidad de Chile), **81**: 25–48.
Born, M. (1953). "Physical reality." *Philosophical Quarterly*, **3**: 139–149.
Born, M. (ed.) (1971) *The Born-Einstein Letters*. Translation of M. Born 1969. New York: Walker.
Bose, S. K. (1995). "The Galilean group in 2+1 space-times and its central extension." *Communications in Mathematical Physics*, **169**: 385–395.
Brading, K. and E. Castellani (2007). "Symmetries and invariances in classical physics." Pp. 1331–1367 in J. Butterfield and J. Earman (eds.), *Philosophy of Physics*. Amsterdam: Elsevier.
Brading, K., E. Castellani, and N. Teh (2023). "Symmetry and symmetry breaking." in E. N. Zalta and U. Nodelman (eds.), *The Stanford Encyclopedia of Philosophy* (Fall 2023 Edition), URL: https://plato.stanford.edu/archives/fall2023/entries/symmetry-breaking/
Brassard, G. and P. Raymond-Robichaud (2013). "Can free will emerge from determinism in quantum theory?" Pp. 41–61 in A. Suarez and P. Adams (eds.), *Is Science Compatible with Free Will?* New York: Springer.
Brown, H. (1999). "Aspects of objectivity in quantum mechanics." Pp. 45–70 in J. Butterfield and C. Pagonis (eds.), *From Physics to Philosophy*. Cambridge: Cambridge University Press.
Brown, H. and P. Holland (1999). "The Galilean covariance of quantum mechanics in the case of external fields." *American Journal of Physics*, **67**: 204–214.
Brown, H., M. Suárez, and G. Bacciagaluppi (1998). "Are 'sharp values' of observables always objective elements of reality?" Pp. 289–306 in D. Dieks and P. E. Vermaas (eds.), *The Modal Interpretation of Quantum Mechanics*. Dordrecht: Kluwer Academic Publishers.
Brukner, Č. (2018). "A no-go theorem for observer-independent facts." *Entropy*, **20**: #350.
Brukner, Č. (2020). "Facts are relative." *Nature Physics*, **16**: 1172–1174.
Bub, J. (1979). "Conditional probabilities in non-Boolean possibility structures." Pp. 209–226 in C. A. Hooker (ed.), *The Logic-Algebraic Approach to Quantum Mechanics*. Dordrecht: Reidel.
Bub, J. (1992). "Quantum mechanics without the projection postulate." *Foundations of Physics*, **22**: 737–754.

Bub, J. (1994). "On the structure of quantal proposition systems." *Foundations of Physics*, **24**: 1261–1279.

Bub, J. (1997). *Interpreting the Quantum World*, Cambridge: Cambridge University Press.

Bub, J. (2020a). "In defense of a «single-world» interpretation of quantum mechanics." *Studies in History and Philosophy of Modern Physics*, **72**: 251–255.

Bub, J. (2020b). "'Two dogmas' redux." Pp. 199–215 in M. Hemmo and O. Shenker (eds.), *Quantum, Probability, Logic*. Jerusalem Studies in Philosophy and History of Science. Cham: Springer.

Bub, J. and R. Clifton (1996). "A uniqueness theorem for interpretations of quantum mechanics." *Studies in History and Philosophy of Modern Physics*, **27**: 181–219.

Bub, J., R. Clifton, and S. Goldstein (2000). "Revised proof of the uniqueness theorem for 'no collapse' interpretations of quantum mechanics." *Studies in History and Philosophy of Modern Physics*, **31**: 95–98.

Bueno, O. (2014). "Constructive empiricism, partial structures, and the modal interpretation of quantum mechanics." *Quanta*, **3**: 1–15.

Bueno, O. and S. A. Shalkowski (2015). "Modalism and theoretical virtues: toward an epistemology of modality." *Philosophical Studies*, **172**: 671–689.

Bunge, M. (1977). *Treatise on Basic Philosophy, Vol.3: Ontology I*. Dordrecht-Boston: Reidel Publishing Company.

Busch P. (1990a). "On the energy-time uncertainty relation. Part I: Dynamical time and time indeterminacy." *Foundations of Physics*, **20**: 1–32.

Busch P. (1990b). "On the energy-time uncertainty relation. Part II: Pragmatic time versus energy indeterminacy." *Foundations of Physics*, **20**: 33–43.

Busch P. (2008). "The time-energy uncertainty relation." Pp. 73–105 in J. Muga, R. S. Mayato, and I. Egusquiza (eds.), *Time in Quantum Mechanics*. Lecture Notes in Physics, *Vol 734*. Berlin-Heidelberg: Springer.

Busch, P. and P. Lahti (2009). "Lüders Rule." Pp. 356–357 in D. Greenberger, K. Hentschel, and F. Weinert (eds.), *Compendium of Quantum Physics*. Berlin: Springer.

Butterfield, J. (1993). "Interpretation and identity in quantum theory." *Studies in History and Philosophy of Science*, **24**: 443–476.

Butterfield, J. (2007). "On symplectic reduction in classical mechanics." Pp. 1–131 in J. Butterfield and J. Earman (eds.), *Philosophy of Physics*. Amsterdam: Elsevier.

Cabello, A. (2017). "Interpretations of quantum theory: A map of madness." Pp. 138–143 in O. Lombardi, S. Fortin, F. Holik, and C. López (eds.), *What is Quantum Information?* Cambridge: Cambridge University Press.

Calosi, C. (2022). "Quantum modal indeterminacy." *Studies in History and Philosophy of Science*, **95**: 177–184.

Calosi, C. and C. Mariani (2020). "Quantum relational indeterminacy." *Studies in History and Philosophy of Modern Physics*, **71**: 158–169.

Calosi, C. and C. Mariani (2021). "Quantum indeterminacy." *Philosophy Compass*. **2021**: 1–15.

Calosi, C. and J. M. Wilson (2019). "Quantum metaphysical indeterminacy." *Philosophical Studies*, **176**: 2599–2627.

Calosi, C. and J. Wilson (2021). "Quantum indeterminacy and the double-slit experiment." *Philosophical Studies*, **178**: 3291–3317.

Campbell, K. (1990). *Abstract Particulars*. Oxford: Basil Blackwell.

Carnap, R. (1950). *Logical Foundation of Probability* 2nd ed. 1962. Chicago: University of Chicago Press.

Carroll, S. and A. Singh (2021). "Quantum mereology: Factorizing Hilbert space into subsystems with quasi-classical dynamics." *Physical Review A*, **103**: #022213.

Cartwright, N. (1983). *How the Laws of Physics Lie*. Oxford: Clarendon Press.

Cartwright, N., T. Shomar, and M. Suarez (1996). "The tool box of science: Tools for building of models with a superconductivity example." Pp. 137–149 in W. E. Herfel, W. Krajewski, I. Niiniluoto, and R. Wójcicki (eds.), *Theories and Models in Scientific Processes*. Amsterdam: Editions Rodopi.

Casati, G. and T. Prosen (2005). "Quantum chaos and the double-slit experiment." *Physical Review A*, **72**: #032111.

Cassinelli, G. and N. Zanghí (1983). "Conditional probabilities in quantum mechanics I. Conditioning with respect to a single event." *Il Nuovo Cimento B*, **73**: 237–245.

Cassirer, E. (1938) [1944]. "Le concept de groupe et la théorie de la perception." *Journal de Psychologie Normale et Pathologique*, **1938** (Juillet–Décembre): 368–414. Translated in 1944 as "The concept of group and the theory of perception." *Philosophy and Phenomenological Research*, **5**: 1–36.

Castagnino, M., S. Fortin, and O. Lombardi (2010a). "Suppression of decoherence in a generalization of the spin-bath model." *Journal of Physics A: Mathematical and Theoretical*, **43**: #065304.

Castagnino, M., S. Fortin, and O. Lombardi (2010b). "Is the decoherence of a system the result of its interaction with the environment?" *Modern Physics Letters* A, **25**: 1431–1439.

Castagnino, M., L. Lara, and O. Lombardi (2003). "The cosmological origin of time-asymmetry." *Classical and Quantum Gravity*, **20**: 369–391.

Castagnino, M., R. Laura, and O. Lombardi (2007). "A general conceptual framework for decoherence in closed and open systems." *Philosophy of Science*, **74**: 968–980.

Castagnino, M. and O. Lombardi (2004). "Self-induced decoherence: a new approach." *Studies in History and Philosophy of Modern Physics*, **35**: 73–107.

Castagnino, M. and O. Lombardi (2005a). "Self-induced decoherence and the classical limit of quantum mechanics." *Philosophy of Science*, **72**: 764–776.

Castagnino, M. and O. Lombardi (2005b). "Decoherence time in self-induced decoherence." *Physical Review A*, **72**: #012102.

Castagnino, M. and O. Lombardi (2008). "The role of the Hamiltonian in the interpretation of quantum mechanics." *Journal of Physics. Conferences Series*, **28**: #012014.

Castagnino, M. and O. Lombardi (2009). "The global non-entropic arrow of time: from global geometrical asymmetry to local energy flow." *Synthese*, **169**: 1–25.

Castagnino, M., O. Lombardi, and L. Lara (2003). "The global arrow of time as a geometrical property of the universe." *Foundations of Physics*, **33**: 877–912.

Caulton A. (2011). *Issues of Identity and Individuality in Quantum Mechanics*. PhD Thesis, Cambridge: Darwin College, University of Cambridge.

Caves, C. M., C. A. Fuchs, and R. Schack, (2002a). "Quantum probabilities as Bayesian probabilities." *Physical Review A*, **65**: #022305.

Caves, C. M., C. A. Fuchs, and R. Schack, (2002b). "Unknown quantum states: The quantum de Finetti representation." *Journal of Mathematical Physics*, **43**: 4537–4559.

Chang, H. (2009). "Ontological principles and the intelligibility of epistemic activities." Pp. 64–82 in H. De Regt, S. Leonelli, and K. Eigner (eds.), *Scientific Understanding: Philosophical Perspectives*. Pittsburgh: Pittsburgh University Press.

Chang, H. (2015). "Reductionism and the relation between chemistry and physics." Pp. 193–210 in T. Arabatzis, J. Renn, and A. Simoes (eds.), *Relocating the History of Science: Essays in Honor of Kostas Gavroglu*. New York: Springer.

Chiang, T. (2002). *Stories of Your Life and Others*. New York: Tor Books.

Claverie, P. and S. Diner (1980). "The concept of molecular structure in quantum theory: interpretation problems." *Israel Journal of Chemistry*, **19**: 54–81.

Clifton, R. (1995a). "Independently motivating the Kochen-Dieks modal interpretation of quantum mechanics." *The British Journal for the Philosophy of Science*, **46**: 33–57.

Clifton, R. (1995b). "Making sense of the Kochen-Dieks 'no-collapse' interpretation of quantum mechanics independent of the measurement problem." *Annals of the New York Academy of Science*, **755**: 570–578.

Clifton, R. (1995c). "Why modal interpretations of quantum mechanics must abandon classical reasoning about the physical properties." *International Journal of Theoretical Physics*, **34**: 1302–1312.

Clifton, R. (1996). "The properties of modal interpretations of quantum mechanics." *The British Journal for the Philosophy of Science*, **47**: 371–398.

Clifton, R. (2000). "The modal interpretation of algebraic quantum field theory." *Physics Letters A*, **271**: 167–177.

Cohen, D. W. (1989). *An Introduction to Hilbert Space and Quantum Logic*. New York: Springer-Verlag.

Cohen-Tannoudji, C., B. Diu, and F. Laloë (1977). *Quantum Mechanics*. New York: John Wiley & Sons.

Colussi, V. and S. Wickramasekara (2008). "Galilean and U(1)-gauge symmetry of the Schrödinger field." *Annals of Physics*, **323**: 3020–3036.

Cover, T. and J. A. Thomas (1991). *Elements of Information Theory*. New York: John Wiley & Sons.

Cramer, J. G. (1986). "The transactional interpretation of quantum mechanics." *Reviews of Modern Physics*, **58**: 647–688.

Cramer, J. G. (1988). "An overview of the transactional interpretation." *Internacional Journal of Theoretical Physics*, **27**: 227–236.

Cuffaro, M. and S. Hartmann (2024). "The open systems view." *Philosophy of Physics*, **2**: #6.

Curiel, E. (2014). "Classical mechanics is Lagrangian; it is not Hamiltonian." *The British Journal for the Philosophy of Science*, **65**: 269–321.

Cushing, J. (1994). *Quantum Mechanics. Historical Contingency and the Copenhagen Hegemony*. Chicago: The University of Chicago Press.

da Costa, N., S. French, and D. Krause (1992). "The Schrödinger problem." Pp. 445–460 in M. Bitbol and O. Darrigol (eds.), *Erwin Schrödinger: Philosophy and the Birth of Quantum Mechanics*. Paris: Editions Frontières.

da Costa, N. and D. Krause (1994). "Schrödinger logics." *Studia Logica*, **53**: 533–550.

da Costa, N. and D. Krause (1997). "An intensional Schrödinger logic." *Notre Dame Journal of Formal Logic*, **38**: 179–194.

da Costa, N. and D. Krause (1999). "Set-theoretical models for quantum systems." Pp. 171–181 in M. L. Dalla Chiara, R. Giuntini, and F. Laudisa (eds.), *Language, Quantum, Music*. Dordrecht: Kluwer.

da Costa, N. and O. Lombardi (2014). "Quantum mechanics: ontology without individuals." *Foundations of Physics*, **44**: 1246–1257.

da Costa, N., O. Lombardi, and M. Lastiri (2013). "A modal ontology of properties for quantum mechanics." *Synthese*, **190**: 3671–3693.

Dalla Chiara, M. L. and G. Toraldo di Francia (1993). "Individuals, kinds and names in physics." Pp. 261–283 in G. Corsi, M. L. Dalla Chiara, and G. C. Ghirardi (eds.), *Bridging the Gap: Philosophy, Mathematics and Physics*. Dordrecht: Kluwer.

Dalla Chiara, M. L. and G. Toraldo di Francia (1995). "Identity questions from quantum theory." Pp. 39–46 in K. Gavroglu, J. Stachel and M. W. Wartofski (eds.), *Physics, Philosophy and the Scientific Community*. Dordrecht: Kluwer.

Daneri, A., A. Loinger, and G. Prosperi (1962). "Quantum theory of measurement and ergodicity conditions." *Nuclear Physics*, **33**: 297–319.

Das, S. and D. Dürr (2019). "Arrival time distributions of spin-1/2 particles." *Scientific Reports*, **9**: #2242.

Davidson, D. (1967), "Truth and meaning." *Synthese*, **17**: 304–323.

de Broglie, L. (1928). "La nouvelle dynamique des quanta." Pp. 105–132 in *Electrons et photons. Rapports et discussions du cinquième Conseil de Physique tenu à Bruxelles du 24 au 29 octobre 1927 sous les auspices de l'Institut International de Physique Solvay*. Paris: Gauthier-Villars. Translated in 2009, pp. 341–371 in G. Bacciagaluppi and A. Valentini (eds.), *Quantum Theory at the Crossroads. Reconsidering the 1927 Solvay Conference*. Cambridge: Cambridge University Press.

de Finetti, B. (1937). "La prévision: ses lois logiques, ses sources subjectives." *Annales de l'Institut Henri Poincaré*, **7**: 1–68. Translated in 1980 as "Foresight. its logical laws, its subjective sources." in H. Kyburg Jr. and H. Smokler (eds)., *Studies in Subjective Probability*, New York: Krieger Publishing Company.

De Haro, S. (2021). "Theoretical equivalence and duality." *Synthese*, **198**: 5139–5177.

de Regt, H. W. (2017). Understanding Scientific Understanding. Oxford: Oxford University Press.

de Regt, H. W. (2004). "Making sense of understanding." *Philosophy of Science*, **71**: 98–109.

d'Espagnat, B. (1976). *Conceptual Foundations of Quantum Mechanics*. Reading MA: Benjamin.

d'Espagnat, B. (1995). *Veiled Reality. An Analysis of Present-Day Quantum Mechanical Concepts*. Reading MA: Addison-Wesley.

DeWitt, B. S. M. (1970). "Quantum mechanics and reality." *Physics Today*, **23**: 30–35.

Dickson, M. (1994). "Wavefunction tails in the modal interpretation." Pp. 366–376 in D. Hull, M. Forbes, and R. Burian (eds.), *Proceedings of the PSA 1994, Vol. 1*. East Lansing, Michigan: Philosophy of Science Association.

Dickson, M. (1995a). "Faux-Boolean algebras, classical probability, and determinism." *Foundations of Physics Letters*, **8**: 231–242.

Dickson, M. (1995b). "Faux-Boolean algebras and classical models." *Foundations of Physics Letters*, **8**: 401–415.

Dickson, M. and R. Clifton (1998). "Lorentz-invariance in modal interpretations." Pp. 9–48 in D. Dieks and P. Vermaas (eds.), *The Modal Interpretation of Quantum Mechanics*. Dordrecht: Kluwer Academic Publishers.

Dieks, D. (1988). "The formalism of quantum theory: an objective description of reality?" *Annalen der Physik*, **7**: 174–190.

Dieks, D. (1989). "Resolution of the measurement problem through decoherence of the quantum state." *Physics Letters A*, **142**: 439–446.

Dieks, D. (1998). "Preferred factorizations and consistent property attribution." Pp. 144–160 in R. Healey and G. Hellman (eds.), *Quantum Measurement: Beyond Paradox*. Minneapolis: University of Minnesota Press.

Dieks, D. (2005). "Quantum mechanics: an intelligible description of objective reality?" *Foundations of Physics*, **35**: 399–415.

Dieks, D. (2007). "Probability in modal interpretations of quantum mechanics." *Studies in History and Philosophy of Modern Physics*, **19**: 292–310.
Dieks, D. (2009). "Objectivity in perspective: relationism in the interpretation of quantum mechanics." *Foundations of Physics*, **39**: 760–775.
Dieks, D. (2017). "Niels Bohr and the formalism of quantum mechanics." Pp. 303–333 in J. Faye and H. J. Folse (eds.), *Niels Bohr and the Philosophy of Physics: Twenty-First-Century Perspectives*. London: Bloomsbury Academic.
Dieks, D. (2019). "Quantum mechanics and perspectivalism." Pp. 51–70 in O. Lombardi, S. Fortin, C. López, and F. Holik (eds.), *Quantum Worlds: Perspectives on the Ontology of Quantum Mechanics*. Cambridge: Cambridge University Press.
Dieks, D. (2020). "Identical quantum particles, entanglement, and individuality." *Entropy*, **22**: #134.
Dieks, D. (2021). "Modal interpretations of quantum mechanics." Pp. 1155–1174 in O. Freire (ed.), *Oxford Handbook of the History of Interpretations of Quantum Mechanics*. Oxford: Oxford University Press.
Dieks, D. and A. Lubberdink (2011). "How classical particles emerge from the quantum world." *Foundations of Physics*, **41**: 1051–1064.
Dieks, D. and A. Lubberdink (2022). "Identical quantum particles as distinguishable objects." *Journal of General Philosophy of Science*, **53**: 259–274.
Dieks, D. and P. Vermaas (eds.) (1998). *The Modal Interpretation of Quantum Mechanics*. Dordrecht: Kluwer Academic Publishers.
Dieks, D. and M. Versteegh (2008). "Identical quantum particles and weak discernibility." *Foundations of Physics*, **38**: 923–934.
Dirac, P. A. M. (1958). *The Principles of Quantum Mechanics*. Oxford: Clarendon Press.
Dretske, F. I. (1977). "Laws of nature." *Philosophy of Science*, **44**: 248–268.
Dugić, M. and J. Jeknić-Dugić (2008). "What is «system»: The information-theoretic arguments." *International Journal of Theoretical Physics*, **47**: 805–813.
Dummett, M. (1975). "Wang's paradox." *Synthese*, **30**: 201–232.
Dürr, D., S. Goldstein, and N. Zanghì (1992). "Quantum equilibrium and the origin of absolute uncertainty." *Journal of Statistical Physics*, **67**: 843–907.
Dürr, D., S. Goldstein, and N. Zanghì (2013). *Quantum Physics without Quantum Philosophy*. Berlin: Springer.
Earman, J. (2004a). "Laws, symmetry, and symmetry breaking: invariance, conservation principles, and objectivity." *Philosophy of Science*, **71**: 1227–1241.
Earman, J. (2004b). "Curie's principle and spontaneous symmetry breaking." *International Studies in the Philosophy of Science*, **18**: 173–198.
Earman, J. (2015). "Some puzzles and unresolved issues about quantum entanglement." *Erkenntnis*, **80**: 303–337.
Earman, J. (2019) "Lüders conditionalization: Conditional probability, transition probability, and updating in quantum probability theory." *PhilSci Archive*, URL: http://philsci-archive.pitt.edu/15914/
Earman, J. and L. Ruetsche (2005). "Relativistic invariance and modal interpretations." *Philosophy of Science*, **72**: 557–583.
Egg, M. (2021). "Quantum ontology without speculation." *European Journal for Philosophy of Science*, **11**: #32.
Einstein, A. (1948). "Quantenmechanik und Wirklichkeit." *Dialectica*, **2**: 320–324.
Einstein, A., B. Podolsky, and N. Rosen (1935). "Can quantum-mechanical description of physical reality be considered complete?" *Physical Review*, **47**: 777–780.
Elby, A. (1993). "Why 'modal' interpretations of quantum mechanics don't solve the measurement problem." *Foundations of Physics Letters*, **6**: 5–19.
Elby, A. (1994). "The 'decoherence' approach to the measurement problem in quantum mechanics." Pp. 355–365 in D. Hull, M. Forbes, and R. M. Burian (eds.), *Proceedings of the of the 1994 Biennial Meeting of the Philosophy of Science Association, Vol. 1*. Cambridge: Cambridge University Press.
Elitzur, A. and L. Vaidman (1993). "Quantum mechanical interaction-free measurements." *Foundations of Physics*, **23**: 987–997.
Esfeld, M. (2019). "Individuality and the account of nonlocality: The case for the particle ontology in quantum physics." Pp. 222–244 in O. Lombardi, S. Fortin, C. López, and F. Holik (eds.), *Quantum Worlds. Perspectives on the Ontology of Quantum Mechanics*. Cambridge: Cambridge University Press.
Evans, G. (1978). "Can there be vague objects?" *Analysis*, **38**: 208.
Everett, H. (1957). "'Relative state' formulation of quantum mechanics." *Reviews of Modern Physics*, **29**: 454–462.

Everett, H. (1973). "The theory of universal wave function." Pp. 3–140 in B. S. DeWitt and N. Graham (eds.), *The Many-Worlds Interpretation of Quantum Mechanics*. Princeton: Princeton University Press.

Faye, J. (2019). "Copenhagen Interpretation of quantum mechanics." in E. N. Zalta and U. Nodelman (eds.), *The Stanford Encyclopedia of Philosophy* (Winter 2019 Edition). URL: https://plato.stanford.edu/archives/win2019/entries/qm-copenhagen/

Fletcher, S. C. and D. E. Taylor (2024). "The representation and determinable structure of quantum properties." *Synthese*, **204**: #44.

Fock, V. (1935). "Zur Theorie des Wasserstoffatoms." *Zeitschrift für Physik*, **98**: 145–154.

Ford, G. and T. O'Connell (2001). "Decoherence without dissipation." *Physics Letters A*, **286**: 87–90.

Fortin, S. and O. Lombardi (2014), "Partial traces in decoherence and in interpretation: What do reduced states refer to?" *Foundations of Physics*, **44**: 426–446.

Fortin, S. and O. Lombardi (2016). "A top-down view of the classical limit of quantum mechanics." Pp. 435–468 in R. E. Kastner, J. Jeknić-Dugić, and G. Jaroszkiewicz (eds.), *Quantum Structural Studies: Classical Emergence from the Quantum Level*. Singapore: World Scientific.

Fortin, S. and O. Lombardi (2018). "Understanding decoherence as an irreversible process." *International Journal of Quantum Foundations*, **4**: 247–267.

Fortin, S. and O. Lombardi (2020). "The Frauchiger-Renner argument: A new no-go result?" *Studies in History and Philosophy of Modern Physics*, **70**: 1–7.

Fortin, S. and O. Lombardi (2021). "Is the problem of molecular structure just the quantum measurement problem?" *Foundations of Chemistry*, **23**: 379–395.

Fortin, S. and O. Lombardi (2022). "Entanglement and indistinguishability in a quantum ontology of properties." *Studies in History and Philosophy of Science*, **91**: 234–243.

Fortin, S., O. Lombardi, and M. Castagnino (2014). "Decoherence: A closed-system approach." *Brazilian Journal of Physics*, **44**: 138–153.

Fortin, S., O. Lombardi, and J. C. Martínez González (2016). "Isomerism and decoherence." *Foundations of Chemistry*, **18**: 225–240.

Fortin, S., O. Lombardi, and J. C. Martínez González (2018). "A new application of the modal-Hamiltonian interpretation of quantum mechanics: the problem of optical isomerism." *Studies in History and Philosophy of Modern Physics*, **62**: 123–135.

Fortin, S., O. Lombardi, and M. Pasqualini (2022). "Relational event-time in quantum mechanics." *Foundations of Physics*, **52**: #10.

Franco, M. and E. Calzetta (2011). "Decoherence in the cosmic background radiation." *Classical and Quantum Gravity*, **28**: #145024.

Frasca, M. (2003). "General theorems on decoherence in the thermodynamic limit." *Physics Letters A*, **308**: 135–139.

Frauchiger, D. and R. Renner (2016). "Single-world interpretations of quantum theory cannot be self-consistent." URL: https://arxiv.org/abs/1604.07422v1

Frauchiger, D. and R. Renner (2018). "Quantum theory cannot consistently describe the use of itself." *Nature Communications*, **9**: #3711.

French, S. (1989). "Identity and individuality in classical and quantum physics." *Australasian Journal of Philosophy*, **67**: 432–446.

French, S. (1998). "On the withering away of physical objects." Pp. 93–113 in E. Castellani (ed.), *Interpreting Bodies. Classical and Quantum Objects in Modern Physics*. Princeton: Princeton University Press.

French, S. (2006). "Structure as a weapon of the realist." *Proceedings of the Aristotelian Society*, **106**: 167–185.

French, S. (2018). "Between factualism and substantialism: Structuralism as a third way." *International Journal of Philosophical Studies*, **26**: 701–721.

French, S. (2020). "What is this thing called structure? (Rummaging in the toolbox of metaphysics for an answer)." URL: http://philsci-archive.pitt.edu/id/eprint/16921.

French, S. and T. Bigaj (2024). "Identity and individuality in quantum theory." in E. N. Zalta and U. Nodelman (eds.), *The Stanford Encyclopedia of Philosophy* (Spring 2024 Edition), URL: https://plato.stanford.edu/archives/spr2024/entries/qt-idind/.

French, S. and D. Krause (2006). *Identity in Physics a Historical, Philosophical, and Formal Analysis*. Oxford: Clarendon Press.

French, S. and J. Ladyman (2003). "Remodelling structural realism: quantum physics and the metaphysics of structure." *Synthese*, **136**: 31–56.

Friebe, C. (2014). "Individuality, distinguishability, and (non-)entanglement: A defense of Leibniz's principle." *Studies in History and Philosophy of Modern Physics*, **48**: 89–98.

Friedman, M. (1974). "Explanation and scientific understanding." *The Journal of Philosophy*, **71**: 5–19.
Frigg, R. (2006). "Scientific representation and the semantic view of theories." *Theoria*, **55**: 37–53.
Frigg, R. (2007). "A field guide to recent work on the foundations of thermodynamics and statistical mechanics." Pp. 99–196 in D. Rickles (ed.), *The Ashgate Companion to the New Philosophy of Physics*. London: Ashgate.
Frigg, R. and Hoefer, C. (2007). "Probability in GRW theory." *Studies in History and Philosophy of Modern Physics*, **38**: 371–389.
Frigg, R. and C. Werndl (2024). "Philosophy of statistical mechanics." in E. N. Zalta and U. Nodelman (eds.), *The Stanford Encyclopedia of Philosophy* (Spring 2024 Edition), URL: https://plato.stanford.edu/archives/spr2024/entries/statphys-statmech/.
Fuchs, C., D. Mermin, and R. Schack (2014). "An introduction to QBism with an application to the locality of quantum mechanics." *American Journal of Physics*, **82**: 749–754.
Gallois, A. (2016). "Identity over time." in E. N. Zalta (ed.), *The Stanford Encyclopedia of Philosophy* (Winter 2016 Edition), URL: https://plato.stanford.edu/archives/win2016/entries/identity-time/.
Gambini, R., R. Porto, and J. Pullin (2007). "Fundamental decoherence from quantum gravity: a pedagogical review." *General Relativity and Gravitation*, **39**: 1143–1156.
Gambini, R. and J. Pullin (2007). "Relational physics with real rods and clocks and the measurement problem of quantum mechanics." *Foundations of Physics*, **37**: 1074–1092.
Gambini, R. and J. Pullin (2010). "Modern space-time and undecidability." Pp. 149–162 in V. Petkov (ed.), *Fundamental Theories of Physics (Minkowski Spacetime: A Hundred Years Later)*. Heidelberg: Springer.
Gelfand, I. M. and M. A. Naimark (1943). "On the imbedding of normed rings into the ring of operators in Hilbert space." *Matematiceskij Sbornik*, **12**: 197–213.
Gell-Mann, M. and J. B. Hartle (1990). "Quantum mechanics in the light of quantum cosmology." Pp. 425–458 in W. H. Zurek (ed.), *Complexity, Entropy and the Physics of Information*. Reading, MA: Addison-Wesley.
Gell-Mann, M. and J. B. Hartle (1993). "Classical equations for quantum systems." *Physical Review D*, **47**: 3345–3382.
Ghirardi, GC. (2005). *Sneaking a Look at God's Cards: Unraveling the Mysteries of Quantum Mechanics*. Princeton NJ: Princeton University Press.
Ghirardi, GC and A. Bassi (2024). "Collapse theories." in E. N. Zalta and U. Nodelman (eds.), *The Stanford Encyclopedia of Philosophy* (Fall 2024 Edition). URL: https://plato.stanford.edu/archives/sum2020/entries/qm-collapse/
Ghirardi, GC., R. Grassi, and F. Benatti (1995). "Describing the macroscopic world: Closing the circle within the dynamical reduction program." *Foundations of Physics*, **25**: 5–38.
Ghirardi, GC. and L. Marinatto (2003). "Entanglement and properties." *Fortschritteder Physik*, **51**: 379–387.
Ghirardi, GC. and L. Marinatto (2004). "General criterion for the entanglement of two indistinguishable particles." *Physical Review A*, **70**: 12109.
Ghirardi, GC. and L. Marinatto (2005). "Identical particles and entanglement." *Optics and Spectroscopy*, **99**: 386–390.
Ghirardi, GC., L. Marinatto, and T. Weber (2002). "Entanglement and properties of composite quantum systems: A conceptual and mathematical analysis." *Journal of Statistical Physics*, **108**: 49–122.
Ghirardi, GC., P. Pearle, and A. Rimini (1990). "Markov Processes in Hilbert Space and Continuous Spontaneous Localization of Systems of Identical Particles." *Physical Review A*, **42**: 78–89.
Ghirardi, GC., A. Rimini, and T. Weber (1986). "Unified dynamics for microscopic and macroscopic systems." *Physical Review D*, **34**: 470–491.
Glick, D. (2017). "Against quantum indeterminacy." *Thought: Journal of Philosophy*, **6**: 204–213.
Glick, D. (2022). "Quantum mechanics without indeterminacy." Pp. 319–335 in V. Allori (ed.), *Quantum Mechanics and Fundamentality*. Synthese Library, *Vol. 460*. Cham: Springer.
Glick, D. (2023). "The principle of least action and teleological explanation in physics." *Synthese*, **202**: #25.
Glymour, C. (1980). *Theory and Evidence*. Princeton: Princeton University Press.
Goldstein, S. (2021). "Bohmian mechanics." in E. N. Zalta (ed.), *The Stanford Encyclopedia of Philosophy* (Fall 2021 Edition), URL: https://plato.stanford.edu/archives/fall2021/entries/qm-bohm/
Goodstein, D. and J. Goodstein (1996). *Feynman's Lost Lecture*. New York: W. W. Norton & Company.
Gordon, E. C. (2017). "Understanding in epistemology." in J. Fieser and B. Dowden (eds.), *Internet Encyclopedia of Philosophy*. URL: https://iep.utm.edu/understa/
Gould, J. B. (1970). *The Philosophy of Chrysippus*. Albany: Suny Press.
Griffiths, R. B. (1984). "Consistent histories and the interpretation of quantum mechanics." *Journal of Statistical Physics*, **36**: 219–272.

Griffiths, R. B. (2002). *Consistent Quantum Theory.* Cambridge: Cambridge University Press.
Griffiths, R. B. (2013). "A consistent quantum ontology." *Studies in History and Philosophy of Modern Physics*, **44**: 93–114.
Grimm, S. (2021). "Understanding." in E. N. Zalta (ed.), *The Stanford Encyclopedia of Philosophy* (Summer 2021 Edition). URL: https://plato.stanford.edu/archives/sum2021/entries/understanding/.
Gryb, S. and K. Thébault (2016). "Time remains." *The British Journal for the Philosophy of Science*, **67**: 663–705.
Guerra Bobo, I. (2013). "On quantum conditional probability." *Theoria*, **76**: 115–137.
Haack, S. (1974). *Deviant Logic.* Cambridge: Cambridge University Press.
Haack, S. (1978). *Philosophy of Logics.* Cambridge: Cambridge University Press.
Haag, R. (1992). *Local Quantum Physics: Fields, Particles, Algebras.* Heidelberg: Springer-Verlag.
Haag, R. and D. Kastler (1964). "An algebraic approach to quantum field theory." *Journal of Mathematical Physics*, **7**: 848–861.
Hacking, I. (1975). *The Emergence of Probability.* Cambridge: Cambridge University Press.
Hacking, I. (2002). *Historical Ontology.* Cambridge MA: Harvard University Press.
Hájek, A. (2023). "Interpretations of probability." in E. N. Zalta and U. Nodelman (eds.), *The Stanford Encyclopedia of Philosophy* (Winter 2023 Edition), URL:https://plato.stanford.edu/archives/win2023/entries/probability-interpret/
Halvorson, H. (2012). "What scientific theories could not be." *Philosophy of Science*, **79**: 183–206.
Halvorson, H. (2016). "Scientific theories." Pp. 585–608 in P. Humphreys (ed.), *The Oxford Handbook of Philosophy of Science.* Oxford: Oxford University Press.
Hamermesh, M. (1962). *Group Theory and its Applications to Physical Problems.* Reading Mass: Addison-Wesley.
Harshman, N. L. (2012). "Observables and entanglement in the two-body system." *AIP Conference Proceedings*, 1508: 386–390.
Harshman, N. L. (2016). "Symmetry and natural quantum structures for three-particles in one-dimension." Pp. 373–399 in R. E. Kastner, J. Jeknić-Dugić, and G. Jaroszkiewicz (eds.), *Quantum Structural Studies: Classical Emergence from the Quantum Level.* Singapore: World Scientific.
Harshman, N. L. and K. S. Ranade (2011). "Observables can be tailored to change the entanglement of any pure state." *Physical Review A*, **84**: #012303.
Harshman, N. L. and S. Wickramasekara (2007a). "Galilean and dynamical invariance of entanglement in particle scattering." *Physical Review Letters*, **98**: #080406.
Harshman, N. L. and S. Wickramasekara (2007b). "Tensor product structures, entanglement, and particle scattering." *Open Systems and Information Dynamics*, **14**: 341–351.
Healey, R. (1995). "Dissipating the quantum measurement problem." *Topoi*, **14**: 55–65.
Healey, R. (2018). "Quantum theory and the limits of objectivity." *Foundations of Physics*, **48**: 1568–1589.
Healey, R. (2022). "Quantum-Bayesian and pragmatist views of quantum theory." in E. N. Zalta (ed.), *The Stanford Encyclopedia of Philosophy* (Summer 2022 Edition), URL: https://plato.stanford.edu/archives/sum2022/entries/quantum-bayesian/
Healey, R. and H. Gomes (2022). "Holism and nonseparability in physics." in E. N. Zalta and U. Nodelman (eds.), *The Stanford Encyclopedia of Philosophy* (Winter 2022 Edition), URL: https://plato.stanford.edu/archives/win2022/entries/physics-holism/
Heisenberg, W. (1958). *Physics and Philosophy.* London: George Allen and Unwin.
Hempel, C. (1966). *Philosophy of Natural Science.* Englewood Cliffs, NJ: Prentice Hall.
Hendry, R. F. (1998). "Models and approximations in quantum chemistry." Pp. 123–142 in N. Shanks (ed.), *Idealization in Contemporary Physics.* Amsterdam-Atlanta: Rodopi.
Hendry, R. F. (2004). "The physicists, the chemists, and the pragmatics of explanation." *Philosophy of Science*, **71**: 1048–1059.
Hendry, R. F. (2010). "Ontological reduction and molecular structure." *Studies in History and Philosophy of Modern Physics*, **41**: 183–191.
Hensen, B., H. Bernien, A. E. Dréau, A. Reiserer, N. Kalb, M. S. Blok, J. Ruitenberg, R. F. L. Vermeulen, R. N. Schouten, C. Abellán, et al. (2015). "Loophole-free Bell inequality violation using electron spins separated by 1.3 kilometres." *Nature*, **526**: 682–686.
Hettema, H. (2012). *Reducing Chemistry to Physics. Limits, Models, Consecuences.* Groningen: University of Groningen.
Hoefer, C. (2019). *Chance in the World: A Humean Guide to Objective Chance.* Oxford: Oxford University Press.
Holik, F., J. P. Jorge, D. Krause, and O. Lombardi (2022). "Quasi-set theory for a quantum ontology of properties." *Synthese*, **200**: #401.

Howard, D. (1985). "Einstein on locality and separability." *Studies in History and Philosophy of Science*, **16**: 171–201.

Howard, D. (1989). "Holism, separability and the metaphysical implications of the Bell experiments." Pp. 224–253 in J. Cushing and E. McMullin (eds.), *Philosophical Consequences of Quantum Theories: Reflections on Bell's Theorem*. Notre Dame: University of Notre Dame Press.

Howard, D. (2004). "Who invented the 'Copenhagen Interpretation?' A study in mythology." *Philosophy of Science*, **71**: 669–682.

Huggett, N. and T. Imbo (2009). "Indistinguishability." Pp. 311–317 in D. Greenberger, K. Hentschel, and F. Weinert (eds.), *Compendium of Quantum Physics*. Berlin-New York: Springer.

Hughes, R. I. G. (1989). *The Structure and Interpretation of Quantum Mechanics*. Cambridge MA: Harvard University Press.

Hund. F. (1927). "Zur Deutung der Molekelspektren. III." *Zeitschrift für Physik*, **43**: 805–826.

Hunt, J., G. Carcassi, and C. Aidala (2025). "Hamiltonian privilege." *Erkenntnis*, **90**: 443–466.

Iguri, S. and M. Castagnino (1999). "The formulation of quantum mechanics in terms of nuclear algebras." *International Journal of Theoretical Physics*, **38**: 143–164.

Isham, C. J. (1993) "Canonical quantum gravity and the problem of time." Pp 157–287 in L. A. Ibort and M. A. Rodríguez (eds.), *Integrable Systems, Quantum Groups, and Quantum Field Theories*, NATO ASI Series (Series C: Mathematical and Physical Sciences), *Vol 409*. Dordrecht: Springer.

James, W. (1897) [1956]. "The dilemma of determinism." Pp. 145–183 in *The Will to Believe*. New York: Dover Publications.

Jauch, J. M. and C. Piron (1969). "On the structure of quantal propositional systems." *Helvetica Physica Acta*, **42**: 842–848.

Joos, E. (2000). "Elements of environmental decoherence." Pp. 1–17 in P. Blanchard, D. Giulini, E. Joos, C. Kiefer and I.-O. Stamatescu (eds.), *Decoherence: Theoretical, Experimental, and Conceptual Problems*, Lecture Notes in Physics, *Vol. 538*. Heidelberg-Berlin: Springer

Kant, I. (1902-). *Gesammelte Schriften*. Berlin: Herausgegeben von der Preußischen Akademie der Wissenschaften (Bde. 1–22), der Deutschen Akademie der Wissenschaften zu Berlin (Bd. 23), und der Akademie der Wissenschaften zu Göttingen (Bde. 24, 25, 27–29).

Kaplan, D. (1975). "How to Russell a Frege-Church." *The Journal of Philosophy*, **72**: 716–729.

Kastner, R. E. (2013). *The Transactional Interpretation of Quantum Mechanics. The Reality of Possibility*. Cambridge: Cambridge University Press.

Kastner, R. E., J. Jeknić-Dugić, and G. Jaroszkiewicz (2016). *Quantum Structural Studies: Classical Emergence from the Quantum Level*. Singapore: World Scientific.

Kastner, R. E., S. Kauffman, and M. Epperson (2018). "Taking Heisenberg's potentia seriously." *International Journal of Quantum Foundations*, **4**: 158–172.

Keinänen, M. and J. Hakkarinen (2014). "The problem of trope individuation: A reply to Lowe." *Erkenntnis*, **79**: 65–79.

Keränen, J. (2001). "The identity problem for realist structuralism." *Philosophia Mathematica*, **9**: 308–330.

Keynes, J. M. (1921). *A Treatise on Probability*. London: Macmillan.

Kiefer, C. and D. Polarski (2009). "Why do cosmological perturbations look classical to us?" *Advanced Science Letters*, **2**: 164–173.

King, P. (1992). "Duns Scotus on the common nature." *Philosophical Topics*, **20**: 50–76.

Kitajima, Y. (2004). "A remark on the modal interpretation of algebraic quantum field theory." *Physics Letters A*, **331**: 181–186.

Kment, B. (2021). "Varieties of modality." in E. N. Zalta (ed.), *The Stanford Encyclopedia of Philosophy* (Spring 2021 Edition), URL: https://plato.stanford.edu/archives/spr2021/entries/modality-varieties/

Kneale, W. and M. Kneale (1962). *The Development of Logic*. Oxford: Clarendon Press.

Knuuttila, S. (2021). "Medieval theories of modality." in E. N. Zalta (ed.), *The Stanford Encyclopedia of Philosophy* (Summer 2021 Edition), URL: https://plato.stanford.edu/archives/sum2021/entries/modality-medieval

Kochen, S. (1985). "A new interpretation of quantum mechanics." Pp. 151–169 in P. Mittelstaedt and P. Lahti (eds.), *Symposium on the Foundations of Modern Physics 1985*. Singapore: World Scientific.

Kochen, S. and E. Specker (1967). "The problem of hidden variables in quantum mechanics." *Journal of Mathematics and Mechanics*, **17**: 59–87.

Koopman, B. O. (1931). "Hamiltonian systems and transformations in Hilbert space." *Proceedings of the National Academy of Sciences*, **18**: 315–318.

Kramer, E. (1970). *The Nature and Growth of Modern Mathematics*. Princeton: Princeton University Press.

Krause, D. (1992). "On a quasi-set theory." *Notre Dame Journal of Formal Logic*, **33**: 402–411.

Krause, D. and A. M. N. Coelho (2005). "Identity, indiscernibility, and philosophical claims." *Axiomathes*, **15**: 191–210.

Krause, D. and J. P. Jorge (2025). "Sobre una teoría 'pura' de casi-conjuntos y su aplicación a una ontología cuántica de propiedades." *Principia*, forthcoming.

Krause, D., J. P. Jorge, and O. Lombardi (2025). "A quasi-set theory without atoms and its application to a quantum ontology of properties," *Synthese*, forthcoming.

Kuchař, K. (1991). "The problem of time in canonical quantization." Pp. 141–171 in A. Ashtekar and J. Stachel (eds.), *Conceptual Problems of Quantum Gravity*. Boston: Birkhäuser.

Kuhlmann, M. (2010). *The Ultimate Constituents of the Material World – In Search of an Ontology for Fundamental Physics*. Frankfurt: Ontos-Verlag.

Kuhlmann, M. (2023). "Quantum field theory." in E. N. Zalta and U. Nodelman (eds.), *The Stanford Encyclopedia of Philosophy* (Summer 2023 Edition). URL: https://plato.stanford.edu/archives/sum2023/entries/quantum-field-theory/

Ladyman, J. (1998). "What is structural realism?" *Studies in History and Philosophy of Science*, **29**: 409–424.

Ladyman, J. (2001). "Science, metaphysics and structural realism." *Philosophica*, **67**: 57–76.

Ladyman J., Ø. Linnebo, and T. Bigaj (2013). "Entanglement and non-factorizability." *Studies in the History and Philosophy of Modern Physics*, **44**: 215–221.

Lagerlund, H. (2021). "Medieval theories of the syllogism." in E. N. Zalta (ed.), *The Stanford Encyclopedia of Philosophy* (Summer 2021 Edition), URL: https://plato.stanford.edu/archives/sum2021/entries/medieval-syllogism/

Laloë, F. (2018). "Can quantum mechanics be considered consistent? A discussion of Frauchiger and Renner's argument." URL: https://arxiv.org/abs/1802.06396v3

Landsman, N. P. (2007). "Between classical and quantum." Pp. 417–554 in J. Butterfield and J. Earman (eds.), *Handbook of the Philosophy of Science Vol. 2: Philosophy of Physics*. North-Holland-Elsevier, Amsterdam.

Landsman, N. P. (2009). "Algebraic quantum mechanics." Pp. 6–9 in D. Greenberger, K. Hentschel, and F. Weinert (eds.), *Compendium of Quantum Physics*. Berlin-Heidelberg: Springer.

Laudisa, F. and C. Rovelli (2021). "Relational quantum mechanics." in E. N. Zalta (ed.), *The Stanford Encyclopedia of Philosophy* (Spring 2021 Edition), URL: https://plato.stanford.edu/entries/qm-relational/

Laue, H. (1996). "Space and time translations commute, don't they?" *American Journal of Physics*, **64**: 1203–1205.

Laura R. and L. Vanni (2008). "Conditional probabilities and collapse in quantum measurements." *International Journal of Theoretical Physics*, **47**: 2382–2392.

Laycock, H. (2010), "Object." in E. N. Zalta (ed.), *The Stanford Encyclopedia of Philosophy* (Fall 2010 Edition), URL: http://plato.stanford.edu/archives/fall2010/entries/object/

Lazarovici, D. and M. Hubert (2019). "How quantum mechanics can consistently describe the use of itself." *Scientific Reports*, **9**: #470.

Lee, H.-W. and J. Kim (2000). "Quantum teleportation and Bell's inequality using single-particle entanglement." *Physical Review A*, **63**: 012305.

Leggett, A. (1987). "Reflections on the quantum measurement paradox." Pp. 85–104 in B. J. Hiley and F. D. Peat (eds.), *Quantum Implications*. London: Routledge and Kegan Paul.

Leinaas, J. M. and J. Myrheim (1977). "On the theory of identical particles." *Nuovo Cimento B*, **37**: 1–23.

Lévi-Leblond, J. M. (1974). "The pedagogical role and epistemological significance of group theory in quantum mechanics." *Nuovo Cimento*, **4**: 99–143.

Lewis, D. (1986a). *Philosophical Papers*, Volume II. Oxford: Oxford University Press.

Lewis, D. (1986b). *On the Plurality of Worlds*. Oxford: Blackwell.

Lewis, D. (1994). "Humean supervenience debugged." *Mind*, **103**: 473–490.

Lewowicz, L. (2005). *Del Relativismo Lingüístico al Relativismo Ontológico en el Último Kuhn*. Montevideo: Departamento de Publicaciones de la Facultad de Humanidades y Ciencias de la Educación, Universidad de la República.

Lewowicz, L. and O. Lombardi (2013). "Stuff versus individuals." *Foundations of Chemistry*, **15**: 65–77.

Lo Franco, R. and G. Compagno (2018). "Indistinguishability of elementary systems as a resource for quantum information processing." *Physical Review Letters*, **120**: #240403.

Lombardi, O. (2010). "The central role of the Hamiltonian in quantum mechanics: decoherence and interpretation." *Manuscrito*, **33**: 307–349.

Lombardi, O. (2023a). "Entanglement and indistinguishability: facing some challenges from a new perspective." *Philosophical Transactions of the Royal Society A*, **381**: #20220101.

Lombardi, O. (2023b). "A Kantian-rooted pluralist realism for science." Pp. 81–101 in C. Soto (ed.), *Current Debates in Philosophy of Science: In Honor of Roberto Torretti*, Cham: Springer-Synthese Library.

Lombardi, O. (2024). "Pluralist realism: where onticity and practice meet." Pp. 93–120 in A. Mets, E. Lõhkivi, P. Müürsepp, and J. Eigi-Watkin (eds.), *Practical Realist Philosophy of Science, Reflecting on Rein Vihalemm's Ideas*. London: Lexington Books.

Lombardi, O. (2025). "The relative nature of open quantum systems." Pp. 53–67 in M. Cuffaro and S. Hartmann (eds.), *The Open Systems View: Physics, Metaphysics and Methodology*. Oxford: Oxford University Press.

Lombardi, O. and J. S. Ardenghi (2022). "How different interpretations of quantum mechanics can enrich each other: the case of the Relational Quantum Mechanics and the Modal-Hamiltonian Interpretation." *Foundations of Physics*, **52**: #64.

Lombardi, O., and M. Castagnino (2008). "A modal-Hamiltonian interpretation of quantum mechanics." *Studies in History and Philosophy of Modern Physics*, **39**: 380–443.

Lombardi, O., and M. Castagnino (2010). "Matters are not so clear on the physical side." *Foundations of Chemistry*, **12**: 159–166.

Lombardi, O., M. Castagnino, and J. S. Ardenghi (2010). "The modal-Hamiltonian interpretation and the Galilean covariance of quantum mechanics." *Studies in History and Philosophy of Modern Physics*, **41**: 93–103.

Lombardi, O. and D. Dieks (2016). "Particles in a quantum ontology of properties." Pp. 123–143 in T. Bigaj and C. Wüthrich (eds.), *Metaphysics in Contemporary Physics* (Poznan Studies in the Philosophy of the Sciences and the Humanities). Leiden: Brill-Rodopi.

Lombardi, O. and D. Dieks (2024). "Modal interpretations of quantum mechanics." in E. N. Zalta and U. Nodelman (eds.), *The Stanford Encyclopedia of Philosophy* (Summer 2024 Edition). URL: https://plato.stanford.edu/archives/sum2024/entries/qm-modal/

Lombardi, O., S. Fortin, and M. Castagnino (2012). "The problem of identifying the system and the environment in the phenomenon of decoherence." Pp. 161–174 in H. W. de Regt, S. Hartmann, and S. Okasha (eds.), *EPSA Philosophy of Science: Amsterdam 2009*, Berlin: Springer.

Lombardi, O., S. Fortin, and C. López (2015). "Measurement, interpretation and information." *Entropy*, **17**: 7310–7330.

Lombardi, O., S. Fortin, and M. Pasqualini (2022). "Possibility and time in quantum mechanics." *Entropy*, **24**: #249.

Lombardi, O. and M. Labarca (2005). "The ontological autonomy of the chemical world." *Foundations of Chemistry*, **7**: 125–148.

Lombardi, O. and M. Labarca (2006). "The ontological autonomy of the chemical world: A response to Needham." *Foundations of Chemistry*, **8**: 81–92.

Lombardi, O. and C. Matta (2022). "Coarse graining and the Quantum Theory of Atoms in Molecules." Pp. 217–241, in O. Lombardi, J. C. Martínez González, and S. Fortin (eds.), *Philosophical Perspectives in Quantum Chemistry*. Cham: Springer Nature, Synthese Library.

Lombardi, O. and A. R. Pérez Ransanz (2012). *Los Múltiples Mundos de la Ciencia*. México: UNAM-Siglo XXI.

Lombardo, F. and D. Mazzitelli (1996). "Coarse graining and decoherence in quantum field theory." *Physical Review D*, **53**: #2001.

López, C. (2024). "Quantum ontology without textbooks. Nor overlapping." *European Journal for Philosophy of Science*, **14**: #11.

Losada, M., R. Laura, and O. Lombardi (2019), "The Frauchiger-Renner argument and quantum histories." *Physical Review A*, **100**: #052114.

Lutz, S. (2012). "On a straw man in the philosophy of science: A defense of the received view." *HOPOS*, **2**: 77–120.

Lutz, S. (2017). "What was the syntax-semantics debate in the philosophy of science about?" *Philosophy and Phenomenological Research*, **95**: 319–352.

Lychkovskiy, O. (2013). "Dependence of decoherence-assisted classicality on the way a system is partitioned into subsystems." *Physical Review A*, **87**: #022112.

Mackey, M. C. (1989). "The dynamic origin of increasing entropy." *Review of Modern Physics*, **61**: 981–1015.

MacLeod, M. and E. Rubenstein (2006). "Universals." in J. Fieser and B. Dowden (eds.), *The Internet Encyclopedia of Philosophy*, URL: https://iep.utm.edu/universa/

Manzano, D. (2020). "A short introduction to the Lindblad master equation." *AIP Advances*, **10**: #025106.

Margenau, H. (1954). "Advantages and disadvantages of various interpretations of the quantum theory." *Physics Today*, **7**: 6–13.

Mariani, C. (2022). "Non-accessible mass and the ontology of GRW." *Studies in History and Philosophy of Modern Science*, **91**: 270–279

Mariani, C. and G. Torrengo (2021). "The indeterminate present and the open future." *Synthese*, **199**: 3923–3944.

Martínez González, J. C., S. Fortin, and O. Lombardi (2019). "Why molecular structure cannot be strictly reduced to quantum mechanics." *Foundations of Chemistry*, **21**: 31–45

Marx, K. (1845). "Theses on Feuerbach." Pp. 13–15 in 1969, *Marx/Engels Selected Works*. Moscow: Progress Publishers.

Masillo, F., G. Scolarici, and S. Sozzo (2009). "Proper versus improper mixtures: towards a quaternionic quantum mechanics." *Theoretical and Mathematical Physics*, **160**: 1006–1013.

Matarese, V. (2023). *The Metaphysics of Bohmian Mechanics. A Comprehensive Guide to the Different Interpretations of Bohmian Ontology*. Berlin-Boston: De Gruyter.

Matta, C. F., O. Lombardi, and J. Jaimes Arriaga (2020). "Two-step emergence: the quantum theory of atoms in molecules as a bridge between quantum mechanics and molecular chemistry." *Foundations of Chemistry*, **22**: 107–129

Maudlin, T. (1995). "Three measurement problems." *Topoi*, **14**: 7–15.

Maudlin, T. (2007). *The Metaphysics within Physics*. New York: Oxford University Press.

Maudlin, T. (2013). "The nature of the quantum state." Pp. 126–153 in A. Ney and D. Albert (eds.), *The Wave Function*. New York: Oxford University Press.

Maurin, A.-S. (2018). "Tropes." in E. N. Zalta (ed.), *The Stanford Encyclopedia of Philosophy* (Summer 2018 Edition), URL: https://plato.stanford.edu/archives/sum2018/entries/tropes/

Maxwell, N. (1988). "Quantum propensiton theory: A testable resolution to the wave/particle dilemma." *British Journal for the Philosophy of Science*, **39**: 1–50.

McKay, T. (2008). "Critical notice of *Words without Objects*." *Canadian Journal of Philosophy*, **38**: 301–323.

Meijer, P. and E. Bauer (2004). *Group Theory. The Application to Quantum Mechanics*. New York: Dover

Menzel, C. (2021). "Actualism." in E. N. Zalta (ed.), *The Stanford Encyclopedia of Philosophy* (Fall 2021 Edition), URL: https://plato.stanford.edu/archives/fall2021/entries/actualism/

Menzel, C. (2024). "The possibilism-actualism debate." in E. N. Zalta and U. Nodelman (eds.), *The Stanford Encyclopedia of Philosophy* (Summer 2024 Edition), URL: https://plato.stanford.edu/archives/sum2024/entries/possibilism-actualism/

Messiah, A. (1961). *Quantum Mechanics, Vol. 1*. Amsterdam: North-Holland.

Messiah, A. M. L. and O. W. Greenberg (1964). "Symmetrization postulate and its experimental foundation." *Physical Review B*, **136**: 248–267.

Minkowski, H. (1923). "Space and time." Pp. 75–91 in W. Perrett and G. B. Jeffrey (eds.), *The Principle of Relativity. A Collection of Original Memoirs on the Special and General Theory of Relativity*. New York: Dover.

Misner, C. W., K. S. Thorne, and J. A. Wheeler (1973). *Gravitation*. New York: W.H. Freeman and Company.

Mittelstaedt, P. (1998). *The Interpretation of Quantum Mechanics and the Measurement Process*. Cambridge: Cambridge University Press.

Montague, R. (1960). "Logical necessity, physical necessity, ethics and quantifiers." *Inquiry*, 3: 259–269. Reprinted in (1974), *Formal Philosophy. Selected Papers of Richard Montague*. Thomason NH: Yale University Press.

Monton, B. (1999). "Van Fraassen and Ruetsche on preparation and measurement." *Philosophy of Science*, **66**: S82–S91.

Morris, B., B. Yadin, M. Fadel, T. Zibold, P. Treutlein, and G. Adesso (2020). "Entanglement between identical particles is a useful and consistent resource." *Physical Review X*, **10**: #041012.

Morrison, M. (1999). "Models as autonomous agents." Pp. 38–65 in M. Morgan and M. Morrison (eds.), *Models as Mediators*. Cambridge: Cambridge University Press.

Muciño, R., E. Okon, and D. Sudarsky (2022). "Assessing Relational Quantum Mechanics." *Synthese*, **200**: #399.

Muller, F. A. (2015). "Circumveiloped by Obscuritads: The nature of interpretation in quantum mechanics, hermeneutic circles and physical reality, with cameos of James Joyce and Jacques Derrida." Pp. 107–136 in J.-Y. Béziau, D. Krause, and J. R. Becker Arenhart (eds.), *Conceptual Clarifications. Tributes to Patrick Suppes (1922–2014)*. London: College Publications.

Muller, F. A. and G. Leegwater (2022). "The case against factorism: On the labels of $\otimes$-factor Hilbert-spaces of similar particles in quantum mechanics." *Journal for General Philosophy of Science*, **53**: 291–306.

Muller, F. A. and S. Saunders (2008). "Discerning fermions." *The British Journal for the Philosophy of Science*, **59**: 499–548.

Muller, F. A. and M. Seevinck (2009). "Discerning elementary particles." *Philosophy of Science*, **76**: 179–200.

Myrvold, W. (2022a). "Philosophical issues in quantum theory." in E. N. Zalta and U. Nodelman (eds.), *The Stanford Encyclopedia of Philosophy* (Fall 2022 Edition). URL: https://plato.stanford.edu/archives/fall2022/entries/qt-issues/

Myrvold, W. (2002b). "Modal interpretations and relativity." *Foundations of Physics*, **32**: 1773–1784.

Narvaja, M., M. Córdoba, and O. Lombardi (2012). "Different domains, the same problems." Pp. 67–87 in S. Pintuck and C. Reynolds (eds.), *Philosophy of Science*. New York: Nova Science.

Needham, P. (2006). "Ontological reduction: A comment on Lombardi-Labarca." *Foundations of Chemistry*, **8**: 73–80.

Neurath, O. (1935). "Pseudorationalism of falsification." Pp. 121–131 in R. Cohen and M. Neurath (eds.) (1983), *Philosophical Papers 1913–1946*. Dordrecht: Reidel.

Ney, A. (2013), "Ontological reduction and the wave function ontology." Pp. 168–183 in A. Ney and D. Albert (eds.), *The Wave Function. Essays on the Metaphysics of Quantum Mechanics*. Oxford: Oxford University Press.

Ney, A. and D. Albert (2013). *The Wave Function*. New York: Oxford University Press.

Noonan, H. and B. Curtis (2022). "Identity." in E. N. Zalta and U. Nodelman (eds.), *Stanford Encyclopedia of Philosophy* (Winter 2022 Edition), URL: https://plato.stanford.edu/archives/win2022/entries/identity/

North, J. (2009). "The 'Structure' of Physics: A Case Study." *Journal of Philosophy*, **106**: 57–88.

Nosrati, F., A. Castellini, G. Compagno, R. Lo Franco (2020). "Robust entanglement preparation against noise by controlling spatial indistinguishability." *npj Quantum Information*, **6**: #39.

Nozick, R. (2001). *Invariances: The Structure of the Objective World*. Harvard: Harvard University Press.

Ohanian, H. and R. Ruffini (1994). *Gravitation and Spacetime*, 2nd ed. London: W. W. Norton and Company.

Øhrstrøm, P. and P. Hasle (2020). "Future contingents." in E. N. Zalta (ed.), *The Stanford Encyclopedia of Philosophy* (Summer 2020 Edition), URL: https://plato.stanford.edu/archives/sum2021/entries/future-contingents/

Oldofredi, A. (2021). "The bundle theory approach to Relational Quantum Mechanics." *Foundations of Physics*, **51**: #18.

O'Leary-Hawthorne, J. (1995). "The bundle theory of substance and the identity of indiscernibles." *Analysis*, **55**: 191–196.

Omnès, R. (1987). "Interpretation of quantum mechanics." *Physics Letters A*, **125**: 169–172.

Omnès, R. (1994). *The Interpretation of Quantum Mechanics*. Princeton: Princeton University Press.

Omnès, R. (1999). *Understanding Quantum Mechanics*. Princeton: Princeton University Press.

Omnès, R. (2001). "Decoherence: An irreversible process." arXiv:quant-ph/0106006.

Omnès, R. (2002). "Decoherence, irreversibility and the selection by decoherence of quantum states with definite probabilities." *Physical Review A*, **65**: #052119.

Orilia, F. and M. Paolini Paoletti (2022). "Properties." in E. N. Zalta and U. Nodelman (eds.), *The Stanford Encyclopedia of Philosophy* (Winter 2022 Edition), URL: https://plato.stanford.edu/archives/win2022/entries/properties/

Page, D. and W. Wootters (1983). "Evolution without evolution." *Physical Review D*, **27**: 2885–2892.

Pan, J.-W., D. Bouwmeester, H. Weinfurter, and A. Zeilinger (1998). "Experimental entanglement swapping: Entangling photons that never interacted." *Physical Review Letters*, **80**: 3891–3894.

Pauli, W. (1933). "Die allgemeinen Prinzipien der Wellenmechanik." Pp. 83–272 in H Geiger and K Scheel (eds.), *Handbuch der Physik, Vol. 24*. Berlin: Springer. English translation: *General Principles of Quantum Mechanics*, 1980. Berlin: Springer.

Paz, J. P. (1994). "Decoherence in quantum Brownian motion." Pp. 213–220 in J. J. Halliwell, J. Pérez-Mercader, and W. H. Zurek (eds.), *Physical Origins of Time Asymmetry*. Cambridge: Cambridge University Press.

Paz, J. P. and A. J. Roncaglia (2009). "Redundancy of classical and quantum correlations during decoherence." *Physical Review A*, **80**: 042111.

Paz, J. P. and W. H. Zurek (1999). "Quantum limit of decoherence: environment induced superselection of energy eigenstates." *Physical Review Letters*, **82**: 5181–5185.

Paz, J. P. and W. H. Zurek (2002). "Environment-induced decoherence and the transition from quantum to classical." Pp. 77–148 in D. Heiss (ed.), *Lecture Notes in Physics, Vol. 587*. Heidelberg-Berlin: Springer.

Pearle, P. (1989). "Combining stochastic dynamical state-vector reduction with spontaneous localization." *Physical Review A*, **39**: 2277–2289.

Pelletier, F. J. (ed.) (1979). *Mass Terms: Some Philosophical Problems.* Dordrecht: Reidel Publishing Company.

Peres, A. (1993). *Quantum Theory: Concepts and Methods.* New York: Kluwer.

Pessoa Jr., O. (1998). "Can the decoherence approach help to solve the measurement problem?" *Synthese*, **113**: 323–346.

Pihlström, S. (2008). "How (not) to write the history of pragmatist philosophy of science?" *Perspectives on Science*, **16**: 26–69.

Pihlström, S. (2012). "Toward pragmatically naturalized transcendental philosophy of scientific inquiry and pragmatic scientific realism." *Studia Philosophica Estonica*, **5**: 79–94.

Piron, C. (1976). *Foundations of Quantum Physics.* Reading: W. A. Benjamin.

Polarski, D. and A. A. Starobinsky (1996). "Semiclassicality and decoherence of cosmological perturbations." *Classical and Quantum Gravity*, **13**: 377–392.

Popper, K. (1959). "The propensity interpretation of probability." *The British Journal for the Philosophy of Science*, **10**: 25–42.

Popper, K. (1982). *The Open Universe: An Argument for Indeterminism.* London: Hutchinson.

Post, H. (1963). "Individuality and physics." *The Listener*, **70**: 534–537.

Primas, H. (1983). *Chemistry, Quantum Mechanics and Reductionism.* Berlin: Springer.

Primas, H. (1994). "Hierarchic quantum descriptions and their associated ontologies." Pp. 210–220 in K. V. Laurikainen, C. Montonen and K. Sunnarborg (eds.), *Symposium on the Foundations of Modern Physics 1994.* Gif-sur-Yvette: Editions Frontières.

Primas, H. (1998). "Emergence in exact natural sciences." *Acta Polytechnica Scandinavica*, **91**: 83–98.

Prior, A. (1967). *Past, Present, and Future.* Oxford: Clarendon Press.

Putnam, H. (1962). "What theories are not." Pp. 240–251 in E. Nagel, P. Suppes, and A. Tarski (eds.), *Logic, Methodology, and Philosophy of Science: Proceedings of the 1960 International Congress*, Stanford, CA: Stanford University Press.

Putnam, H. (1981). *Reason, Truth and History.* Cambridge: Cambridge University Press.

Quine, W. V. O. (1953). "On what there is." Pp. 1–19 in *From a Logical Point of View.* New York: Harper.

Quine, W. V. O. (1960). *Word and Object.* Cambridge, MA: MIT Press.

Quine, W. V. O. (1976). "Whither physical objects?" Pp. 497–504 in R. S. Cohen, P. Feyerabend, and M. Wartofsky (eds.), *Essays in Memory of Imre Lakatos.* Dordrecht: Reidel.

Quine, W. V. O. (1969). *Ontological Relativity and Other Essays.* New York: Columbia University Press.

Ramsey, F. P. (1931). "Truth and probability." Pp. 156–198 in R. B. Braithwaite (ed.), *The Foundations of Mathematics and other Logical Essays.* London: Routledge and Kegan Paul.

Redhead, M. (1980). "Models in physics." *The British Journal for the Philosophy of Science*, **31**: 145–163

Redhead, M. and P. Teller (1992). "Particle labels and the theory of indistinguishable particles in quantum mechanics." *British Journal for the Philosophy of Science*, **43**: 201–218.

Reichenbach, H. (1949). *The Theory of Probability.* Berkeley: University of California Press.

Rettler, B. and A. M. Bailey (2022). "Object." in E. N. Zalta and U. Nodelman (eds.), *The Stanford Encyclopedia of Philosophy* (Winter 2022 Edition), URL: https://plato.stanford.edu/archives/win2022/entries/object/

Robinson, H. (2021). "Substance." in E. N. Zalta (ed.) *The Stanford Encyclopedia of Philosophy* (Fall 2021 Edition), URL: https://plato.stanford.edu/archives/fall2021/entries/substance/

Roch, N., M. E. Schwartz, F. Motzoi, C. Macklin, R. Vijay, A. W. Eddins, A. N. Korotkov, K. B. Whaley, M. Sarovar, and I. Siddiqi (2014). "Observation of measurement-induced entanglement and quantum trajectories of remote superconducting qubits." *Physical Review Letters*, **112**: #170501.

Rovelli, C. (1991). "Is there incompatibility between the ways time is treated in general relativity and in standard quantum mechanics?" Pp. 126–136 in A. Ashtekar and J. Stachel (eds.), *Conceptual Problems of Quantum Gravity.* New York: Birkhauser.

Rovelli, C. (1996). "Relational quantum mechanics." *International Journal of Theoretical Physics*, **35**: 1637–1678.

Rovelli, C. (2002). "Partial observables." *Physical Review D*, **65**: #124013.

Rovelli, C. (2006). "The disappearance of space and time." Pp. 25–35 in D. Dieks (ed.), *The Ontology of Spacetime*, Amsterdam: Elsevier.

Ruetsche, L. (1995). "Measurement error and the Albert-Loewer problem." *Foundations of Physics Letters*, **8**: 327–344.

Ruetsche, L. (1996). "Van Fraassen on preparation and measurement." *Philosophy of Science*, **63**: S338–S346.

Ruetsche, L. (2003). "Modal semantics, modal dynamics and the problem of state preparation." *International Studies in the Philosophy of Science*, **17**: 25–41.

Ruetsche, L. (2011). *Interpreting Quantum Theories.* Oxford: Oxford University Press.
Russell, B. (1914). *Our Knowledge of the External World.* London: Allen and Unwin.
Russell, B. (1919). *Introduction to Mathematical Philosophy.* London: George Allen and Unwin.
Russell, R. (1923). "Vagueness." *Australasian Journal of Philosophy*, **1**: 84–92.
Russell, B. (1940). *An Inquiry into Meaning and Truth.* London: Allen and Unwin.
Russell, B. (1953). "On the notion of cause with applications to the free-will problem." Pp. 387–418 in H. Feigl and G. Brodbeck (eds.), *Readings in the Philosophy of Science.* New York: Appleton.
Salmon, N. (1981). *Reference and Essence.* Princeton: Princeton University Press.
Saunders, S. (2003). "Physics and Leibniz's principles." Pp. 289–307 in K. Brading and E. Castellani (eds.), *Symmetries in Physics: Philosophical Reflections.* Cambridge: Cambridge University Press.
Scerri, E. R. (2011). "Editorial 37." *Foundations of Chemistry*, **13**: 1–7.
Scerri, E. R. (2013). "Philosophy of chemistry: where has it been and where is it going." Pp. 208–225 in J.-P. Llored (ed.), *The Philosophy of Chemistry: Practices, Methodologies, and Concepts.* Newcastle: Cambridge Scholars Publishing.
Schaffer, J. (2001). "The individuation of tropes." *Australasian Journal of Philosophy*, **79**: 247–259.
Schaffer, J. (2010). "Monism: The priority of the whole." *Philosophical Review*, **119**: 31–76.
Schlosshauer, M. (2004). "Decoherence, the measurement problem, and interpretations of quantum mechanics." *Reviews of Modern Physics*, **76**: 1267–1305.
Schlosshauer, M. (2007). *Decoherence and the Quantum-to-Classical Transition.* Heidelberg-Berlin: Springer.
Schroeren, D. (2021). "Quantum metaphysical indeterminacy and the ontological foundations of orthodoxy." *Studies In History and Philosophy of Science Part A*, **90**: 235–246.
Schrödinger, E. (1935). "Discussion of probability relations between separated systems." *Proceedings of the Cambridge Philosophical Society*, **31**: 555–563.
Segal, I. E. (1947a). "Irreducible representations of operator algebras." *Bulletin* of the *American Mathematical Society*, **53**: 73–88.
Segal, I. E. (1947b). "Postulates for general quantum mechanics." *Annals of Mathematics*, **48**: 930–948.
Shannon, C. (1948). "The mathematical theory of communication." *Bell System Technical Journal*, **27**: 379–423.
Shao, J. and P. Hänggi (1997). "Control of molecular chirality." *The Journal of Chemical Physics*, **107**: 9935–9941.
Shapiro, S. (1997). *Philosophy of Mathematics: Structure and Ontology.* New York-Oxford: Oxford University Press.
Sklar, L. (1974). *Space, Time, and Spacetime.* Berkeley-Los Angeles: University of California Press.
Solé, A. (2013). "Bohmian mechanics without wave function ontology." *Studies in History and Philosophy of Modern Physics*, **44**: 365–378.
Suárez, M. (2004a). "On quantum propensities: two arguments revisited." *Erkenntnis*, **61**: 1–16.
Suárez, M. (2004b). "Quantum selections, propensities, and the problem of measurement." *The British Journal for the Philosophy of Science*, **55**: 219–255.
Suárez, M. (2007). "Quantum propensities." *Studies in History and Philosophy of Modern Physics*, **38**: 418–438.
Sudbery, A. (2017). "Single-world theory of the extended Wigner's friend experiment." *Foundations of Physics*, **47**: 658–669.
Suppe, F. (1974). "The Search for Philosophic Understanding of Scientific Theories." Pp. 1–241 in F. Suppe, F. (ed.), *The Structure of Scientific Theories.* Urbana: University of Illinois Press.
Suppes, P. (2000). "Invariance, symmetry and meaning." *Foundations of Physics*, **30**: 1569–1585.
Sutcliffe, B. T. and R. G. Woolley (2011). "A comment on Editorial 37." *Foundations of Chemistry*, **13**: 93–95.
Sutcliffe, B. T. and R. G. Woolley (2012). "Atoms and molecules in classical chemistry and quantum mechanics." Pp. 387–426 in R. F. Hendry, A. Woody, and P. Needham (eds.), *Handbook of Philosophy of Science. Vol. 6, Philosophy of Chemistry.* Oxford: Elsevier.
Sutcliffe, B. T. and R. G. Woolley (2015). "The position of the clamped nuclei electronic Hamiltonian in quantum mechanics." Pp. 13–54 in J. Leszczynski (ed.), *Handbook of Computational Chemistry.* Cham: Springer.
Takesaki, M. (2003). *Theory of Operator Algebras. Vols. I–III.* New York: Springer-Verlag.
Tanona, S. (2013). "Decoherence and the Copenhagen cut." *Synthese*, **190**, 3625–3649.
Tarski, A. (1941). "On the calculus of relations." *The Journal of Symbolic Logic*, **6**: 73–89.
Tegtmeier, E. (2000). "Events as facts." Pp. 219–228 in J. Faye, U. Scheffler, and M. Urchs (eds.), *Things, Facts, and Events.* Amsterdam, Rodopi.

Teller, P. (1998). "Quantum mechanics and haecceities." Pp. 114–141 in E. Castellani (ed.), *Interpreting Bodies: Classical and Quantum Objects in Modern Physics*. Princeton: Princeton University Press.

Terra Cunha, M. O., J. A. Dunningham, and V. Vedral (2007). "Entanglement in single-particle systems." *Proceedings of the Royal Society A*, **463**: 2277–2286.

Thomason, S. K. (1989). "Free construction of time from events." *Journal of Philosophical Logic*, **18**: 43–67.

Tichy, M. C., F. Mintert, and A. Buchleitner (2011). "Essential entanglement for atomic and molecular physics." *Journal of Physics B: Atomic, Molecular and Optical Physics*, **44**: #192001.

Tinkham, M. (1964). *Group Theory and Quantum Mechanics*. New York: McGraw-Hill.

Tolman, R. (1938). *The Principles of Statistical Mechanics*. Oxford: Clarendon Press.

Tooley, M. (1987). *Causation: A Realist Approach*. Oxford: Oxford University Press.

Torretti, R. (2000). "'Scientific realism' and scientific practice." Pp. 113–122 in E. Agazzi and M. Pauri (eds.). *The Reality of the Unobservable*. Dordrecht: Springer.

Torretti, R. (2005a). *Manuel Kant*. Santiago de Chile: Ediciones Universidad Diego Portales.

Torretti, R. (2005b). "Conocimiento discursivo." *Revista UDP*, **2**: 67–72.

Torza, A. (2023). *Indeterminacy in the World*. Cambridge: Cambridge University Press

Trost, J. and K. Hornberger (2009). "Hund's paradox and the collisional stabilization of chiral molecules." *Physical Review Letters*, **103**: 023202.

Tugendhat, E. (1982). *Traditional and Analytical Philosophy: Lectures on the Philosophy of Language*. Cambridge: Cambridge University Press.

Tumulka, R. (2006). "A relativistic version of the Ghirardi-Rimini-Weber model." *Journal of Statistical Physics*, **125**: 821–840.

Tumulka, R. (2022). "Bohmian mechanics." Pp. 257–271 in E. Knox and A. Wilson (eds.), *The Routledge Companion to the Philosophy of Physics*. Abingdon: Routledge.

Tung, W. K. (1985). *Group Theory in Physics*. Singapore: World Scientific.

Uffink, J. (2007). "Compendium of the foundations of classical statistical physics." Pp. 923–1074 in J. Butterfield and J. Earman (eds.), *Philosophy of Physics*. Amsterdam: Elsevier.

Vaidman, L. (2021). "Many-Worlds Interpretation of quantum mechanics." in E. N. Zalta (ed.), *The Stanford Encyclopedia of Philosophy* (Fall 2021 Edition). URL: https://plato.stanford.edu/archives/fall2021/entries/qm-manyworlds/

van Fraassen, B. (1970). "On the extension of Beth's semantics of physical theories." *Philosophy of Science*, **37**: 325–334.

van Fraassen, B. (1972). "A formal approach to the philosophy of science." Pp. 303–366 in R. Colodny (ed.), *Paradigms and Paradoxes: The Philosophical Challenge of the Quantum Domain*. Pittsburgh: University of Pittsburgh Press.

van Fraassen, B. (1974). "The Einstein-Podolsky-Rosen paradox." *Synthese*, **29**: 291–309.

van Fraassen, B. (1980). *The Scientific Image*. Oxford: Oxford University Press.

van Fraassen, B. (1985). "Statistical behaviour of indistinguishable particles: Problems of interpretation." Pp. 161–187 in P. Mittelstaedt and E.-W. Stachow (eds.), *Recent Developments in Quantum Logic*. Cologne: Mannheim.

van Fraassen, B. (1989). *Laws and Symmetry*. Oxford: Oxford University Press.

van Fraassen, B. (1991), *Quantum Mechanics*. Oxford: Clarendon Press.

Van Kampen, N. G. (1954). "Quantum statistics of irreversible processes." *Physica*, **20**: 603–622.

Veldman, M. (2024). "Mathematizing metaphysics: The case of the principle of least action." *Philosophy of Science*, **91**: 351–369.

Vermaas, P. (1996). "Unique transition probabilities in the modal interpretation." *Studies in History and Philosophy of Modern Physics*, **27**: 133–159.

Vermaas, P. (1997). "A no-go theorem for joint property ascriptions in modal interpretations of quantum mechanics." *Physical Review Letters*, **78**: 2033–2037.

Vermaas, P. (1998). "The pros and cons of the Kochen-Dieks and the atomic modal interpretation." Pp. 103–148 in D. Dieks and P. Vermaas (eds.), *The Modal Interpretation of Quantum Mechanics*. Dordrecht: Kluwer Academic Publishers

Vermaas, P. (1999). *A Philosopher's Understanding of Quantum Mechanics: Possibilities and Impossibilities of a Modal Interpretation*. Cambridge: Cambridge University Press.

Vermaas, P. and D. Dieks (1995). "The modal interpretation of quantum mechanics and its generalization to density operators." *Foundations of Physics*, **25**: 145–158.

Viola, L. and H. Barnum (2010). "Entanglement and subsystems, entanglement beyond subsystems, and all that." Pp. 16–43 in A. Bokulich and G. Jaeger (eds.), *Philosophy of Quantum Information and Entanglement*. Cambridge: Cambridge University Press.

von Mises R. (1957). *Probability, Statistics and Truth.* New York: Macmillan.

von Neumann, J. (1932). *Mathematische Grundlagen der Quantenmechanik*, Berlin: Springer-Verlag. English translation: (1955). *Mathematical Foundations of Quantum Mechanics.* Princeton: Princeton University Press.

Wallace, D. (2008). "Philosophy of quantum mechanics." Pp. 16–98 in D. Rickles (ed.), *The Ashgate Companion to Contemporary Philosophy of Physics.* New York: Routledge.

Wallace, D. (2012). *The Emergent Multiverse: Quantum Theory according to the Everett Interpretation.* Oxford: Oxford University Press.

Wallace, D. (2019). "What is orthodox quantum mechanics?" Pp. 285–312 in A. Cordero (ed.), *Philosophers Look at Quantum Mechanics.* Synthese Library, *Vol 406.* Cham: Springer.

Wallace, D. (2022). "Stating structural realism: mathematics-first approaches to physics and metaphysics." *Philosophical Perspectives,* **36**: 345–378.

Weatherall, J. O. (2016). "Are Newtonian gravitation and geometrized Newtonian gravitation theoretically equivalent?" *Erkenntnis,* **81**: 1073–1091.

Weatherall, J. O. (2019). "Theoretical equivalence in physics. Parts 1 and 2." *Philosophy Compass,* 14: #e12591and#e12592.

Weinberg, S. (1972). *Gravitation and Cosmology: Principles and Applications of the General Theory of Relativity.* New York: John Wiley and Sons.

Weinberg, S. (1995). *The Quantum Theory of Fields, Volume I: Foundations.* Cambridge: Cambridge University Press.

Weyl, H. (1950). *The Theory of Groups and Quantum Mechanics.* New York: Dover.

Weyl, H. (1952). *Symmetry.* Princeton: Princeton University Press.

Whitehead, A. N. (1929). *Process and Reality. An Essay in Cosmology.* New York: Macmillan.

Wigner, E. (1961). "Remarks on the mind-body question." Pp. 284–301 in I. J. Good (ed.), *The Scientist Speculates.* London: Heinemann.

Williams, R. (2008). "Multiple actualities and ontically vague identity." *Philosophical Quarterly,* **58**: 134–154.

Williams, T. (2022). "John Duns Scotus." in E. N. Zalta and U. Nodelman (eds.), *The Stanford Encyclopedia of Philosophy* (Fall 2022 Edition). URL: https://plato.stanford.edu/archives/fall2022/entries/duns-scotus/

Wilson, A. (2020). *The Nature of Contingency: Quantum physics as Modal Realism.* Oxford: Oxford University Press.

Wilson, J. M. (2013). "A determinable-based account of metaphysical indeterminacy." *Inquiry: An Interdisciplinary Journal of Philosophy,* **56**: 359–385.

Wilson, J. M. (2016). "Are there indeterminate states of affairs? Yes." Pp. 105–125 in E. Barnes (ed.), *Current Controversies in Metaphysics.* Abingdon: Routledge.

Wilson, J. M. (2013). "Determinables and determinates." in E. N. Zalta and U. Nodelman (eds.), *The Stanford Encyclopedia of Philosophy* (Spring 2023 Edition), URL: https://plato.stanford.edu/archives/spr2023/entries/determinate-determinables/

Wilson, N. L. (1974). "Facts, events, and their identity conditions." *Philosophical Studies,* **25**: 303–321.

Wittgenstein, L. (1921). "Logisch-Philosophische Abhandlung." *Annalen der Naturphilosophische,* **XIV** (3/4). English translation: C. K. Ogden (1922). *Tractatus Logico-Philosophicus.* London: Routledge & Kegan Paul.

Wolff, Ch. (1728). *Philosophia Rationalis Sive Logica.* Reprint of the 1740 edition with introduction, notes and index in (1980) by Jean École (ed.). New York: Georg Olms.

Wolff, J. (2015). "Spin as a determinable." *Topoi,* **34**: 379–386.

Woolley, R. G. (1976). "Quantum theory and molecular structure." *Advances in Physics,* **25**: 27–52.

Woolley, R. G. (1978). "Must a molecule have a shape?" *Journal of the American Chemical Society,* **100**: 1073–1078.

Woolley, R. G. (1982). "Natural optical activity and the molecular hypothesis." *Structure and Bonding,* **52**: 1–35.

Woolley, R. G. and B. T. Sutcliffe (1977). "Molecular structure and the Born-Oppenheimer approximation." *Chemical Physics Letters,* **45**: 393–398.

Wootters, W. (1984). "'Time' replaced by quantum correlations." *International Journal of Theoretical Physics,* **23**: 701–711.

Worrall, J. (1989). "Structural realism: the best of both worlds?" *Dialectica,* **43**: 99–124.

Yagisawa, T. (2023). "Possible objects." in E. N. Zalta and U. Nodelman (eds.), *The Stanford Encyclopedia of Philosophy* (Summer 2023 Edition), URL: https://plato.stanford.edu/archives/sum2023/entries/possible-objects/

Zanardi, P. (2001). "Virtual quantum systems." *Physical Review Letters*, **87**: #077901.

Zanardi, P., D. A. Lidar, and S. Lloyd (2004). "Quantum tensor product structures are observable induced." *Physical Review Letters*, **92**: 06042.

Zeh, H.-D. (2003). "Basic concepts and their interpretation." Pp. 7–40 in E. Joos, H.-D. Zeh, C. Kiefer, D. Giulini, J. Kupsch, and I.-O. Stamatescu (eds.), *Decoherence and the Appearance of a Classical World in Quantum Theory*. Berlin: Springer.

Zeh, H. D. (2005). "Roots and fruits of decoherence." *Séminaire Poincaré*, **2**: 1–19.

Żukowski, M., A. Zeilinger, M. A. Horne, and A. K. Ekert (1993). "Event-ready-detectors. Bell experiment via entanglement swapping." *Physical Review Letters*, **71**: 4287–4290.

Zurek, W. H. (1981). "Pointer basis of quantum apparatus: into what mixtures does the wave packet collapse?" *Physical Review D*, **24**: 1516–1525.

Zurek, W. H. (1982). "Environment-induced superselection rules." *Physical Review D*, **26**: 1862–1880.

Zurek, W. H. (1991). "Decoherence and the transition from quantum to classical," *Physics Today*, **44**: 36–44.

Zurek, W. H. (1993). "Preferred states, predictability, classicality and the environment-induced decoherence," *Progress of Theoretical Physics*, **89**: 281–312.

Zurek, W. H. (1994). "Preferred sets of states, predictability, classicality and environment-induced decoherence." Pp. 175–207 in J. J. Halliwell, J. Pérez-Mercader, and W. H. Zurek (eds.), *Physical Origins of Time Asymmetry*. Cambridge: Cambridge University Press.

Zurek, W. H. (1998). "Decoherence, einselection and the existential interpretation (the rough guide)." *Philosophical Transactions: Mathematical, Physical and Engineering Sciences*, **356**: 1793–1821.

Zurek, W. H. (2000). "Decoherence and einselection." Pp. 309–342 in P. Blanchard, D. Giulini, E. Joos, C. Kiefer, and I.-O. Stamatescu (eds.), *Decoherence: Theoretical, Experimental, and Conceptual Problems*. Berlin-Heidelberg: Springer-Verlag.

Zurek, W. H. (2003). "Decoherence, einselection, and the quantum origins of the classical." *Reviews of Modern Physics*, **75**: 715–776.

Zwanzig, R. (1961). "Statistical mechanics of irreversibility." Pp. 106–141 in W. E. Britten, B. W. Downs, and J. Downs (eds.), *Lectures in Theoretical Physics* (Boulder), *Vol. 3*. New York: Inter-Science.

Index